Common and Scientific
Names of Aquatic Invertebrates
from the United States and Canada:

Mollusks

Cover design by F. H. Roberts

Publication of this book was made possible by a grant from the

Shell Companies Foundation

Travel and institutional support was extended to the Committee by the

National Marine Fisheries Service
National Oceanic and Atmospheric Administration
U.S. Department of Commerce

Additional travel support was provided by the

Fish and Wildlife Service
U.S. Department of the Interior

The American Fisheries Society is indebted to

F. H. Roberts
Creative Design International

for his design of the cover,
his photographic contributions, and
his arrangement of the color portfolio
featured in this volume

American Fisheries Society Special Publication 16

Common and Scientific
Names of Aquatic Invertebrates
from the United States and Canada:

Mollusks

Donna D. Turgeon, Chair
Arthur E. Bogan, Eugene V. Coan, William K. Emerson,
William G. Lyons, William L. Pratt, Clyde F. E. Roper,
Amelie Scheltema, Fred G. Thompson,
and James D. Williams

*Committee on Scientific and Vernacular
Names of Mollusks of the
Council of Systematic Malacologists*

American Malacological Union

Bethesda, Maryland
1988

QL
411
.C65x

The American Fisheries Society Special Publication series is a registered serial. A suggested citation format for this book follows.

Turgeon, D. D., A. E. Bogan, E. V. Coan, W. K. Emerson, W. G. Lyons, W. L. Pratt, C. F. E. Roper, A. Scheltema, F. G. Thompson, and J. D. Williams. 1988. Common and scientific names of aquatic invertebrates from the United States and Canada: mollusks. American Fisheries Society Special Publication 16.

© Copyright by the American Fisheries Society, 1988

Library of Congress Catalog Card Number: 88-70617

ISBN 0-913235-47-4 (cloth)   ISSN 0097-0638

ISBN 0-913235-48-2 (paper)

Address orders to
American Fisheries Society
5410 Grosvenor Lane, Suite 110
Bethesda, Maryland 20814, USA

CONTENTS

Foreword	vii
Alphabetical List of Families	1
Phylogenetic List by Class, Order, and Family	5
Introduction	9
History and Background	9
List of Molluscan Names	10
Area of Coverage	11
Vernacular versus Common Names	11
Plan of the List	11
AFS Principles Governing the Selection of Common Names	12
Resolutions from the Committee on Scientific and Vernacular Names of Mollusks	14
Future of the Molluscan List	15
Acknowledgments	15
Dedication	19
Names of Mollusks	20
Class Aplacophora—Solenogasters	20
Class Bivalvia—Clams or Bivalves	21
Class Scaphopoda—Tuskshells and Toothshells	50
Class Gastropoda—Gastropods	52
Class Polyplacophora—Chitons	148
Class Cephalopoda—Squids and Octopuses	151
Appendix 1: Endangered and Threatened Mollusks of North America	153
Appendix 2: Possibly Extinct Mollusks of the United States	156
Index	158
Portfolio of Mollusk Diversity	

FOREWORD

The American Fisheries Society (AFS) Committee on Names of Aquatic Invertebrates (CNAI) was established by AFS President John J. Magnuson on September 30, 1981. The main goal of this Committee is to achieve uniformity and avoid confusion in vernacular nomenclature of aquatic invertebrates. The present charge by the Society to this Committee is as follows.

> The Committee shall be responsible for studying and reporting on matters concerning common and scientific names of aquatic invertebrates, and shall prepare checklists of names to achieve uniformity and avoid confusion in nomenclature. The Chairman shall be custodian of the master checklists. The Committee shall coordinate its activities with those of other societies and organizations throughout the world. The Committee shall be composed of up to 12 members who are outstanding specialists in invertebrate taxonomy and nomenclature.

The Names of Aquatic Invertebrates Committee has benefited substantially from the long experience and decisions reached by the AFS Names of Fishes Committee. The Names of Fishes Committee was originally appointed in 1933 as the result of a resolution adopted by the Society to form a permanent committee of experts in the field of ichthyology "to prepare and submit for publication a list of common names of fishes corresponding to the accepted scientific names." Because the AFS membership does not include much of the invertebrate taxonomic expertise needed to develop a comprehensive list of common and scientific names of aquatic invertebrates from the United States and Canada, it was decided to enlist the cooperation of other professional societies to accomplish the job.

The American Fisheries Society acknowledges with considerable gratitude both the Committee on Scientific and Vernacular Names of Mollusks of the Council of Systematic Malacologists and the American Malacological Union for developing this comprehensive list of common and scientific names of mollusks. Two CNAI members, Dr. Donna D. Turgeon and Dr. Fred G. Thompson, in particular, were instrumental in bringing this task to fruition. The AFS Committee on Names of Aquatic Invertebrates has approved the list for publication.

Committee on Names of Aquatic Invertebrates

Donna D. Turgeon, *Chair*

Edward L. Bousfield	Frank Maturo
Steven Cairns	David Pawson
Larry L. Eng	Lynn Starnes
Mark Holliday	Fred G. Thompson
Kenneth Manuel	Austin B. Williams

ALPHABETICAL LIST OF FAMILIES

CLASS/Family	PAGE
APLACOPHORA	20
Chaetodermatidae	20
Dondersiidae	20
Limifossoridae	20
Meiomeniidae	20
Prochaetodermatidae	20
Proneomeniidae	20
BIVALVIA (PELECYPODA)	21
Anomiidae	27
Arcidae	24
Arcticidae	43
Astartidae	38
Bernardinidae	43
Cardiidae	39
Carditidae	38
Chamidae	36
Chlamydoconchidae	37
Condylocardiidae	38
Cooperellidae	45
Corbiculidae	44
Corbulidae	46
Crassatellidae	39
Cuspidariidae	49
Cyrenoididae	36
Dimyidae	27
Donacidae	42
Dreissenidae	43
Entoliidae	26
Galeommatidae	37
Gastrochaenidae	46
Glossidae	43
Glycymerididae	24
Gryphaeidae	28
Hiatellidae	46
Isognomonidae	25
Kelliidae	36
Limidae	25
Limopsidae	24
Lucinidae	34
Lyonsiidae	47
Mactridae	40
Malleidae	25
Malletiidae	21
Manzanellidae	25
Mesodesmatidae	40
Montacutidae	37
Myidae	45
Mytilidae	23
Nuculanidae	21

CLASS/Family	PAGE
Nuculidae	21
Ostreidae	28
Pandoridae	48
Pectinidae	26
Periplomatidae	48
Petricolidae	45
Philobryidae	25
Pholadidae	46
Pholadomyidae	47
Pinnidae	25
Plicatulidae	27
Poromyidae	49
Propeamussiidae	27
Psammobiidae	42
Pteriidae	25
Semelidae	42
Solecurtidae	43
Solemyidae	22
Solenidae	40
Sphaeriidae	34
Spheniopsidae	46
Spondylidae	27
Sportellidae	38
Tellinidae	40
Teredinidae	47
Thraciidae	48
Thyasiridae	35
Trapeziidae	43
Turtoniidae	38
Ungulinidae	36
Unionidae	28
Veneridae	44
Verticordiidae	49
Vesicomyidae	43
SCAPHOPODA	50
Dentaliidae	50
Entalinidae	51
Gadilidae	51
Gadilinidae	50
Laevidentaliidae	50
Pulsellidae	51
Siphonodentaliidae	51
GASTROPODA	52
Achatinidae	129
Aclididae	76
Acmaeidae	53
Acroloxidae	122
Acteonidae	107

CLASS/Family	PAGE
Actinocyclidae	117
Addisoniidae	54
Aegiretidae	116
Aeolidiidae	121
Aglajidae	108
Akeridae	111
Aldisidae	117
Ammonitellidae	141
Ancylidae	125
Annulariidae	66
Aplysiidae	112
Aporrhaididae	78
Archidorididae	117
Architectonicidae	74
Arionidae	131
Arminidae	119
Assimineidae	68
Asteronotidae	117
Atlantidae	81
Atyidae	109
Babakinidae	121
Barleeiidae	68
Bithyniidae	59
Boselliidae	112
Bradybaenidae	142
Buccinidae	87
Bulimulidae	130
Bullidae	109
Bullinidae	107
Bursidae	83
Cadlinidae	116
Caecidae	71
Caliphyllidae	112
Camaenidae	141
Calycidorididae	117
Calyptraeidae	79
Cancellariidae	94
Capulidae	78
Carinariidae	81
Carychiidae	125
Cassidae	82
Cavoliniidae	110
Ceriidae	128
Cerithiidae	72
Cerithiopsidae	73
Charopidae	130
Chromodorididae	116
Clionidae	111
Cliopsidae	111
Cocculinidae	54
Cochlicopidae	126
Colubrariidae	90

CLASS/Family	PAGE
Columbellidae	86
Conidae	94
Conualeviidae	117
Coralliophilidae	86
Corambidae	114
Coryphellidae	120
Costasiellidae	112
Costellariidae	92
Cumanotidae	120
Cyclostrematidae	57
Cylichnidae	108
Cylindrobullidae	112
Cymbuliidae	110
Cypraeidae	80
Dendrodorididae	117
Dendronotidae	118
Desmopteridae	111
Diaphanidae	109
Discidae	131
Discodorididae	118
Dironidae	119
Dorididae	117
Dotoidae	119
Elachisinidae	69
Elysiidae	113
Entoconchidae	78
Epitoniidae	75
Eubranchidae	120
Eulimidae	77
Facelinidae	121
Falsicingulidae	69
Fasciolariidae	91
Ferussaciidae	128
Ficidae	83
Fionidae	121
Fissurellidae	52
Fossaridae	78
Gastropteridae	108
Glaucidae	122
Goniodorididae	114
Gymnodorididae	116
Haliotididae	52
Hancockiidae	118
Haplotrematidae	129
Harpidae	92
Helicarionidae	133
Helicellidae	146
Helicidae	146
Helicinidae	58
Helicodiscidae	130
Helminthoglyptidae	143
Hermaeidae	113

CLASS/Family	PAGE
Heterodorididae	115
Hipponicidae	78
Hydatinidae	107
Hydrobiidae	59
Hydromylidae	111
Janolidae	120
Janthinidae	75
Juliidae	113
Kentrodorididae	118
Lacunidae	66
Lamellariidae	79
Lepetidae	54
Limacidae	136
Limacinidae	110
Littorinidae	66
Lobigeridae	113
Lomanotidae	119
Lymnaeidae	122
Marginellidae	93
Mathildidae	74
Melampodidae	122
Melongenidae	90
Microhedylidae	110
Milacidae	137
Mitridae	92
Modulidae	72
Muricidae	84
Nassariidae	91
Naticidae	82
Neritidae	58
Notobranchaeidae	111
Olividae	92
Omalogyridae	69
Onchidiidae	147
Onchidorididae	115
Oocorythidae	83
Oreohelicidae	141
Ovulidae	80
Oxynoidae	113
Pelycidiidae	69
Peraclididae	110
Phasianellidae	58
Phenacolepadidae	58
Philinidae	108
Philomycidae	132
Phyllidiidae	117
Phylliroidae	119
Pilidae	59
Planaxidae	72
Platydorididae	118
Pleurobranchaeidae	114
Pleurobranchidae	114

CLASS/Family	PAGE
Pleuroceridae	63
Pleurotomariidae	52
Pneumodermatidae	111
Physidae	123
Planorbidae	124
Polyceridae	116
Polygyridae	137
Pomatiopsidae	63
Potamididae	72
Pterotracheidae	81
Punctidae	130
Pupillidae	126
Pyramidellidae	101
Ranellidae	83
Retusidae	109
Ringiculidae	107
Rissoellidae	69
Rissoidae	67
Rostangidae	117
Runcinidae	109
Sagdidae	141
Scaphandridae	107
Scissurellidae	52
Scyllaeidae	119
Seguenziidae	56
Siliquariidae	72
Siphonariidae	126
Skeneidae	57
Skeneopsidae	69
Spiraxidae	129
Spurillidae	121
Stiligeridae	113
Streptaxidae	129
Strobilopsidae	128
Strombidae	78
Subulinidae	128
Succineidae	132
Terebridae	95
Tergipedidae	120
Testacellidae	137
Tethyidae	119
Thiaridae	63
Thliptodontidae	111
Thysanophoridae	141
Tonnidae	83
Tornidae	71
Trichotropidae	79
Trimusculidae	126
Triophidae	115
Triphoridae	74
Tritoniidae	118
Triviidae	80

FAMILY LIST

CLASS/Family	PAGE
Trochidae	54
Truncatellidae	69
Turbinellidae	93
Turbinidae	57
Turridae	95
Turritellidae	71
Tylodinidae	114
Umbraculidae	114
Urocoptidae	129
Valloniidae	128
Valvatidae	58
Vanikoroidae	78
Vermetidae	72
Veronicellidae	147
Vitrinellidae	69
Vitrinidae	136
Viviparidae	59
Volutidae	93
Volutomitridae	93
Xenophoridae	79
Zonitidae	134

CLASS/Family	PAGE
POLYPLACOPHORA	148
Acanthochitonidae	150
Chaetopleuridae	149
Chitonidae	150
Hanleyidae	148
Ischnochitonidae	148
Katherinidae	150
Lepidopleuridae	148
Mopaliidae	149
CEPHALOPODA	151
Argonautidae	152
Gonatidae	151
Loliginidae	151
Octopodidae	152
Ommastrephidae	151
Onychoteuthididae	151
Sepiolidae	151
Spirulidae	151
Thysanoteuthidae	152
Tremoctopodidae	152

PHYLOGENETIC LIST BY CLASS, ORDER, AND FAMILY

CLASS/Order/Family	PAGE
APLACOPHORA	20
CHAETODERMOMORPHA	20
Chaetodermatidae	20
Prochaetodermatidae	20
Limifossoridae	20
NEOMENIOMORPHA	20
Dondersiidae	20
Meiomeniidae	20
Proneomeniidae	20
BIVALVIA (PELECYPODA)	21
NUCULOIDA	21
Nuculidae	21
Malletiidae	21
Nuculanidae	21
SOLEMYOIDA	22
Solemyidae	22
MYTILOIDA	23
Mytilidae	23
ARCOIDA	24
Arcidae	24
Limopsidae	24
Glycymerididae	24
Manzanellidae	25
Philobryidae	25
PTERIOIDA	25
Pinnidae	25
Pteriidae	25
Isognomonidae	25
Malleidae	25
LIMOIDA	25
Limidae	25
OSTREOIDA	26
Entoliidae	26
Pectinidae	26
Propeamussiidae	27
Plicatulidae	27
Spondylidae	27
Dimyidae	27
Anomiidae	27
Ostreidae	28
Gryphaeidae	28
UNIONOIDA	28
Unionidae	28
VENEROIDA	34
Sphaeriidae	34
Lucinidae	34
Thyasiridae	35
Ungulinidae	36
Cyrenoididae	36

CLASS/Order/Family	PAGE
Chamidae	36
Kelliidae	36
Montacutidae	37
Galeommatidae	37
Chlamydoconchidae	37
Turtoniidae	38
Sportellidae	38
Carditidae	38
Condylocardiidae	38
Astartidae	38
Crassatellidae	39
Cardiidae	39
Mactridae	40
Mesodesmatidae	40
Solenidae	40
Tellinidae	40
Donacidae	42
Psammobiidae	42
Semelidae	42
Solecurtidae	43
Dreissenidae	43
Arcticidae	43
Bernardinidae	43
Trapeziidae	43
Vesicomyidae	43
Glossidae	43
Corbiculidae	44
Veneridae	44
Petricolidae	45
Cooperellidae	45
MYOIDA	45
Myidae	45
Corbulidae	46
Spheniopsidae	46
Gastrochaenidae	46
Hiatellidae	46
Pholadidae	46
Teredinidae	47
PHOLADOMYOIDA	47
Pholadomyidae	47
Lyonsiidae	47
Pandoridae	48
Thraciidae	48
Periplomatidae	48
Poromyidae	49
Verticordiidae	49
Cuspidariidae	49
SCAPHOPODA	50
DENTALIIDA	50

CLASS/ORDER/Family	PAGE
Dentaliidae	50
Laevidentaliidae	50
Gadilinidae	50
GADILIDA	51
Entalinidae	51
Pulsellidae	51
Siphonodentaliidae	51
Gadilidae	51
GASTROPODA	52
ARCHAEOGASTROPODA	52
Pleurotomariidae	52
Scissurellidae	52
Haliotididae	52
Fissurellidae	52
Acmaeidae	53
Lepetidae	54
Cocculinidae	54
Addisoniidae	54
Trochidae	54
Seguenziidae	56
Cyclostrematidae	57
Skeneidae	57
Turbinidae	57
Phasianellidae	58
Neritidae	58
Phenacolepadidae	58
Helicinidae	58
MESOGASTROPODA	58
Valvatidae	58
Viviparidae	59
Pilidae	59
Bithyniidae	59
Hydrobiidae	59
Pomatiopsidae	63
Thiaridae	63
Pleuroceridae	63
Annulariidae	66
Lacunidae	66
Littorinidae	66
Rissoidae	67
Barleeiidae	68
Assimineidae	68
Falsicingulidae	69
Pelycidiidae	69
Elachisinidae	69
Truncatellidae	69
Rissoellidae	69
Skeneopsidae	69
Omalogyridae	69
Vitrinellidae	69
Tornidae	71

CLASS/ORDER/Family	PAGE
Caecidae	71
Turritellidae	71
Siliquariidae	72
Vermetidae	72
Planaxidae	72
Modulidae	72
Potamididae	72
Cerithiidae	72
Cerithiopsidae	73
Mathildidae	74
Architectonicidae	74
Triphoridae	74
Janthinidae	75
Epitoniidae	75
Aclididae	76
Eulimidae	77
Entoconchidae	78
Aporrhaididae	78
Strombidae	78
Hipponicidae	78
Fossaridae	78
Vanikoroidae	78
Capulidae	78
Trichotropidae	79
Calyptraeidae	79
Xenophoridae	79
Lamellariidae	79
Triviidae	80
Cypraeidae	80
Ovulidae	80
Atlantidae	81
Carinariidae	81
Pterotracheidae	81
Naticidae	82
Cassidae	82
Ranellidae	83
Bursidae	83
Tonnidae	83
Oocorythidae	83
Ficidae	83
NEOGASTROPODA	84
Muricidae	84
Coralliophilidae	86
Columbellidae	86
Buccinidae	87
Colubrariidae	90
Melongenidae	90
Nassariidae	91
Fasciolariidae	91
Olividae	92
Harpidae	92
Mitridae	92

CLASS/ORDER/Family	PAGE	CLASS/ORDER/Family	PAGE
Costellariidae	92	Lobigeridae	113
Volutomitridae	93	Oxynoidae	113
Turbinellidae	93	Stiligeridae	113
Volutidae	93	NOTASPIDEA	114
Marginellidae	93	Tylodinidae	114
Cancellariidae	94	Umbraculidae	114
Conidae	94	Pleurobranchidae	114
Terebridae	95	Pleurobranchaeidae	114
Turridae	95	NUDIBRANCHIA	114
PYRAMIDELLOIDA	101	Corambidae	114
Pyramidellidae	101	Goniodorididae	114
CEPHALASPIDEA	107	Onchidorididae	115
Acteonidae	107	Triophidae	115
Bullinidae	107	Heterodorididae	115
Hydatinidae	107	Aegiretidae	116
Ringiculidae	107	Gymnodorididae	116
Scaphandridae	107	Polyceridae	116
Cylichnidae	108	Cadlinidae	116
Aglajidae	108	Chromodorididae	116
Philinidae	108	Asteronotidae	117
Gastropteridae	108	Actinocyclidae	117
Diaphanidae	109	Conualeviidae	117
Runcinidae	109	Calycidorididae	117
Bullidae	109	Aldisidae	117
Atyidae	109	Rostangidae	117
Retusidae	109	Dorididae	117
ACOCHLIDIOIDEA	110	Dendrodorididae	117
Microhedylidae	110	Phyllidiidae	117
THECOSOMATA	110	Archidorididae	117
Limacinidae	110	Discodorididae	118
Cavoliniidae	110	Kentrodorididae	118
Peraclididae	110	Platydorididae	118
Cymbuliidae	110	Tritoniidae	118
Desmopteridae	111	Hancockiidae	118
GYMNOSOMATA	111	Dendronotidae	118
Clionidae	111	Tethyidae	119
Cliopsidae	111	Lomanotidae	119
Hydromylidae	111	Scyllaeidae	119
Notobranchaeidae	111	Phylliroidae	119
Pneumodermatidae	111	Dotoidae	119
Thliptodontidae	111	Arminidae	119
ANASPIDEA	111	Dironidae	119
Akeridae	111	Janolidae	120
Aplysiidae	112	Coryphellidae	120
SACOGLOSSA	112	Eubranchidae	120
Boselliidae	112	Cumanotidae	120
Caliphyllidae	112	Tergipedidae	120
Costasiellidae	112	Fionidae	121
Cylindrobullidae	112	Babakinidae	121
Elysiidae	113	Facelinidae	121
Hermaeidae	113	Aeolidiidae	121
Juliidae	113	Spurillidae	121

CLASS/Order/Family	PAGE
Glaucidae	122
ARCHAEOPULMONATA	122
Melampodidae	122
BASOMMATOPHORA	122
Acroloxidae	122
Lymnaeidae	122
Physidae	123
Planorbidae	124
Ancylidae	125
Carychiidae	125
Siphonariidae	126
Trimusculidae	126
STYLOMMATOPHORA	126
Cochlicopidae	126
Pupillidae	126
Valloniidae	128
Strobilopsidae	128
Ceriidae	128
Ferussaciidae	128
Subulinidae	128
Spiraxidae	129
Achatinidae	129
Streptaxidae	129
Haplotrematidae	129
Urocoptidae	129
Bulimulidae	130
Punctidae	130
Charopidae	130
Helicodiscidae	130
Discidae	131
Arionidae	131
Philomycidae	132
Succineidae	132
Helicarionidae	133
Zonitidae	134
Vitrinidae	136
Limacidae	136
Milacidae	137
Testacellidae	137

CLASS/Order/Family	PAGE
Polygyridae	137
Sagdidae	141
Thysanophoridae	141
Camaenidae	141
Ammonitellidae	141
Oreohelicidae	141
Bradybaenidae	142
Helminthoglyptidae	143
Helicellidae	146
Helicidae	146
SYSTELLOMMATOPHORA	147
Onchidiidae	147
Veronicellidae	147
POLYPLACOPHORA	148
NEOLORICATA	148
Lepidopleuridae	148
Hanleyidae	148
Ischnochitonidae	148
Chaetopleuridae	149
Mopaliidae	149
Katherinidae	150
Chitonidae	150
Acanthochitonidae	150
CEPHALOPODA	151
SEPIOIDEA	151
Spirulidae	151
Sepiolidae	151
TEUTHOIDEA	151
Loliginidae	151
Gonatidae	151
Onychoteuthididae	151
Ommastrephidae	151
Thysanoteuthidae	152
OCTOPODA	152
Octopodidae	152
Argonautidae	152
Tremoctopodidae	152

INTRODUCTION

The purpose of this list is to provide a checklist of species and to recommend selected common names for North American mollusks, thereby achieving uniformity and avoiding confusion in molluscan nomenclature. It is not to impose scientific names. Common names, we believe, can be stabilized by general agreement. Scientific names, on the other hand, will inevitably change with advancing knowledge. The Committee on Scientific and Vernacular Names of Mollusks (Committee) realizes that many users of this list are not completely aware of the literature and are not interested in systematics. Therefore, the scientific nomenclature involved has been edited carefully with regard to spelling and authority; this nomenclature reflects the majority opinion of the Committee.

On September 7, 1985, at the American Fisheries Society (AFS) annual meeting, the Committee as a unit within the Council of Systematic Malacologists (CSM), in concert with the American Malacological Union (AMU), formally submitted this list of North American mollusks to the AFS Chairman of the Committee on Names of Aquatic Invertebrates (CNAI) for its review and subsequent publication.

History and Background

In April 1983, responsibility for preparing a list of scientific and vernacular names of mollusks was transferred from AMU to CSM. A committee was formed of 10 molluscan systematists, each with oversight responsibility: R. Tucker Abbott (marine bivalves), Arthur E. Bogan (freshwater bivalves), William K. Emerson (scaphopods), William G. Lyons (marine gastropods and polyplacophorans), William L. Pratt (helicacean terrestrial gastropods), Clyde F. E. Roper (cephalopods), Amelie Scheltema (aplacophorans), Fred G. Thompson (freshwater, land, and nonhelicacean gastropods), Donna D. Turgeon (Chair), and James D. Williams (extinct, endangered, and threatened species). Later, Dr. Abbott withdrew from the Committee and was replaced by Eugene V. Coan. The Committee assumed the name "Committee on Scientific and Vernacular Names of Mollusks" to reflect its strong feeling that most of the names selected would be descriptive regional or local names, some of which may have been in use for well over 100 years. The Committee recognized that once a single name was selected for a species and published, that name would assume "common" usage with time.

The Committee was charged with providing an initial draft list for a Committee meeting held prior to the CSM business meeting in July 1983. Those members with responsibility for the larger phylogenetic groups invited other systematists to join them as coauthors; authors are listed in the Acknowledgments section.

Preparing the "strawman" list identified a large set of problems, some of which were unique to a taxonomic unit. These problems were discussed and resolutions were proposed by the Committee. The Committee also voted to join the larger AFS endeavor to prepare a list of the aquatic invertebrates of North America and agreed to accept, without modification, 17 principles prepared by AFS to help select appropriate names (pages 12–14). The scope of the AFS endeavor included aquatic species that occur on the American continent north of Mexico and was restricted to species living at 200 meter depths or less, including coastal islands but not the West Indies. The Committee modified this scope to include terrestrial mollusks. The Committee also agreed to use the same style for the list as that of AFS Special Publication 12 (Robins et al., 1980. A List of Common and Scientific Names of Fishes from the United States and Canada, 4th edition, American Fisheries Society, Bethesda, Maryland).

The resolutions listed on pages 14–15, the proposals and recommendations cited above, and a status report from the Committee were presented to CSM and AMU. All were discussed, then approved unanimously by both organizations.

By the 1984 annual meeting of AMU, each specific taxonomic group had been reviewed by the designated specialist's peers, and larger phylogenetic units had been reviewed by more general systematists. This resulted in a more refined list. Again, the Committee met prior to the CSM business meeting to discuss problems, formulate resolutions, and agree on the Committee status report. The Committee report was accepted unanimously as presented again at both the CSM and AMU business meetings. Copies of that draft list were made available to all AMU members during the meeting for further review and suggestions. Mr. Steven J. Long, Editor of *Shells and Sea Life*, volunteered to publish gratis the draft list for his approximately 2,000 subscribers to review. The AMU accepted that offer to publish on a one-time basis the draft list current at that time.

The AFS principles governing selection of common names and a general statement from the CSM Committee Chair requesting review of the draft list were published in the September 1984 issue of *Shells and Sea Life*. The list of terrestrial gastropods was published in the October (1984) issue; the list of marine gastropods in the November (1984), December (1984), and January (1985) issues; and the lists of freshwater gastropods, scaphopods, cephalopods, aplacophorans, polyplacophorans, and marine and freshwater bivalves in the February (1985) issue. Reviewer comments were incorporated in a revised list of terrestrial mollusks and shared with readers in the February (1985) issue. Infrequent editorials by readers and publisher and further comments appeared in subsequent issues of that magazine.

By the 1985 CSM–AMU meeting, the molluscan list of scientific and vernacular names had been reviewed by over 100 experts. The list was formally presented to and accepted by CSM and AMU at their business meetings. The Committee, and by association the authors of the list, was voted by CSM to be retained as a standing committee for the next 5 years to allow this body of specialists to follow through with publication and revisions of the large project.

List of Molluscan Names

The goal of this list is to achieve uniformity and to avoid confusion in nomenclature of the mollusks of North America. Approximately 5,700 species are listed from 332 families in 31 orders (the two subclasses of Aplacophora are additional) within the six classes of mollusks. The Committee has endeavored to include all native and introduced species in the region of coverage.

Common names were assigned to most but not all species. Future editions of this list probably will increase the number of assigned common names. A common name was not assigned to a species if (1) the species is known only from a few specimens or from non-living shells, (2) the species is very difficult to identify, (3) the species belongs to a taxonomic group in need of revision, or (4) the species is represented by very small, rare, or poorly known forms. Species from the periphery of the study area yet may be found to have populations that extend into the study area; the Arctic molluscan fauna, for example, is relatively unknown. Continued systematic studies not only will add new species, but also will identify previously included species as synonyms and thus eliminate them from the list. The Committee hopes that the list will serve as a vehicle to identify taxonomic groups in need of systematic revision; foster more interest in studying molluscan populations in such poorly studied areas as Alaska, the Arctic, and offshore shelf waters; and perhaps even stimulate malacologists to compile and publish systematic and distributional data that they already have but have not yet published.

This list reflects a nearly complete survey of the literature as of July 1985. Errors and omissions inevitably have occurred, but future revisions will resolve such deficiencies.

Area of Coverage

The list is intended to include all species of mollusks, even though not all have common names assigned, from the United States and Canada that live in (1) fresh waters, (2) marine waters from shoreline habitats out to a depth of 200 m on the continental shelf, (3) estuaries, and (4) terrestrial habitats (e.g., gardens, woodlands, mountains, deserts, and caves). Mollusks from the Arctic Ocean, eastern Canada and eastern United States, and the northern Gulf of Mexico to the mouth of the Rio Grande River are included. Mollusks from Mexico and from offshore islands (e.g., Greenland, Iceland, Bermuda, the Bahamas, and the West Indies) are excluded unless they occur also in the region covered. The lack of workers and the absence of a good compilation of literature on the bivalves of the Atlantic and Gulf coasts may have resulted in the inclusion of taxa that only occur deeper than the 200-m depth selected for this list.

For the eastern Pacific region, published records of mollusks from waters along the western Canadian and U.S. shores and coastal islands, from the Aleutian Islands to the Mexico–United States boundary, are listed. Hawaii is excluded from the study area for two major reasons: (1) its fauna is of Indo-Pacific origin, and (2) the AFS endeavor excluded Hawaii from its study area. However, Hawaiian mollusks may be included in later editions.

Vernacular versus Common Names

The Committee, CSM, AMU, and AFS recognize the need to establish a body of common names that reflects broad current usage, to adopt appropriate names from the rich and colorful vernacular names that exist, and to develop descriptive names when desirable, based on AFS guidelines and principles. The Committee maintains that the list is best described as vernacular. By electing to adopt the style previously published by AFS for its list of fishes and designating *common names* for publication, the Committee expects that the selected vernacular names and newly designated names of this list will be widely used after publication and then will become "common."

Standardized and uniform vernacular names of mollusks are needed not only for commercial shellfish, but for amateur collections and the industrial trade, for aquarium specimens, for the import or export trade of mollusks or molluscan products, for legal documents, for governmental listings of threatened and endangered species, for shell crafts, and for popular or scientific writing. (Common names should be referenced *with* appropriate scientific names and not as substitutes for them.) Regional variations in common names and contradictory taxonomic listings directly affect governmental resource agencies by confounding the reporting of commercial and recreational catch statistics for invertebrates (B. G. Thompson, personal communication). Proper identification and recording of these species are important because of their high economic values, which generally exceed those of finfish. The existence of different names in local areas of a species' geographic range creates difficulties. Conversely, a given name may be employed in several places for diverse species. Listings by the Committee are not expected to change local usage quickly, and in some cases maybe never.

Plan of the List

The list of mollusks is presented in a natural or phyletic sequence of classes, orders, and families of mollusks, with the genera and species within each family arranged alphabetically. Two appendixes list extinct species or those thought to be extinct after having been cited in the study area, and species listed by the U.S. government as endangered or threatened. An index of common and scientific names concludes the book.

AFS Principles Governing the Selection of Common Names

The following principles were prepared by the CNAI and used by the CSM Committee and authors in selecting common names.

1. *A primary vernacular name shall be accepted for each species or taxonomic unit included.* Alternative published names may be listed in order of prominence. Rationale for selection of the primary name and etymologies may be indicated.

2. *No two species on a list shall have the same primary name.* Commonly used names of extralimital species should be avoided wherever possible.

3. *The expression "common" as part of an invertebrate's name shall be avoided if possible.* Use of adjectives that also describe age or size and thus may have dual meanings shall be avoided as part of an invertebrate's name wherever possible (e.g., little, small, big, fat).

4. *Simplicity in names is favored.* Hyphens and suffixes shall be omitted from names (e.g., coonstripe shrimp) except where they are orthographically essential (e.g., cup-and-saucer, brown-banded), or are necessary to avoid possible misunderstanding (e.g., half-slippershell). Compounded modifying words, including paired structures, usually should be treated as singular nouns in apposition with a group name (e.g., hawkwing conch, roostertail conch), but a plural modifier should usually be placed in adjectival form (e.g., fingered limpet, checkered pheasant) unless its plural nature is obvious (e.g., lineolate periwinkle). Preference shall be given to names that are short and euphonious.

The compounding of brief, familiar words into a single name, written without a hyphen, may in some cases promote clarity and simplicity (slippersnail, moonsnail, hairysnail, seaslug, dovesnail), but the habitual practice of combining words, especially those that are lengthy, awkward, or unfamiliar, is to be avoided. Spelling of compound names should follow rules set forth in the CBE Style Manual, 5th Edition, Council of Biological Editors, Bethesda, Maryland, USA.

5. *Common names shall not be capitalized in text use except for those elements that are proper names* (e.g., channeled whelk but San Diego doris).

6. *Names intended to honor persons* (e.g., Marshall mussel, Jay river snail) *are discouraged in that they are without descriptive value.* In some large groups, identical specific patronymics in scientific names (sometimes honoring different persons) exist in related genera, and use of a patronymic in the common name is confusing. However, some patryonymics are already well established in literature, agency regulations, and industry (e.g., Tanner crab). Apostrophes should be deleted.

7. *Only clearly defined and well-marked taxonomic entities* (usually species) *shall be assigned common names.* Most subspecies are not suitable subjects for common names, but those forms that are so different in appearance (not just in geographic distribution) as to be distinguished readily by laymen or for which a common name constitutes a significant aid in communication may merit separate names. There is a wide divergence of opinion concerning the criteria for recognition of subspecies. We have usually not named subspecies. Exceptions are those of certain invertebrates such as

the yellow-blossom *Epioblasma florentina florentina,* which is listed by the U.S. Department of the Interior as endangered. Subspecies have importance in evolutionary inquiry but are rarely of significance to laymen or in those aspects of biological endeavor in which common names are of concern. The common name for the species should apply to all subspecies of a taxon and may be appropriately modified by those treating subspecies. The practice of adding geographic modifiers to designate regional populations makes for a cumbersome terminology.

Hybrids in general are not named. The established common name of a hybrid, if important, is indicated by a footnote. Cultured varieties, phases, and morphological variants also are not named even though they are important in commercial trade of aquarium animals.

8. *The common name shall not be intimately tied to the scientific name.* Thus, the vagaries of scientific nomenclature do not entail constant changing of common names. The practice of applying a name to each genus, a modifying name for each species, and still another modifier for each subspecies, while appealing in its simplicity, has the defect of inflexibility. If an invertebrate is transferred from genus to genus, or shifted from species to subspecies or vice versa, the common name should nevertheless remain unaffected. It is not a primary function of common names to indicate relationship. When two or more taxonomic groups (e.g., nominal species) are found to be identical, one name shall be adopted for the combined group.

This principle is regarded not only as fundamental to the achievement of stability, but as essential to the development of a true vernacular nomenclature.

9. *Names shall not violate the tenets of good taste.*

The foregoing principles are largely in the name of procedural precepts. Those given below are criteria regarded as aids in the selection of suitable names.

10. *Colorful, romantic, fanciful, metaphorical, and otherwise distinctive and original names are especially appropriate.* Such terminology adds to the richness and breadth of nomenclature and yields a harvest of satisfaction to the user. Examples of such names include arrow crab, batwing seaslug, blackberry drupe, marsh walker, rugose nutmeg, and scotch bonnet.

11. *American Indian or other truly vernacular names are welcome for adoption as common names.* Indian names in current use include the cayuse, physa, geoduck, and quahog. In addition to aboriginal names, names of American invertebrates have been derived from foreign languages such as Spanish (abalone, aglaja). Although too little genuine originality is evident, excellent names have been developed by American immigrants. Most of these conform to principles 13 and 14 below.

12. *Commonly employed names adopted from traditional English usage* (e.g., crab, crayfish, limpet, mussel, periwinkle, prawn, shrimp) *are given considerable latitude in taxonomic placement.* Adherence to customary English practice is to be preferred if this does not conflict with the broad general usage of another name. Many English names, however, have been applied to similar-appearing but often distantly related invertebrates in America. We find shrimp in use for representatives of such diverse groups as the Cambaridae, Nephropidae, Palaemonidae, and Palinuridae. For widely known species, the Committee believes it preferable to recognize and adopt general use than to adopt bookish or pedantic substitutes. Thus, established practice should outweigh consistency with original English usage. This may not be well understood by some zoologists who may suggest strict adherence to the former usage.

13. *Structural attributes, color, and color pattern are desirable and are in common use in forming names.* Beaded, channeled, copper, fluted, giant, hairy, keeled, mottled, shouldered, splendid, and a multitude of other descriptors decorate invertebrate names. Efforts should be made to select terms that are descriptively accurate, and to hold repetition of those most frequently employed (e.g., white, black, spotted, banded) to a minimum.

Following tradition in American invertebrate zoology, we have attempted to restrict use of the terms line or stripe to longitudinal marks that parallel the body axis and the terms bar or band to vertical or transverse marks.

14. *Ecological characteristics are useful in making good names.* They, too, should be properly descriptive. Terms such as reef, pond, coral, sand, rock, riffle, freshwater, and mountain are well known in invertebrate names.

15. *Geographic distribution provides suitable adjectival modifiers.* Poorly descriptive or misleading geographic characterizations should be corrected unless they are too deeply entrenched in current usage. In the interest of brevity, it is usually possible to delete words such as lake, river, or ocean in the names of species (e.g., Ohio pebblesnail, not Ohio River pebblesnail).

16. *Generic names may be employed outright* (e.g., elimia, valvata) *or in modified form* (e.g., diplodon, from *Diplodonta;* nerite, from *Nerita*) *as common names.* Once adopted, such names should be maintained even if the generic name is changed. These vernaculars should be written in Roman and without capitalization. Brevity and euphony are of especial importance for names of this type.

17. *The duplication of common names of invertebrates and other organisms should be avoided if possible, but names in wide general use need not be rejected on this basis alone.* The name tulip is commonly applied to bulbous herbs of the genus *Tulipa* and also to certain gastropods. Similarly, olive is employed for the fruit of a tree, *Olea europaea,* and for various gastropods. On the basis of prevailing use, these names are admissible as invertebrate names.

Resolutions from the Committee on Scientific and Vernacular Names of Mollusks

In addition to adopting AFS principles, the CSM Committee prepared the following specific resolutions in 1983, which were unanimously accepted at the annual CSM, AMU, and AFS business meetings.

• Taxonomic classification should indicate class, order, family, and genus without author.

• Names of authors and dates of scientific names should be used only for the taxonomic levels of species and (when used) subspecies.

• The following descriptors should be used: F (freshwater), T (terrestrial), E (estuarine, used only for species restricted to estuaries), P (Pacific Ocean), A (Atlantic Ocean, Gulf of Mexico, etc.), Ac (Arctic Ocean), [I] (introduced and established), and [X] (extinct or thought to be extinct).

• "Linnaeus" will be used, not "Linne."

• Authors' initials will be used consistently if more than one author has the same surname (e.g., A. E. Verrill and A. H. Verrill).

• Prefixes of authors' names should be used consistently (e.g., von Ihering, da Costa, de Folin, d'Orbigny).

• The fauna of Hawaii is not included but may be considered later.

- Synonyms clearly established in the taxonomic literature should be omitted from the lists. Unpublished opinion of synonymy is not a valid criterion for exclusion of a scientific name from the list.

- Common names should not be designated for genera or higher taxa, except for aplacophorans for which common names for class and subclass currently are designated.

- Only species with documented occurrence within the study area will be listed.

- Species will not be included that are mentioned only in undocumented records (naked) or on check lists that lack diagnostic comment or illustration. Monographs should be used whenever possible, rather than less extensive, noncomparative literature notations.

- Any species described by an author which was listed in another author's publication has been listed here without the cumbersome "in" reference sometimes used (i.e., the original author and date is used).

- Species without published scientific names will not be included.

- Species of uncertain taxonomic status (*incertae sedis*) should not be given vernacular names.

- Species known or thought to have become extinct since their description should be designated.

- Introduced species should be designated and accompanied by appropriate comment.

- The most current literature on North American fauna should be used for systematic classification.

- Previously used vernacular names are not binding; prior publication has no priority.

- Vernacular names with the term "shell" should be changed to a whole-animal descriptor (e.g., "snail," "clam," etc.) to reflect that the shell is only a portion of the animal.

- Common names for groups may cross generic and familial lines (e.g., "crown conch" and "queen conch") and may also differ within a genus (e.g., "spike" and "lance" in the genus *Elliptio*).

- Justification should be presented when necessary to explain inclusion or deletion of a scientific or common name. (This will be a procedural requirement in future editions.)

Future of the Molluscan List

Subsequent to publication, proposed changes to the published molluscan list will be collected, the list will be revised where necessary, and the second edition will be published 5 years after the first. Future editions are planned for every 10 years after the second edition.

Readers of the molluscan list who wish to recommend changes in listed scientific or common names, reorganization of the phyletic arrangement, or additions or deletions of species should (1) clearly identify the desired change or changes, (2) briefly and specifically justify the change with reference to literature sources, and (3) send the request to the Chair, Committee on the Scientific and Vernacular Names of Mollusks, c/o American Fisheries Society, 5410 Grosvenor Lane, Suite 110, Bethesda, Maryland 20814, USA. All such requests will be considered by the Committee for the second edition.

Acknowledgments

This list has been prepared by many molluscan taxonomic specialists who are members of the CSM, the AMU, or both. Authors for the list of terrestrial gastropods include William L. Pratt, Arthur E. Bogan,

Kenneth C. Emberton, Jane E. Deisler, Dorothea Franzen, Fred G. Thompson, and Amy S. Van Devender. Fresh-water gastropods were listed by Fred G. Thompson and Charlotte M. Porter. The voluminous marine gastropod list was prepared by William G. Lyons and James F. Quinn, Jr. William G. Lyons compiled the list of Polyplacophora. The marine bivalve list was prepared by Eugene V. Coan, R. Tucker Abbott, and Donna D. Turgeon. The freshwater bivalve list was prepared by Arthur E. Bogan and David H. Stansbery. The scaphopod list was prepared by William K. Emerson, Margaret McFadien Carter, and John N. Kraeuter. The Aplacophora were compiled by Amelie H. Scheltema and the Cephalopoda by Clyde F. E. Roper.

To all reviewers of the draft lists in various stages of development, we are indeed grateful. Without the help of those who spent so much time reviewing these lists, we could not have such a complete product. Acknowledgment and thanks are due the following reviewers:

Steven A. Ahlstedt, M. James Allen, Herbert D. Athearn, Rae Baxter, Frank R. Bernard, Hans Bertsch, Constance E. Boone, Kathy G. Borror, Kenneth J. Boss, Branley Allan Branson, Twila Bratcher, Joseph C. Britton, Jack W. Brookshire, Alan C. Buchanan, John G. Burch, Sam M. Call, Lyle Campbell, Kerry B. Clark, Arthur H. Clarke, William J. Clench, George M. Davis, Ralph W. Dexter, Helen DuShane, Larry L. Eng, Glen J. Fallo, Antonio J. Ferreira, Thomas M. Freitag, Samuel L. H. Fuller, Mark E. Gordon, Terrence M. Gosliner, Peter L. Haaker, M. G. Harasewych, Harold W. Harry, Marian E. Havlik, David J. Heath, K. Elaine Hoagland, F. G. Hochberg, Michael A. Hoggarth, Ellet Hoke, Richard S. Houbrick, Leslie Hubricht, Billy G. Isom, John J. Jenkinson, Paul R. Jennewein, Sally D. Kaicher, William N. Kasson, Eugene P. Keferl, Bretton W. Kent, Alan J. Kohn, Frank Kokai, Silvard Kool, Louise R. Kraemer, Harry G. Lee, A. Byron Leonard, David R. Lindberg, Steven J. Long, Gerald L. Mackie, Virginia O. Maes, Glen K. Mahoney, Lawrence Master, Harold A. Mathiak, Gary R. McDonald, James H. McLean, Albert R. Mead, Arthur S. Merrill, Paula M. Mikkelsen, Paul S. Mikkelsen, Donald R. Moore, Joseph P. E. Morrison, Harold D. Murray, Raymond W. Neck, Richard J. Neves, David Nicol, Christopher Noe, James Nybakken, D. Ronald Oesch, Paul W. Parmalee, Juan J. Parodiz, Timothy A. Pearce, Richard E. Petit, Hugh J. Porter, Robert S. Prezant, Thomas E. Pulley, Warren F. Rathjen, Gary Rosenberg, Joseph Rosewater, Barry Roth, Walter Sage, John E. Schmidt, Guenter A. Schuster, Paul Scott, Ronald L. Shimek, James B. Sickel, Scott E. Siddall, Douglas G. Smith, Lynn B. Starnes, Carol B. Stein, Michael J. Sweeney, Ralph W. Taylor, Jane Topping, Ruth D. Turner, Henry van der Schalie, Emily H. Vokes, Gilbert L. Voss, Thomas R. Waller, Anders Waren, Thomas G. Watters, Susan M. Wells, Kirk E. Wright, Paul Yokley, Jr., and Michael A. Zeto.

Additionally, many people have commented verbally on the list, adding, subtracting, revising, and generally improving its quality. Countless others have encouraged the effort to establish a checklist of molluscan species and have lent assistance in innumerable ways when needed. Copies of a draft list were made available at the 1984 annual AMU meeting to all in attendance for review and comment. The majority of AMU commenters were anonymous, but we want each to know that his or her efforts were heeded and appreciated.

Steven J. Long, Editor of *Shells and Sea Life*, published the draft list for his subscribers' review in the September 1984 through February 1985 issues. Comments from his readers were useful and appreciated but, more importantly, an enlarged audience was offered the opportunity to review and comment upon the Committee's draft list.

The task of typing and reordering the draft lists on the word processor over the 4 years of this project was done by Gwendolyn Jones, Alma Taylor, Elizabeth Reed, and Elizabeth Haynes, Regulations Division of the U.S. National Marine Fisheries Service (NMFS). Mrs. Haynes did the bulk of the formatting of the lists on the word processor and acted as technical advisor for data processing.

George C. Steyskal of the Systematic Entomology Laboratory, U.S. Department of Agriculture, checked the list for compliance with the International Code of Zoological Nomenclature and with the rules of Latin and Greek grammar. With his assistance, the spellings of several family and species names have been researched and corrected from their traditional forms.

We are indebted to F. H. Roberts, President of Creative Design International, for furnishing the color photographs and designing the color insert for all of the 4-color molluscan subjects.

Mark Holliday, a member of the CNAI and senior staff for the Fishery Statistics Office, NMFS, developed the computer program and created the index for this volume. Mary Guinn, of his office, assisted him.

Travel funds for committee members to attend AFS work sessions have been provided in part by AFS through cooperative agreements with NMFS and the U.S. Fish and Wildlife Service. We thank our home institutions for subsidizing our efforts on this project, for secretarial help, and for word-processing and duplicating facilities. The NMFS, especially, has provided extensive institutional support to the efforts of the CNAI Chair.

The AFS CNAI membership has changed since this project began. Fred A. Mangum, Richard Neves, James R. Sedell, Scott E. Siddall, and Elizabeth L. Wenner, although no longer members, are acknowledged for their oversight reviews of this volume.

The CSM Committee and CNAI are deeply grateful to Shell Companies Foundation for its generous grant to fund publication of this volume.

DEDICATION

Dr. Robert F. Hutton

This first volume is dedicated to Dr. Robert Hutton, without whose perseverence and advocacy this series on scientific and common names of aquatic invertebrates would not have materialized. Dr. Hutton conceived this series before his tenure as the first chairman of CNAI. He has remained closely associated with the CSM Committee since its initiation in 1983 through the final publication stages of its manuscript on mollusks.

Bob Hutton is internationally well known as a scientist and fisheries manager. He earned his Ph.D. in marine biology at the University of London in 1954 while on a Fulbright Scholarship. His areas of research include estuarine ecology, fisheries, marine biology, and parasitology. Two trematodes, *Gigantobilharzia huttoni* and *Neostictodora huttoni*, were named in his honor.

His fisheries experience includes research, teaching, directing state and U.S. federal programs, and administration of AFS. In 1971, he joined the National Marine Fisheries Service as Associate Director for Resource Management and filled a variety of key positions until his retirement in June 1986. Over the years, he has managed or initiated most of the major programs of that agency: habitat conservation, national recreational fisheries, marine mammals and endangered species, fishery conservation and management, and state–federal coordination.

In 24 years of active membership in AFS, Bob has served in many capacities. He was appointed the first full-time Executive Secretary in 1966, elected President in 1976, honored as a life member in 1981, and awarded the first AFS Meritorious Award in 1986.

In tribute, Bob Hutton epitomizes the senior fishery professional—well educated with broad expertise and experience, highly respected and liked, always supportive of his many affiliated organizations. He willingly gives his time to anyone who asks.

NAMES OF MOLLUSKS

SCIENTIFIC NAME	OCCURRENCE[1]	COMMON NAME

CLASS APLACOPHORA—SOLENOGASTERS

SUBCLASS CHAETODERMOMORPHA—FOOTLESS SOLENOGASTERS

Chaetodermatidae

Chaetoderma argenteum Heath, 1911............	P	.......glisten-worm solenogaster[2]
Chaetoderma hancocki (Schwabl, 1963)..........	P	.......glisten-worm solenogaster
Chaetoderma marinelli (Schwabl, 1963)..........	P	.......glisten-worm solenogaster
Chaetoderma nanulum Heath, 1911	P	.......glisten-worm solenogaster
Chaetoderma nitidulum Lovén, 1844	A	.......glisten-worm solenogaster
Chaetoderma pacificum (Schwabl, 1963)	P	.......glisten-worm solenogaster
Chaetoderma productum Wiren, 1892	Ac	.......glisten-worm solenogaster
Chaetoderma scabrum Heath, 1911	P	.......glisten-worm solenogaster
Falcidens hartmanae (Schwabl, 1961)	P	..............tailed solenogaster

Prochaetodermatidae

Prochaetoderma californicum Schwabl, 1963......	P	.... jawed solenogaster

Limifossoridae

Limifossor talpoideus Heath, 1904	P	...... mole solenogaster
Scutopus megaradulatus Salvini-Plawen, 1972.....	A	........corkscrew solenogaster

SUBCLASS NEOMENIOMORPHA—FOOTED SOLENOGASTERS

Dondersiidae

Dondersia californica Heath, 1911	P	.......leaf-spiculed solenogaster

Meiomeniidae

Meiomenia swedmarki M. P. Morse, 1979	P	.... psammon solenogaster

Proneomeniidae

Proneomenia acuminata Heath, 1918............	A	.... elongate solenogaster

---

[1] A = Atlantic; P = Pacific; Ac = Arctic; F = freshwater; T = terrestrial; E = restricted estuarine; [I] = introduced and established; [X] = extinct or thought to be extinct.

[2] A single common name has been used for different species with nearly indistinguishable features. The selected common names generally describe the most distinctive feature of the preserved animal.

SCIENTIFIC NAME	OCCURRENCE	COMMON NAME

CLASS BIVALVIA—CLAMS OR BIVALVES

Order Nuculoida

Nuculidae

Acila castrensis (Hinds, 1843)	P	divaricate nutclam
Nucula aegeensis Jeffreys, 1879	A	Aegean nutclam
Nucula atacellana Schenck, 1939	A	cancellate nutclam
Nucula bellotii A. Adams, 1856	P, Ac	Belloti nutclam
Nucula calcicola Moore, 1977	A	reef nutclam
Nucula crenulata A. Adams, 1856	A	crenulate nutclam
Nucula delphinodonta Mighels and C. B. Adams, 1842	A	dolphintooth nutclam
Nucula exigua Sowerby, 1833	P	short nutclam
Nucula groenlandica Posselt, 1898	A	Greenland nutclam
Nucula linki Dall, 1916	P	Link nutclam
Nucula proxima Say, 1822	A	Atlantic nutclam
Nucula quirica Dall, 1916	P	spear nutclam
Nucula tenuis (Montagu, 1808)	A, P, Ac	smooth nutclam

Malletiidae

Neilonela acinula (Dall, 1890)	A	berry malletclam
Neilonela quadrangularis (Dall, 1881)	A	quadrangular malletclam

Nuculanidae

Adrana notabilis Rehder, 1939	A	notable nutclam
Adrana scaphoides Rehder, 1939	A	skiff nutclam
Adrana tellinoides (Sowerby, 1823)	A, P, Ac	tellinoid nutclam
Nuculana acuta (Conrad, 1832)	A	pointed nutclam
Nuculana amiata (Dall, 1916)	P	rostrate nutclam
Nuculana buccata (Möller, 1842)	A, P, Ac	stout nutclam
Nuculana callimene (Dall, 1908)	P	moon nutclam
Nuculana carpenteri (Dall, 1881)	A	Carpenter nutclam
Nuculana caudata (Donovan, 1801)	A	tailed nutclam
Nuculana cellulita (Dall, 1896)	P	fine-lined nutclam
Nuculana cestrota (Dall, 1889)	A	hammer nutclam
Nuculana concentrica (Say, 1824)	A	concentric nutclam
Nuculana fossa (Baird, 1863)	P	trenched nutclam
Nuculana hamata (Carpenter, 1864)	P	hooked nutclam
Nuculana jacksoni (Gould, 1841)	A	Jackson nutclam
Nuculana jamaicensis (d'Orbigny, 1842)	A	Jamaica nutclam
Nuculana messanensis (Seguenza, 1877)	A	Messanean nutclam
Nuculana minuta (Fabricius, 1776)	A, P, Ac	minute nutclam
Nuculana oxia (Dall, 1916)	P	Pacific nutclam
Nuculana parva (Sowerby, 1833)	Ac	dwarf nutclam
Nuculana penderi (Dall and Bartsch, 1910)	P	Pender nutclam
Nuculana pernula (Müller, 1779)	A, P, Ac	northern nutclam

SCIENTIFIC NAME	OCCURRENCE	COMMON NAME
Nuculana radiata (Krause, 1885)	P, Ac	rayed nutclam
Nuculana spargana (Dall, 1916)	P	San Diego nutclam
Nuculana subaequilatera (Jeffreys, 1879)	A	unequal nutclam
Nuculana taphria (Dall, 1896)	P	furrowed nutclam
Nuculana tenuisulcata (Couthouy, 1838)	A, P, Ac	thin nutclam
Nuculana verrilliana (Dall, 1886)	A	Verrill nutclam
Portlandia arctica (J. E. Gray, 1824)	A, P, Ac	Arctic nutclam
Portlandia dalli Krause, 1885	P	Dall nutclam
Yoldia amygdalea (Valenciennes, 1846)	A, P	almond yoldia
Yoldia cooperii Gabb, 1865	P	Cooper yoldia
Yoldia hyperborea Torell, 1859	A, P, Ac	northern yoldia
Yoldia limatula (Say, 1831)	A, P, Ac	file yoldia
Yoldia martyria Dall, 1897	P	witness yoldia
Yoldia myalis (Couthouy, 1838)	A, P, Ac	oval yoldia
Yoldia regularis A. E. Verrill, 1884	A	regular yoldia
Yoldia sapotilla (Gould, 1841)	A, Ac	short yoldia
Yoldia scissurata Dall, 1897	P, Ac	crisscrossed yoldia
Yoldia seminuda Dall, 1871	P	halfsmooth yoldia
Yoldia thraciaeformis (Storer, 1838)	A, P, Ac	broad yoldia
Yoldiella dissimilis (A. E. Verrill and Bush, 1898)	A	different yoldia
Yoldiella fraterna (A. E. Verrill and Bush, 1898)	A	sibling yoldia
Yoldiella frigida (Torell, 1859)	A, Ac	frigid yoldia
Yoldiella inconspicua (A. E. Verrill and Bush, 1898)	A	inconspicuous yoldia
Yoldiella inflata (A. E. Verrill and Bush, 1897)	A	inflated yoldia
Yoldiella intermedia (M. Sars, 1865)	A, P, Ac	intermediate yoldia
Yoldiella iris (A. E. Verrill and Bush, 1897)	A	rainbow yoldia
Yoldiella lenticula (Möller, 1842)	A, P, Ac	lenticulate yoldia
Yoldiella lucida (Lovén, 1846)	A	iridescent yoldia
Yoldiella sanesia (Dall, 1916)	P	sanesia yoldia
Yoldiella subangulata (A. E. Verrill and Bush, 1898)	A	subangular yoldia

ORDER SOLEMYOIDA

Solemyidae

Solemya borealis Totten, 1834	A	boreal awningclam
Solemya grandis A. E. Verrill and Bush, 1898	A	grand awningclam
Solemya occidentalis Deshayes, 1857	A	West Indian awningclam
Solemya reidi Bernard, 1980	P	gutless awningclam
Solemya valvulus Carpenter, 1864	P	Pacific awningclam
Solemya velum Say, 1822	A	Atlantic awningclam

SCIENTIFIC NAME	OCCURRENCE	COMMON NAME
ORDER MYTILOIDA		
Mytilidae		
Adula californiensis (Philippi, 1847)	P	California datemussel
Adula diegensis (Dall, 1911)	P	San Diego datemussel
Adula falcata (Gould, 1851)	P	curved datemussel
Amygdalum pallidulum (Dall, 1916)	P	pallid papermussel
Amygdalum papyrium (Conrad, 1846)	A	Atlantic papermussel
Amygdalum sagittatum Rehder, 1934	A	arrow papermussel
Botula fusca (Gmelin, 1791)	A, P	cinnamon mussel
Brachidontes adamsianus (Dunker, 1857)	P	Adams mussel
Brachidontes domingensis (Lamarck, 1819)	A	Santo Domingo mussel
Brachidontes exustus (Linnaeus, 1758)	A	scorched mussel
Brachidontes modiolus (Linnaeus, 1767)	A	yellow mussel
Crenella decussata (Montagu, 1808)	A, P	cross-sculpture crenella
Crenella divaricata (d'Orbigny, 1842)	A, P	spreading-sculpture crenella
Crenella faba (Müller, 1776)	A	bean crenella
Crenella fragilis A. E. Verrill, 1885	A	fragile crenella
Crenella glandula (Totten, 1834)	A	glandular crenella
Crenella leana Dall, 1897	P	Lea crenella
Crenella pectinula (Gould, 1841)	A	little-comb crenella
Crenella seminuda (Dall, 1897)	P	partly-sculptured crenella
Crenella skomma (Schwengel, 1944)	A	joking crenella
Dacrydium elegantulum Soot-Ryen, 1955	P	elegant mussel
Dacrydium vitreum (Möller, 1842)	A, P, Ac	glassy mussel
Geukensia demissa (Dillwyn, 1817)	A, P [I]	ribbed mussel
Gregariella chenui (Récluz, 1842)	A, P	Chenu mussel
Gregariella coralliophaga (Gmelin, 1791)	A	coral-eating mussel
Ischadium recurvum (Rafinesque, 1820)	A	hooked mussel
Lioberus castaneus (Say, 1822)	A	chestnut mussel
Lithophaga antillarum (d'Orbigny, 1842)	A	giant datemussel
Lithophaga aristata (Dillwyn, 1817)	A, P	scissor datemussel
Lithophaga bisuculata (d'Orbigny, 1842)	A	mahogany datemussel
Lithophaga nigra (d'Orbigny, 1842)	A	black datemussel
Lithophaga plumula kelseyi Hertlein and Strong, 1946	P	Kelsey datemussel
Lithophaga rogersi S. S. Berry, 1957	P	Rogers datemussel
Megacrenella columbiana (Dall, 1897)	P	British Columbia crenella
Modiolus americanus (Leach, 1815)	A	American horsemussel
Modiolus capax (Conrad, 1837)	P	fat horsemussel
Modiolus carpenteri Soot-Ryen, 1963	P	California horsemussel
Modiolus eiseni Strong and Hertlein, 1937	P	Eisen horsemussel
Modiolus flabellatus (Gould, 1850)	P	fan horsemussel
Modiolus kurilensis Bernard, 1983	P	Kurile horsemussel
Modiolus modiolus (Linnaeus, 1758)	A, P, Ac	northern horsemussel
Modiolus neglectus Soot-Ryen, 1955	P	neglected horsemussel
Modiolus rectus (Conrad, 1837)	P	straight horsemussel
Modiolus sacculifer (S. S. Berry, 1953)	P	bag horsemussel
Musculista senhousia (Benson, 1842)	P	Senhouse mussel
Musculus corrugatus (Stimpson, 1851)	A, P, Ac	corrugate mussel
Musculus cultellus (Deshayes, 1839)	A, P, Ac	knife mussel
Musculus discors (Linnaeus, 1767)	A, P, Ac	discordant mussel

SCIENTIFIC NAME	OCCURRENCE	COMMON NAME
Musculus lateralis (Say, 1822)	A	lateral mussel
Musculus marmoratus (Forbes, 1838)	Ac	spotted mussel
Musculus niger (J. E. Gray, 1824)	A, P, Ac	black mussel
Musculus pygmaeus Glynn, 1964	P	pygmy mussel
Musculus taylori (Dall, 1897)	P	Taylor mussel
Musculus vernicosus (Middendorff, 1849)	P	varnished mussel
Mytilus edulis Linnaeus, 1758	A, P, Ac	blue mussel
Mytilus californianus Conrad, 1837	P	California mussel
Septifer bifurcatus (Conrad, 1837)	P	bifurcate mussel

Order Arcoida

Arcidae

Anadara baughmani Hertlein, 1951	A	Baughman ark
Anadara brasiliana (Lamarck, 1819)	A	incongruous ark
Anadara chemnitzii (Philippi, 1851)	A	Chemnitz ark
Anadara floridana (Conrad, 1869)	A	cut-ribbed ark
Anadara multicostata (Sowerby, 1833)	P	many-ribbed ark
Anadara notabilis (Röding, 1798)	A	eared ark
Anadara ovalis (Bruguière, 1789)	A	blood ark
Anadara transversa (Say, 1822)	A	transverse ark
Arca imbricata Bruguière, 1789	A	mossy ark
Arca zebra (Swainson, 1833)	A	turkey wing
Arcopsis adamsi (Dall, 1886)	A	Adams ark
Barbatia bailyi (Bartsch, 1931)	P	Baily miniature ark
Barbatia cancellaria (Lamarck, 1819)	A	red-brown ark
Barbatia candida (Helbling, 1779)	A	white-beard ark
Barbatia domingensis (Lamarck, 1819)	A	white miniature ark
Barbatia tenera (C. B. Adams, 1845)	A	Doc Bales ark
Bathyarca anomala (A. E. Verrill and Bush, 1898)	A	anomalous bathyark
Bathyarca frielei (Friele, 1879)	A, Ac	Friele bathyark
Bathyarca glacialis (J. E. Gray, 1824)	A, Ac	glacial bathyark
Bathyarca glomerula (Dall, 1881)	A	little-ball bathyark
Bathyarca pectunculoides (Scacchi, 1833)	A	comb bathyark
Noetia ponderosa (Say, 1822)	A	ponderous ark

Limopsidae

Limopsis akutanica Dall, 1916	P	Akutan limops
Limopsis antillensis Dall, 1881	A	Antillean limops
Limopsis aurita (Brocchi, 1814)	A	eared limops
Limopsis cristata Jeffreys, 1876	A	crested limops
Limopsis diegensis Dall, 1908	P	San Diego limops
Limopsis minuta Philippi, 1836	A	minute limops
Limopsis sulcata A. E. Verrill and Bush, 1898	A	sulcate limops

Glycymerididae

Glycymeris americana (DeFrance, 1829)	A	giant bittersweet
Glycymeris corteziana Dall, 1916	P	Cortez bittersweet
Glycymeris decussata (Linnaeus, 1758)	A	decussate bittersweet

SCIENTIFIC NAME	OCCURRENCE	COMMON NAME
Glycymeris pectinata (Gmelin, 1791)	A	comb bittersweet
Glycymeris septentrionalis (Middendorff, 1849)	A	northern bittersweet
Glycymeris spectralis Nicol, 1952	A	spectral bittersweet
Glycymeris subobsoleta (Carpenter, 1864)	P	California bittersweet
Glycymeris subtilis (Nicol, 1956)	A	Bermuda bittersweet
Glycymeris undata (Linnaeus, 1758)	A	wavy bittersweet

Manzanellidae

Huxleyia munita (Dall, 1898)	P	minute nucinellid

Philobryidae

Cosa caribaea Abbott, 1958	A	Caribbean philobrya
Philobrya setosa (Carpenter, 1864)	P	hairy philobrya

ORDER PTERIOIDA

Pinnidae

Atrina oldroydii Dall, 1901	P	Oldroyd penshell
Atrina rigida (Lightfoot, 1786)	A	stiff penshell
Atrina seminuda (Lamarck, 1819)	A	half-naked penshell
Atrina serrata (Sowerby, 1825)	A	sawtooth penshell
Pinna carnea Gmelin, 1791	A	amber penshell
Pinna rudis Linnaeus, 1758	A	rough penshell

Pteriidae

Pinctada imbricata Röding, 1798	A	Atlantic pearl-oyster
Pteria colymbus (Röding, 1798)	A	Atlantic wing-oyster
Pteria longisquamosa (Dunker, 1852)	A	scaly wing-oyster
Pteria vitrea (Reeve, 1857)	A	glassy wing-oyster

Isognomonidae

Isognomon alatus (Gmelin, 1791)	A	flat tree-oyster
Isognomon bicolor (C. B. Adams, 1845)	A	bicolor purse-oyster
Isognomon janus Carpenter, 1857	P	thin purse-oyster
Isognomon radiatus (Anton, 1839)	A	Lister purse-oyster

Malleidae

Malleus candeanus (d'Orbigny, 1842)	A	Caribbean hammer-oyster

ORDER LIMOIDA

Limidae

Lima excavata (Fabricius, 1779)	A	excavated fileclam
Lima floridana Olsson and Harbison, 1953	A	smooth fileclam
Lima lima (Linnaeus, 1758)	A	spiny fileclam

SCIENTIFIC NAME	OCCURRENCE	COMMON NAME
Lima locklini McGinty, 1955	A	Locklin fileclam
Lima pellucida C. B. Adams, 1846	A	Antillean fileclam
Lima scabra (Born, 1778)	A	rough fileclam
Lima subovata Jeffreys, 1876	A	subovate fileclam
Limaria hemphilli (Hertlein and Strong, 1946)	P	Hemphill fileclam
Limatula attenuata Dall, 1916	P	attenuate fileclam
Limatula confusa (E. A. Smith, 1885)	A	confusing fileclam
Limatula hendersoni Olsson and McGinty, 1958	A	Henderson fileclam
Limatula hyalina A. E. Verrill and Bush, 1898	A	hyaline fileclam
Limatula hyperborea A. S. Jensen, 1909	A, Ac	boreal fileclam
Limatula regularis A. E. Verrill and Bush, 1898	A	regular fileclam
Limatula saturna Bernard, 1978	P	saturnine fileclam
Limatula setifera Dall, 1886	A	bristly fileclam
Limatula subauriculata (Montagu, 1808)	A, P	small-ear fileclam
Limatula vancouverensis Bernard, 1978	P	Vancouver fileclam
Limea bronniana Dall, 1886	A	Bronn fileclam

ORDER OSTREOIDA

Entoliidae

Pectinella sigsbeei (Dall, 1886)	A	Sigsbee scallop

Pectinidae

Aequipecten acanthodes (Dall, 1925)	A	thistle scallop
Aequipecten glyptus (A. E. Verrill, 1882)	A	red-ribbed scallop
Aequipecten muscosus (W. Wood, 1828)	A	rough scallop
Amusium papyraceum (Gabb, 1873)	A	paper scallop
Arctinula greenlandica (Sowerby, 1842)	Ac	Greenland scallop
Argopecten "*circularis* (Sowerby, 1835)," auct.	P	Pacific calico scallop
Argopecten gibbus (Linnaeus, 1758)	A	Atlantic calico scallop
Argopecten irradians (Lamarck, 1819)	A	bay scallop
Argopecten lineolaris (Lamarck, 1819)	A	lined scallop
Argopecten nucleus (Born, 1778)	A	nucleus scallop
Bractechlamys antillarum (Récluz, 1853)	A	Antillean scallop
Chlamys albida (Arnold, 1906)	P	white scallop
Chlamys behringiana (Middendorff, 1849)	P	Bering scallop
Chlamys benedicti (A. E. Verrill and Bush, 1897)	A	Benedict scallop
Chlamys hastata (Sowerby, 1842)	P	spiny scallop
Chlamys imbricata (Gmelin, 1791)	A	knobby scallop
Chlamys islandica (Müller, 1776)	A, P, Ac	Iceland scallop
Chlamys jordani Arnold, 1903	P	Jordan scallop
Chlamys lowei (Hertlein, 1935)	P	Lowe scallop
Chlamys mildredae (Bayer, 1941)	A	Mildred scallop
Chlamys multisquamata (Dunker, 1864)	A	fine-ribbed scallop
Chlamys ornata (Lamarck, 1819)	A	ornate scallop
Chlamys rubida (Hinds, 1845)	P	reddish scallop
Chlamys sentis (Reeve, 1853)	A	scaly scallop
Crassedoma giganteum (J. E. Gray, 1825)	P	giant rock scallop
Cryptopecten phrygium (Dall, 1886)	A	spathate scallop

SCIENTIFIC NAME	OCCURRENCE	COMMON NAME
Delectopecten randolphi (Dall, 1897)	P	Randolph scallop
Delectopecten vancouverensis (Whiteaves, 1893)	P	Vancouver scallop
Delectopecten vitreus (Gmelin, 1791)	A, P	vitreous scallop
Hyalopecten undatus (A. E. Verrill and S. Smith, 1885)	A	wavy scallop
Leptopecten latiauratus (Conrad, 1837)	P	kelp scallop
Nodipecten nodosus (Linnaeus, 1758)	A	lions-paw scallop
Palliolum striatum (Müller, 1776)	A	striate scallop
Patinopecten caurinus (Gould, 1850)	P	weathervane scallop
Pecten chazaliei Dautzenberg, 1900	A	dwarf zigzag scallop
Pecten diegensis Dall, 1898	P	San Diego scallop
Pecten raveneli Dall, 1898	A	Ravenel scallop
Pecten ziczac (Linnaeus, 1758)	A	zigzag scallop
Placopecten magellanicus (Gmelin, 1791)	A	sea scallop

Propeamussiidae

Cyclopecten alaskensis (Dall, 1871)	P	Alaska glass-scallop
Cyclopecten barbarensis Grau, 1954	P	Santa Barbara glass-scallop
Cyclopecten catalinensis (Willett, 1931)	P	Catalina glass-scallop
Cyclopecten imbrifer (Lovén, 1847)	A, Ac	shingle glass-scallop
Cyclopecten leptaleus (A. E. Verrill, 1884)	A	delicate glass-scallop
Cyclopecten nanus A. E. Verrill and Bush, 1897	A	dwarf glass-scallop
Cyclopecten reticulus (Dall, 1886)	A	netted glass-scallop
Cyclopecten ringnesius (Dall, 1924)	Ac	Ringnes glass-scallop
Cyclopecten thalassinus (Dall, 1886)	A	costate glass-scallop
Propeamussium cancellatum (E. A. Smith, 1885)	A	cancellate glass-scallop
Propeamussium holmesii (Dall, 1886)	A	Holmes glass-scallop
Propeamussium pourtalesianum (Dall, 1886)	A	Pourtales glass-scallop

Plicatulidae

Plicatula gibbosa Lamarck, 1801	A	Atlantic kittenpaw

Spondylidae

Spondylus americanus Hermann, 1781	A	Atlantic thorny-oyster
Spondylus ictericus Reeve, 1856	A	digitate thorny-oyster

Dimyidae

Basiliomya goreaui Bayer, 1971	A	Goreau dimyid
Dimya argentea Dall, 1886	A	silver dimyid
Dimya californiana S. S. Berry, 1937	P	California dimyid
Dimya coralliotis S. S. Berry, 1944	P	coral dimyid
Dimyella starcki Moore, 1969	A	Starck dimyid

Anomiidae

Anomia peruviana d'Orbigny, 1846	P	Peruvian jingle
Anomia simplex d'Orbigny, 1842	A	common jingle
Anomia squamula Linnaeus, 1758	A	prickly jingle
Pododesmus cepio (J. E. Gray, 1850)	P	Pacific falsejingle
Pododesmus macroschisma (Deshayes, 1839)	P	Alaska falsejingle
Pododesmus pernoides (J. E. Gray, 1853)	P	pedestal falsejingle

SCIENTIFIC NAME	OCCURRENCE	COMMON NAME
Pododesmus rudis (Broderip, 1834)	A	Atlantic falsejingle

Ostreidae

Crassostrea gigas (Thunberg, 1793)	P [I]	Pacific oyster
Crassostrea virginica (Gmelin, 1791)	A, P [I]	eastern oyster
Cryptostrea permollis (Sowerby, 1871)	A	sponge oyster
Dendostrea frons (Linnaeus, 1758)	A	frond oyster
Ostrea edulis Linnaeus, 1750	A [I]	edible oyster
Ostreola conchaphila (Carpenter, 1857)	P	Olympia oyster
Ostreola equestris (Say, 1834)	A	crested oyster
Teskeyostrea weberi (Olsson, 1951)	A	Weber oyster

Gryphaeidae

Neopycnodonte cochlear (Poli, 1795)	A	deepsea oyster
Parahyotissa macgintyi Harry, 1985	A	McGinty oyster

ORDER UNIONOIDA

Unionidae

Actinonaias ligamentina (Lamarck, 1819)	F	mucket
Actinonaias pectorosa (Conrad, 1834)	F	pheasantshell
Alasmidonta arcula (I. Lea, 1838)	F	Altamaha arc-mussel
Alasmidonta atropurpurea (Rafinesque, 1831)	F	Cumberland elktoe
Alasmidonta heterodon (I. Lea, 1829)	F	dwarf wedgemussel
Alasmidonta maccordi Athearn, 1964	F [X]	Coosa elktoe
Alasmidonta marginata Say, 1818	F	elktoe
Alasmidonta raveneliana (I. Lea, 1834)	F	Appalachian elktoe
Alasmidonta robusta Clarke, 1981	F	Carolina elktoe
Alasmidonta undulata (Say, 1817)	F	triangle floater
Alasmidonta varicosa (Lamarck, 1819)	F	brook floater
Alasmidonta viridis (Rafinesque, 1820)	F	slippershell mussel
Alasmidonta wrightiana (Walker, 1901)	F [X]	Ochlocknee arc-mussel
Amblema neislerii (I. Lea, 1858)	F	fat threeridge
Amblema plicata perplicata (Conrad, 1841)	F	roundlake
Amblema plicata plicata (Say, 1817)	F	threeridge
Anodonta beringiana Middendorff, 1851	F	Yukon floater
Anodonta californiensis I. Lea, 1852	F	California floater
Anodonta cataracta cataracta Say, 1817	F	eastern floater
Anodonta cataracta fragilis Lamarck, 1819	F	Newfoundland floater
Anodonta cataracta marginata Say, 1817	F	Gaspé floater
Anodonta couperiana I. Lea, 1840	F	barrel floater
Anodonta gibbosa Say, 1824	F	inflated floater
Anodonta grandis Say, 1829	F	giant floater
Anodonta imbecillis Say, 1829	F	paper pondshell
Anodonta implicata Say, 1829	F	alewife floater
Anodonta kennerlyi I. Lea, 1860	F	western floater
Anodonta nuttaliana I. Lea, 1838	F	winged floater
Anodonta oregonensis I. Lea, 1838	F	Oregon floater
Anodonta peggyae R. I. Johnson, 1965	F	Florida floater
Anodonta suborbiculata Say, 1831	F	flat floater
Anodontoides ferussacianus (I. Lea, 1834)	F	cylindrical papershell
Anodontoides radiatus (Conrad, 1834)	F	rayed creekshell

SCIENTIFIC NAME	OCCURRENCE	COMMON NAME
Arcidens confragosus (Say, 1829)	F	rock-pocketbook
Arkansia wheeleri Ortmann and Walker, 1912	F	Ouchita rock-pocketbook
Cumberlandia monodonta (Say, 1829)	F	spectaclecase
Cyclonaias tuberculata (Rafinesque, 1820)	F	purple wartyback
Cyprogenia aberti (Conrad, 1850)	F	western fanshell
Cyprogenia stegaria (Rafinesque, 1820)	F	fanshell
Cyrtonaias tampicoensis (I. Lea, 1838)	F	Tampico pearlymussel
Disconaias salinasensis (Simpson, 1908)	F	Salina mucket
Dromus dromas (I. Lea, 1834)	F	dromedary pearlymussel
Ellipsaria lineolata (Rafinesque, 1820)	F	butterfly
Elliptio ahenea (I. Lea, 1843)	F	southern lance
Elliptio angustata (I. Lea, 1831)	F	Carolina lance
Elliptio arca (Conrad, 1834)	F	Alabama spike
Elliptio arctata (Conrad, 1834)	F	delicate spike
Elliptio buckleyi (I. Lea, 1843)	F	Florida shiny spike
Elliptio chipolaensis Walker, 1905	F	Chipola slabshell
Elliptio cistelliformis (I. Lea, 1863)	F	box spike
Elliptio complanata (Lightfoot, 1786)	F	eastern elliptio
Elliptio congaraea (I. Lea, 1831)	F	Carolina slabshell
Elliptio crassidens (Lamarck, 1819)	F	elephant-ear
Elliptio dariensis (I. Lea, 1842)	F	Georgia elephant-ear
Elliptio dilatata (Rafinesque, 1820)	F	spike
Elliptio downiei (I. Lea, 1859)	F	Satilla elephant-ear
Elliptio fisheriana (I. Lea, 1838)	F	northern lance
Elliptio folliculata (I. Lea, 1838)	F	pod lance
Elliptio fraterna (I. Lea, 1852)	F	brother spike
Elliptio hopetonensis (I. Lea, 1838)	F	Altamaha slabshell
Elliptio icterina (Conrad, 1834)	F	variable spike
Elliptio jayensis (I. Lea, 1838)	F	flat spike
Elliptio lanceolata (I. Lea, 1828)	F	yellow lance
Elliptio macmichaeli Clench and Turner, 1956	F	fluted elephant-ear
Elliptio marsupiobesa Fuller, 1972	F	Cape Fear spike
Elliptio nigella (I. Lea, 1852)	F	winged spike
Elliptio producta (Conrad, 1836)	F	Atlantic spike
Elliptio raveneli (Conrad, 1834)	F	Carolina spike
Elliptio roanokensis (I. Lea, 1836)	F	Roanoke slabshell
Elliptio shepardiana (I. Lea, 1834)	F	Altamaha lance
Elliptio spinosa (I. Lea, 1836)	F	Altamaha spinymussel
Elliptio steinstansana R. I. Johnson and Clarke, 1983	F	Tar spinymussel
Elliptio waccamawensis (I. Lea, 1863)	F	Waccamaw spike
Elliptio waltoni (Wright, 1888)	F	Florida lance
Elliptoideus sloatianus (I. Lea, 1840)	F	purple bankclimber
Epioblasma arcaeformis (I. Lea, 1831)	F [X]	sugarspoon
Epioblasma biemarginata (I. Lea, 1857)	F [X]	angled riffleshell
Epioblasma brevidens (I. Lea, 1831)	F	cumberlandian combshell
Epioblasma capsaeformis (I. Lea, 1834)	F	oyster mussel
Epioblasma flexuosa (Rafinesque, 1820)	F [X]	leafshell
Epioblasma florentina curtisi (Utterback, 1916)	F	Curtis pearlymussel
Epioblasma florentina florentina (I. Lea, 1857)	F	yellow blossom
Epioblasma florentina walkeri (Wilson and H. W. Clark, 1914)	F	tan riffleshell

SCIENTIFIC NAME	OCCURRENCE	COMMON NAME
Epioblasma haysiana (I. Lea, 1833)	F [X]	acornshell
Epioblasma lenior (I. Lea, 1843)	F [X]	narrow catspaw
Epioblasma lewisii (Walker, 1910)	F [X]	forkshell
Epioblasma metastriata (Conrad, 1840)	F	upland combshell
Epioblasma obliquata obliquata (Rafinesque, 1820)	F	catspaw
Epioblasma obliquata perobliqua (Conrad, 1836)	F	white catspaw
Epioblasma othcaloogensis (I. Lea, 1857)	F	southern acornshell
Epioblasma penita (Conrad, 1834)	F	southern combshell
Epioblasma personata (Say, 1829)	F [X]	round combshell
Epioblasma propinqua (I. Lea, 1857)	F [X]	Tennessee riffleshell
Epioblasma sampsonii (I. Lea, 1861)	F [X]	Wabash riffleshell
Epioblasma stewardsoni (I. Lea, 1852)	F [X]	Cumberland leafshell
Epioblasma torulosa gubernaculum (Reeve, 1865)	F	green blossom
Epioblasma torulosa rangiana (I. Lea, 1839)	F	northern riffleshell
Epioblasma torulosa torulosa (Rafinesque, 1820)	F	tubercled blossom
Epioblasma triquetra (Rafinesque, 1820)	F	snuffbox
Epioblasma turgidula (I. Lea, 1858)	F	turgid blossom
Fusconaia askewi (Marsh, 1896)	F	Texas pigtoe
Fusconaia barnesiana (I. Lea, 1838)	F	Tennessee pigtoe
Fusconaia cerina (Conrad, 1838)	F	Gulf pigtoe
Fusconaia cor (Conrad, 1834)	F	shiny pigtoe
Fusconaia cuneolus (I. Lea, 1840)	F	fine-rayed pigtoe
Fusconaia ebena (I. Lea, 1831)	F	ebonyshell
Fusconaia escambia Clench and Turner, 1956	F	narrow pigtoe
Fusconaia flava (Rafinesque, 1820)	F	Wabash pigtoe
Fusconaia lananensis (Frierson, 1900)	F	triangle pigtoe
Fusconaia masoni (Conrad, 1834)	F	Atlantic pigtoe
Fusconaia ozarkensis (Call, 1887)	F	Ozark pigtoe
Fusconaia subrotunda (I. Lea, 1831)	F	long-solid
Fusconaia succissa (I. Lea, 1852)	F	purple pigtoe
Glebula rotundata (Lamarck, 1819)	F	round pearlshell
Gonidea angulata (I. Lea, 1838)	F	western ridgemussel
Hemistena lata (Rafinesque, 1820)	F	cracking pearlymussel
Lampsilis abrupta (Say, 1831)	F	pink mucket
Lampsilis altilis (Conrad, 1834)	F	fine-lined pocketbook
Lampsilis australis Simpson, 1900	F	southern sandshell
Lampsilis binominata Simpson, 1900	F	lined pocketbook
Lampsilis bracteata (Gould, 1855)	F	Texas fatmucket
Lampsilis cardium (Rafinesque, 1820)	F	plain pocketbook
Lampsilis cariosa (Say, 1817)	F	yellow lampmussel
Lampsilis dolabraeformis (I. Lea, 1838)	F	Altamaha pocketbook
Lampsilis fasciola Rafinesque, 1820	F	wavy-rayed lampmussel
Lampsilis fullerkati R. I. Johnson, 1984	F	Waccamaw fatmucket
Lampsilis haddletoni Athearn, 1964	F	Haddleton lampmussel
Lampsilis higginsi (I. Lea, 1857)	F	Higgins eye
Lampsilis hydiana (I. Lea, 1838)	F	Louisiana fatmucket
Lampsilis ornata (Conrad, 1835)	F	southern pocketbook
Lampsilis ovata (Say, 1817)	F	pocketbook
Lampsilis perovalis (Conrad, 1834)	F	orange-nacre mucket
Lampsilis powelli (I. Lea, 1852)	F	Arkansas fatmucket
Lampsilis radiata conspicua (I. Lea, 1872)	F	Carolina fatmucket
Lampsilis radiata radiata (Gmelin, 1791)	F	eastern lampmussel
Lampsilis rafinesqueana Frierson, 1927	F	Neosho mucket

SCIENTIFIC NAME	OCCURRENCE	COMMON NAME
Lampsilis reeviana brevicula (Call, 1887)	F	ozark broken-ray
Lampsilis reeviana brittsi Simpson, 1900	F	northern broken-ray
Lampsilis reeviana reeviana (I. Lea, 1852)	F	Arkansas broken-ray
Lampsilis satura (I. Lea, 1852)	F	sandbank pocketbook
Lampsilis siliquoidea (Barnes, 1823)	F	fatmucket
Lampsilis splendida (I. Lea, 1838)	F	rayed pink fatmucket
Lampsilis straminea claibornensis (I. Lea, 1838)	F	southern fatmucket
Lampsilis straminea straminea (Conrad, 1834)	F	rough fatmucket
Lampsilis streckeri Frierson, 1927	F	speckled pocketbook
Lampsilis subangulata (I. Lea, 1840)	F	shiny-rayed pocketbook
Lampsilis teres (Rafinesque, 1820)	F	yellow sandshell
Lampsilis virescens (I. Lea, 1858)	F	Alabama lampmussel
Lasmigona complanata alabamensis Clarke, 1985	F	Alabama heelsplitter
Lasmigona complanata complanata (Barnes, 1823)	F	white heelsplitter
Lasmigona compressa (I. Lea, 1829)	F	creek heelsplitter
Lasmigona costata (Rafinesque, 1820)	F	fluted-shell
Lasmigona decorata (I. Lea, 1852)	F	Carolina heelsplitter
Lasmigona holstonia (I. Lea, 1838)	F	Tennessee heelsplitter
Lasmigona subviridis (Conrad, 1835)	F	green floater
Lemiox rimosus (Rafinesque, 1820)	F	birdwing pearlymussel
Leptodea fragilis (Rafinesque, 1820)	F	fragile papershell
Leptodea leptodon (Rafinesque, 1820)	F	scaleshell
Leptodea ochracea (Say, 1817)	F	tidewater mucket
Lexingtonia dolabelloides (I. Lea, 1840)	F	slabside pearlymussel
Lexingtonia subplana (Conrad, 1837)	F	Virginia pigtoe
Ligumia nasuta (Say, 1817)	F	eastern pondmussel
Ligumia recta (Lamarck, 1819)	F	black sandshell
Ligumia subrostrata (Say, 1831)	F	pondmussel
Margaritifera falcata (Gould, 1850)	F	western pearlshell
Margaritifera hembeli (Conrad, 1838)	F	Louisiana pearlshell
Margaritifera margaritifera (Linnaeus, 1758)	F	eastern pearlshell
Margaritifera marrianae R. I. Johnson, 1983	F	Alabama pearlshell
Medionidus acutissimus (I. Lea, 1831)	F	Alabama moccasinshell
Medionidus conradicus (I. Lea, 1834)	F	Cumberland moccasinshell
Medionidus macglameriae van der Schalie, 1939	F [X]	Tombigbee moccasinshell
Medionidus parvulus (I. Lea, 1860)	F	Coosa moccasinshell
Medionidus penicillatus (I. Lea, 1857)	F	gulf moccasinshell
Medionidus simpsonianus Walker, 1905	F	Ochlocknee moccasinshell
Medionidus walkeri (Wright, 1897)	F	Suwannee moccasinshell
Megalonaias boykiniana (I. Lea, 1840)	F	round washboard
Megalonaias nervosa (Rafinesque, 1820)	F	washboard
Obliquaria reflexa Rafinesque, 1820	F	threehorn wartyback
Obovaria jacksoniana (Frierson, 1912)	F	southern hickorynut
Obovaria olivaria (Rafinesque, 1820)	F	hickorynut
Obovaria retusa (Lamarck, 1819)	F	ring pink
Obovaria rotulata (Wright, 1899)	F	round ebonyshell
Obovaria subrotunda (Rafinesque, 1820)	F	round hickorynut
Obovaria unicolor (I. Lea, 1845)	F	Alabama hickorynut
Pegias fabula (I. Lea, 1838)	F	little-wing pearlymussel
Plectomerus dombeyanus (Valenciennes, 1827)	F	bankclimber
Plethobasus cicatricosus (Say, 1829)	F	white wartyback
Plethobasus cooperianus (I. Lea, 1834)	F	orange-foot pimpleback
Plethobasus cyphyus (Rafinesque, 1820)	F	sheepnose

SCIENTIFIC NAME	OCCURRENCE	COMMON NAME
Pleurobema altum (Conrad, 1854)	F	highnut
Pleurobema avellanum Simpson, 1900	F	hazel pigtoe
Pleurobema beadleanum (I. Lea, 1861)	F	Mississippi pigtoe
Pleurobema bournianum (I. Lea, 1840)	F	Scioto pigtoe
Pleurobema chattanoogaense (I. Lea, 1858)	F	painted clubshell
Pleurobema clava (Lamarck, 1819)	F	clubshell
Pleurobema coccineum (Conrad, 1834)	F	round pigtoe
Pleurobema collina (Conrad, 1837)	F	James spinymussel
Pleurobema cordatum (Rafinesque, 1820)	F	Ohio pigtoe
Pleurobema curtum (I. Lea, 1859)	F	black clubshell
Pleurobema decisum (I. Lea, 1831)	F	southern clubshell
Pleurobema flavidulum (I. Lea, 1831)	F	yellow pigtoe
Pleurobema furvum (Conrad, 1834)	F	dark pigtoe
Pleurobema georgianum (I. Lea, 1841)	F	southern pigtoe
Pleurobema gibberum (I. Lea, 1838)	F	Cumberland pigtoe
Pleurobema hanleyanum (I. Lea, 1852)	F	Georgia pigtoe
Pleurobema johannis (I. Lea, 1859)	F	Alabama pigtoe
Pleurobema marshalli Frierson, 1927	F	flat pigtoe
Pleurobema murrayense (I. Lea, 1868)	F	Coosa pigtoe
Pleurobema nucleopsis (Conrad, 1849)	F	longnut
Pleurobema oviforme (Conrad, 1834)	F	Tennessee clubshell
Pleurobema perovatum (Conrad, 1834)	F	ovate clubshell
Pleurobema plenum (I. Lea, 1840)	F	rough pigtoe
Pleurobema pyramidatum (I. Lea, 1840)	F	pyramid pigtoe
Pleurobema pyriforme (I. Lea, 1857)	F	oval pigtoe
Pleurobema riddelli (I. Lea, 1861)	F	Louisiana pigtoe
Pleurobema rubellum (Conrad, 1834)	F	Warrior pigtoe
Pleurobema strodeanum (Wright, 1898)	F	fuzzy pigtoe
Pleurobema taitianum (I. Lea, 1834)	F	heavy pigtoe
Pleurobema troschelianum (I. Lea, 1852)	F	Alabama clubshell
Pleurobema verum (I. Lea, 1860)	F	true pigtoe
Popenaias popei (I. Lea, 1857)	F	Texas hornshell
Potamilus alatus (Say, 1817)	F	pink heelsplitter
Potamilus amphichaenus (Frierson, 1898)	F	Texas heelsplitter
Potamilus capax (Green, 1832)	F	fat pocketbook
Potamilus inflatus (I. Lea, 1831)	F	inflated heelsplitter
Potamilus ohiensis (Rafinesque, 1820)	F	pink papershell
Potamilus purpuratus (Lamarck, 1819)	F	bleufer
Ptychobranchus fasciolaris (Rafinesque, 1820)	F	kidneyshell
Ptychobranchus greeni (Conrad, 1834)	F	triangular kidneyshell
Ptychobranchus jonesi (van der Schalie, 1934)	F	southern kidneyshell
Ptychobranchus occidentalis (Conrad, 1836)	F	Ouachita kidneyshell
Ptychobranchus subtentum (Say, 1825)	F	fluted kidneyshell
Quadrula apiculata (Say, 1829)	F	southern mapleleaf
Quadrula asperata (I. Lea, 1861)	F	Alabama orb
Quadrula aurea (I. Lea, 1859)	F	golden orb
Quadrula couchiana (Lea, 1860)	F	Rio Grande monkeyface
Quadrula cylindrica cylindrica (Say, 1817)	F	rabbitsfoot
Quadrula cylindrica strigillata (Wright, 1898)	F	rough rabbitsfoot
Quadrula fragosa (Conrad, 1835)	F	winged mapleleaf
Quadrula houstonensis (I. Lea, 1859)	F	smooth pimpleback
Quadrula intermedia (Conrad, 1836)	F	Cumberland monkeyface
Quadrula metanevra (Rafinesque, 1820)	F	monkeyface
Quadrula nodulata (Rafinesque, 1820)	F	wartyback

SCIENTIFIC NAME	OCCURRENCE	COMMON NAME
Quadrula petrina (Gould, 1855)	F	Texas pimpleback
Quadrula pustulosa mortoni (Conrad, 1835)	F	western pimpleback
Quadrula pustulosa pustulosa (I. Lea, 1831)	F	pimpleback
Quadrula quadrula (Rafinesque, 1820)	F	mapleleaf
Quadrula refulgens (I. Lea, 1868)	F	purple pimpleback
Quadrula rumphiana (I. Lea, 1852)	F	ridged mapleleaf
Quadrula sparsa (I. Lea, 1841)	F	Appalachian monkeyface
Quadrula stapes (I. Lea, 1831)	F	stirrupshell
Quadrula tuberosa (I. Lea, 1840)	F	rough rockshell
Quincuncina burkei Walker, 1922	F	tapered pigtoe
Quincuncina infucata (Conrad, 1834)	F	sculptured pigtoe
Quincuncina mitchelli (Simpson, 1896)	F	false spike
Simpsonaias ambigua (Say, 1825)	F	salamander mussel
Strophitus connasaugaensis (I. Lea, 1857)	F	Alabama creekmussel
Strophitus subvexus (Conrad, 1834)	F	southern creekmussel
Strophitus undulatus (Say, 1817)	F	squawfoot
Toxolasma corvunculus (I. Lea, 1868)	F	southern lilliput
Toxolasma cylindrellus (I. Lea, 1868)	F	pale lilliput
Toxolasma lividus (Rafinesque, 1831)	F	purple lilliput
Toxolasma mearnsi (Simpson, 1900)	F	western lilliput
Toxolasma parvus (Barnes, 1823)	F	lilliput
Toxolasma paulus (I. Lea, 1840)	F	iridescent lilliput
Toxolasma pullus (Conrad, 1838)	F	Savannah lilliput
Toxolasma texasensis (I. Lea, 1857)	F	Texas lilliput
Tritogonia verrucosa (Rafinesque, 1820)	F	pistolgrip
Truncilla cognata (I. Lea, 1860)	F	Mexican fawnsfoot
Truncilla donaciformis (I. Lea, 1828)	F	fawnsfoot
Truncilla macrodon (I. Lea, 1859)	F	Texas fawnsfoot
Truncilla truncata Rafinesque, 1820	F	deertoe
Uniomerus caroliniana (Bosc, 1801)	F	Florida pondhorn
Uniomerus declivis (Say, 1831)	F	tapered pondhorn
Uniomerus excultus (Conrad, 1838)	F	polished pondhorn
Uniomerus obesus (I. Lea, 1831)	F	southern pondhorn
Uniomerus tetralasmus (Say, 1831)	F	pondhorn
Venustaconcha ellipsiformis (Conrad, 1836)	F	ellipse
Venustaconcha pleasi (Marsh, 1891)	F	bleedingtooth mussel
Villosa amygdala (I. Lea, 1843)	F	Florida rainbow
Villosa arkansasensis (I. Lea, 1862)	F	Ouachita creekshell
Villosa choctawensis Athearn, 1964	F	Choctaw bean
Villosa constricta (Conrad, 1838)	F	notched rainbow
Villosa delumbis (Conrad, 1834)	F	eastern creekshell
Villosa fabalis (I. Lea, 1831)	F	rayed bean
Villosa iris (I. Lea, 1829)	F	rainbow
Villosa lienosa (Conrad, 1834)	F	little spectaclecase
Villosa nebulosa (Conrad, 1834)	F	Alabama rainbow
Villosa ortmanni (Walker, 1925)	F	Kentucky creekshell
Villosa perpurpurea (I. Lea, 1861)	F	purple bean
Villosa taeniata (Conrad, 1834)	F	painted creekshell
Villosa trabalis (Conrad, 1834)	F	Cumberland bean
Villosa vanuxemensis umbrans (I. Lea, 1857)	F	Coosa creekshell
Villosa vanuxemensis vanuxemensis (I. Lea, 1838)	F	mountain creekshell
Villosa vaughaniana (I. Lea, 1838)	F	Carolina creekshell
Villosa vibex (Conrad, 1834)	F	southern rainbow

SCIENTIFIC NAME	OCCURRENCE	COMMON NAME
Villosa villosa (Wright, 1898)	F	downy rainbow

Order Veneroida

Sphaeriidae

SCIENTIFIC NAME	OCCURRENCE	COMMON NAME
Eupera cubensis (Prime, 1865)	F	mottled fingernailclam
Musculium lacustre (Müller, 1774)	F	lake fingernailclam
Musculium partumeium (Say, 1822)	F	swamp fingernailclam
Musculium securis (Prime, 1852)	F	pond fingernailclam
Musculium transversum (Say, 1829)	F	long fingernailclam
Pisidium adamsi Stimpson, 1851	F	Adam peaclam
Pisidium amnicum (Müller, 1774)	F [I]	greater European peaclam
Pisidium casertanum (Poli, 1791)	F	ubiquitous peaclam
Pisidium compressum Prime, 1852	F	ridged-beak peaclam
Pisidium conventus Clessin, 1877	F	Alpine peaclam
Pisidium cruciatum Sterki, 1895	F	ornamented peaclam
Pisidium dubium (Say, 1817)	F	greater eastern peaclam
Pisidium equilaterale Prime, 1852	F	round peaclam
Pisidium fallax Sterki, 1896	F	river peaclam
Pisidium ferrugineum Prime, 1852	F	rusty peaclam
Pisidium henslowanum (Sheppard, 1825)	F [I]	Henslow peaclam
Pisidium idahoense E. W. Roper, 1890	F	giant northern peaclam
Pisidium insigne Gabb, 1868	F	tiny peaclam
Pisidium lilljeborgi (Clessin, 1886)	F	Lilljeborg peaclam
Pisidium milium Held, 1836	F	quadrangular pillclam
Pisidium nitidum Jenyns, 1832	F	shiny peaclam
Pisidium punctatum Sterki, 1895	F	perforated peaclam
Pisidium punctiferum (Guppy, 1867)	F	striate peaclam
Pisidium rotundatum Prime, 1852	F	fat peaclam
Pisidium subtruncatum Malm, 1855	F	short-end peaclam
Pisidium supinum Schmidt, 1850	F [I]	humpback peaclam
Pisidium ultramontanum Prime, 1865	F	montane peaclam
Pisidium variabile Prime, 1852	F	triangular peaclam
Pisidium ventricosum Prime, 1851	F	globular peaclam
Pisidium walkeri Sterki, 1895	F	Walker peaclam
Sphaerium corneum (Linnaeus, 1758)	F [I]	European fingernailclam
Sphaerium fabale (Prime, 1852)	F	river fingernailclam
Sphaerium nitidum Westerlund, 1876	F	Arctic fingernailclam
Sphaerium occidentale (J. Lewis, 1856)	F	Herrington fingernailclam
Sphaerium patella (Gould, 1850)	F	Rocky Mountain fingernailclam
Sphaerium rhomboideum (Say, 1822)	F	rhomboid fingernailclam
Sphaerium simile (Say, 1817)	F	grooved fingernailclam
Sphaerium striatinum (Lamarck, 1818)	F	striated fingernailclam

Lucinidae

SCIENTIFIC NAME	OCCURRENCE	COMMON NAME
Anodontia alba Link, 1807	A	buttercup lucine
Anodontia philippiana (Reeve, 1850)	A	chalky buttercup lucine
Codakia costata (d'Orbigny, 1842)	A	costate lucine
Codakia cubana Dall, 1901	A	Cuba lucine
Codakia orbicularis (Linnaeus, 1758)	A	tiger lucine
Codakia orbiculata (Montagu, 1808)	A	dwarf tiger lucine

SCIENTIFIC NAME	OCCURRENCE	COMMON NAME
Codakia pectinella C. B. Adams, 1852	A	little-comb lucine
Divaricella dentata (W. Wood, 1815)	A	dentate lucine
Divaricella quadrisulcata (d'Orbigny, 1842)	A	cross-hatched lucine
Epilucina californica (Conrad, 1837)	P	California lucine
Here richthofeni (Gabb, 1866)	P	pit lucine
Linga amiantus (Dall, 1901)	A	miniature lucine
Linga leucocyma (Dall, 1886)	A	four-rib lucine
Linga pensylvanica (Linnaeus, 1758)	A	Pennsylvania lucine
Linga sombrerensis (Dall, 1886)	A	Sombrero lucine
Lucina bermudensis Dall, 1901	A	Bermuda lucine
Lucina keenae Chavan, 1971	A	Dosinia lucine
Lucina muricata (Spengler, 1798)	A	spinose lucine
Lucina nassula (Conrad, 1846)	A	woven lucine
Lucina nuttalli (Conrad, 1837)	P	Nuttall lucine
Lucina pectinata (Gmelin, 1791)	A	thick lucine
Lucinoma aequizonatum (Stearns, 1890)	P	equal-zone lucine
Lucinoma annulatum (Reeve, 1850)	P	ringed lucine
Lucinoma blakeanum Bush, 1893	A	Blake lucine
Lucinoma filosum (Stimpson, 1851)	A	northeast lucine
Myrtea compressa (Dall, 1881)	A	compressed lucine
Myrtea lens (A. E. Verrill and S. Smith, 1880)	A	lens lucine
Myrtea pristiphora Dall and Simpson, 1901	A	lamellate lucine
Myrtea sagrinata (Dall, 1886)	A	arrow lucine
Parvilucina approximata (Dall, 1901)	P	approximate lucine
Parvilucina blanda (Dall and Simpson, 1901)	A	three-ridge lucine
Parvilucina multilineata (Tuomey and Holmes, 1857)	A	many-line lucine
Parvilucina tenuisculpta (Carpenter, 1864)	P	fine-lined lucine
Pseudomiltha floridana (Conrad, 1833)	A	Florida lucine

Thyasiridae

SCIENTIFIC NAME	OCCURRENCE	COMMON NAME
Adontorhina cyclia S. S. Berry, 1947	P	circle axinopsid
Adontorhina sphaericosa Scott, 1986	P	inflated axinopsid
Axinopsida cordata (A. E. Verrill and Bush, 1898)	A	heart axinopsid
Axinopsida orbiculata (G. O. Sars, 1878)	A, Ac	orbicular axinopsid
Axinopsida serricata (Carpenter, 1864)	P	silky axinopsid
Axinopsida viridis (Dall, 1901)	P, Ac	green axinopsid
Axinulus careyi Bernard, 1979	Ac	Carey axinopsid
Axinulus redondoensis (T. A. Burch, 1941)	P	Redondo axinopsid
Conchocele bisecta (Conrad, 1849)	A, P	giant cleftclam
Conchocele disjuncta Gabb, 1866	P	disjunct cleftclam
Leptaxinus minutus A. E. Verrill and Bush, 1898	A	minute axinopsid
Thyasira barbarensis (Dall, 1890)	P	Santa Barbara cleftclam
Thyasira brevis A. E. Verrill and Bush, 1898	A	short cleftclam
Thyasira cygnus Dall, 1916	P	swan cleftclam
Thyasira equalis A. E. Verrill and Bush, 1898	A	equal cleftclam
Thyasira eumyaria M. Sars, 1870	Ac	Mya cleftclam
Thyasira ferruginea Winckworth, 1932	A, Ac	reddish cleftclam
Thyasira flexuosa (Montagu, 1803)	A, Ac	flexuose cleftclam

SCIENTIFIC NAME	OCCURRENCE	COMMON NAME
Thyasira gouldii (Philippi, 1845)	P, Ac	Gould cleftclam
Thyasira granulosa di Monterosato, 1874	A	granulose cleftclam
Thyasira rotunda Jeffreys, 1881	A	rotund cleftclam
Thyasira simplex A. E. Verrill and Bush, 1898	A	simple cleftclam
Thyasira succisa Jeffreys, 1876	A	cut-off cleftclam
Thyasira trisinuata d'Orbigny, 1842	A	Atlantic cleftclam

Ungulinidae

Diplodonta aleutica Dall, 1901	P	Aleutian diplodon
Diplodonta candeana (d'Orbigny, 1842)	A	Cande diplodon
Diplodonta impolita S. S. Berry, 1953	P	rough diplodon
Diplodonta notata Dall and Simpson, 1901	A	marked diplodon
Diplodonta nucleiformis Wagner, 1838	A	nut diplodon
Diplodonta orbella (Gould, 1851)	P	orb diplodon
Diplodonta punctata (Say, 1822)	A	Atlantic diplodon
Diplodonta semiaspera Philippi, 1836	A	pimpled diplodon
Diplodonta soror C. B. Adams, 1852	A	sister diplodon
Diplodonta subglobosa C. B. Adams, 1852	A	subglobose diplodon
Diplodonta venezuelensis Dunker, 1848	A	Venezuela diplodon
Diplodonta verrilli Dall, 1900	A	Verrill diplodon
Felaniella parilis (Conrad, 1848)	P	equal diplodon

Cyrenoididae

Cyrenoida floridana (Dall, 1896)	A	Florida marshclam

Chamidae

Arcinella arcinella (Linnaeus, 1767)	A	spiny jewelbox
Arcinella cornuta Conrad, 1866	A	Florida spiny jewelbox
Chama arcana Bernard, 1976	P	secret jewelbox
Chama congregata Conrad, 1833	A	corrugate jewelbox
Chama florida Lamarck, 1819	A	pretty jewelbox
Chama lactuca Dall, 1886	A	milky jewelbox
Chama macerophylla Gmelin, 1791	A	leafy jewelbox
Chama sarda Reeve, 1847	A	cherry jewelbox
Chama sinuosa Broderip, 1835	A	smooth-edge jewelbox
Pseudochama exogyra (Conrad, 1837)	P	Pacific jewelbox
Pseudochama granti Strong, 1934	P	Grant jewelbox
Pseudochama inezae Bayer, 1943	A	Inez jewelbox
Pseudochama radians (Lamarck, 1819)	A	Atlantic jewelbox

Kelliidae

Aligena diegoana Hertlein and Grant, 1972	P	San Diego aligena
Aligena elevata (Stimpson, 1851)	A	eastern aligena
Aligena texasiana Harry, 1969	A	Texas aligena
Bornia longipes (Stimpson, 1855)	A	borniaclam
Erycina balliana Dall, 1916	P	Ball erycina
Erycina coronata Dall, 1916	P	crown erycina
Erycina emmonsi Dall, 1899	A	Emmons erycina
Erycina linella Dall, 1899	A	lined erycina

SCIENTIFIC NAME	OCCURRENCE	COMMON NAME
Erycina periscopiana Dall, 1899	A	periscope erycina
Kellia laperousii (Deshayes, 1839)	P	La Perouse kellyclam
Kellia suborbicularis (Montagu, 1803)	A, P, Ac	suborbicular kellyclam
Lasaea adansoni (Gmelin, 1791)	A	Adanson lepton
Lasaea subviridis Dall, 1899	P	Pacific lepton
Odontogena borealis (I. M. Cowan, 1964)	P	boreal kellyclam
Parabornia squillina Boss, 1965	A	squillaclam
Platomysia meroea (Carpenter, 1864)	P	broad lepton
Rhamphidonta retifera (Dall, 1899)	P	netted kellyclam

Montacutidae

SCIENTIFIC NAME	OCCURRENCE	COMMON NAME
Boreacola vadosus Bernard, 1979	Ac	Arctic montacutid
Entovalva perrieri (Malard, 1903)	A	Perrier montacutid
Isorobitella trigonalis (Carpenter, 1857)	P	trigonal montacutid
Lepton lepidum Say, 1826	A	graceful lepton
Montacuta dawsoni Jeffreys, 1864	A, P, Ac	Dawson montacutid
Montacuta limpida Dall, 1899	A	limpid montacutid
Montacuta percompressa Dall, 1899	A	much-compressed montacutid
Mysella aleutica Dall, 1899	P	Aleutian mysella
Mysella beringensis (Dall, 1916)	P	Bering Sea mysella
Mysella casta (A. E. Verrill and Bush, 1898)	A	pure mysella
Mysella compressa (Dall, 1913)	P	compressed mysella
Mysella grippi (Dall, 1912)	P	eliptical mysella
Mysella moelleri (Mörch, 1875)	A	Moeller mysella
Mysella ovata (Jeffreys, 1881)	A	ovate mysella
Mysella pedroana Dall, 1899	P	San Pedro mysella
Mysella planata (Krause, 1885)	P	flat mysella
Mysella planulata (Stimpson, 1851)	A	plate mysella
Mysella striatula A. E. Verrill and Bush, 1898	A	scratched mysella
Mysella triquetra (A. E. Verrill and Bush, 1898)	A	triangular mysella
Mysella tumida (Carpenter, 1864)	P	robust mysella
Neaeromya compressa (Dall, 1899)	P	compressed montacutid
Neaeromya floridana (Dall, 1899)	A	giant montacutid
Neaeromya rugifera (Carpenter, 1864)	P	wrinkled montacutid
Orobitella bakeri (Dall, 1916)	P	Baker montacutid
Pristes oblongus Carpenter, 1864	P	chiton clam
Pythenella cuneata (A. E. Verrill and Bush, 1898)	A	cuneate montacutid

Galeommatidae

SCIENTIFIC NAME	OCCURRENCE	COMMON NAME
Aclistothyra atlantica McGinty, 1955	A	Atlantic galeommatid
Scintillona bellerophon O'Foighil and Gibson, 1984	P	sea-cucumber clam

Chlamydoconchidae

SCIENTIFIC NAME	OCCURRENCE	COMMON NAME
Chlamydoconcha orcutti Dall, 1884	P	Orcutt nakedclam

SCIENTIFIC NAME	OCCURRENCE	COMMON NAME
Turtoniidae		
Turtonia minuta (Fabricius, 1780)	A, P, Ac	minute turton
Turtonia occidentalis Dall, 1871	P	western turton
Sportellidae		
Basterotia corbuloidea Dall, 1899	A	Corbula sportella
Basterotia elliptica (Récluz, 1850)	A	elliptical sportella
Basterotia quadrata (Hinds, 1843)	A	square sportella
Ensitellops pilsbryi (Dall, 1899)	A	Pilsbry sportella
Ensitellops protexta (Conrad, 1841)	A	textured sportella
Ensitellops tabula Olsson and Harbison, 1953	A	tube sportella
Planktomya henseni Simroth, 1896	A	
Carditidae		
Carditamera floridana Conrad, 1838	A	broad-ribbed carditid
Crassicardia crassidens (Broderip and Sowerby, 1829)	P, Ac	thick carditid
Cyclocardia armilla (Dall, 1903)	A	bracelet cyclocardia
Cyclocardia bailyi (J. Q. Burch, 1944)	P	bumpy cyclocardia
Cyclocardia barbarensis (Stearns, 1891)	P	Santa Barbara cyclocardia
Cyclocardia borealis (Conrad, 1831)	A	northern cyclocardia
Cyclocardia crebricostata (Krause, 1885)	P, Ac	many-rib cyclocardia
Cyclocardia incisa (Dall, 1903)	P	cut cyclocardia
Cyclocardia novangliae (E. S. Morse, 1869)	A	New England cyclocardia
Cyclocardia ovata (Rjabinina, 1952)	P	ovate cyclocardia
Cyclocardia rjabininae (Scarlato, 1955)	P	Rjabinina cyclocardia
Cyclocardia umnaka (Willett, 1932)	P	Umnak cyclocardia
Cyclocardia ventricosa (Gould, 1850)	P	stout cyclocardia
Glans carpenteri (Lamy, 1922)	P	Carpenter carditid
Glans dominguensis (d'Orbigny, 1842)	A	Santo Domingo carditid
Milneria kelseyi Dall, 1916	P	Kelsey pouchclam
Milneria minima (Dall, 1871)	P	tiny pouchclam
Miodontiscus prolongatus (Carpenter, 1864)	P	elongate carditid
Pleuromeris armilla (Dall, 1902)	A	armlet carditid
Pleuromeris perplana (Conrad, 1841)	A	flattened carditid
Pleuromeris tridentata (Say, 1826)	A	threetooth carditid
Condylocardiidae		
Carditopsis bernardi (Dall, 1903)	A	Bernard condylclam
Carditopsis smithii (Dall, 1896)	A	tiny condylclam
Cuna dalli Vanatta, 1904	A	Dall condylclam
Astartidae		
Astarte castanea (Say, 1822)	A	smooth astarte
Astarte compacta Carpenter, 1864	P	compact astarte
Astarte crenata (J. E. Gray, 1824)	A, P, Ac	crenulate astarte
Astarte esquimalti (Baird, 1863)	P	Eskimo astarte
Astarte mirabilis (Dall, 1871)	P	strange astarte
Astarte nana Dall, 1886	A	dwarf astarte

SCIENTIFIC NAME	OCCURRENCE	COMMON NAME
Astarte polaris Dall, 1903	P	polar astarte
Astarte quadrans Gould, 1841	A	squarish astarte
Astarte smithii Dall, 1886	A	Smith astarte
Astarte undata Gould, 1841	A, P	wavy astarte
Astarte willetti Dall, 1917	P	Willett astarte
Tridonta alaskensis (Dall, 1903)	P	Alaska tridonta
Tridonta arctica (J. E. Gray, 1824)	A, P, Ac	Arctic tridonta
Tridonta bennettii (Dall, 1903)	A, P, Ac	Bennett tridonta
Tridonta borealis Schumacher, 1817	A, P, Ac	boreal tridonta
Tridonta filatovae (Habe, 1964)	P	Filatova tridonta
Tridonta montagui (Dillwyn, 1817)	A, P, Ac	Montagu tridonta
Tridonta rollandi (Bernardi, 1859)	P	Rolland tridonta
Tridonta vernicosa (Dall, 1903)	P, Ac	varnished tridonta

Crassatellidae

SCIENTIFIC NAME	OCCURRENCE	COMMON NAME
Crassinella lunulata (Conrad, 1834)	A	lunate crassinella
Crassinella martinicensis (d'Orbigny, 1842)	A	Martinique crassinella
Crassinella pacifica (C. B. Adams, 1852)	P	Pacific crassinella
Eucrassatella fluctuata (Carpenter, 1864)	P	wavy crassatella
Eucrassatella speciosa (A. Adams, 1852)	A	beautiful crassatella

Cardiidae

SCIENTIFIC NAME	OCCURRENCE	COMMON NAME
Americardia biangulata (Broderip and Sowerby, 1829)	P	western strawberry-cockle
Americardia guppyi (Thiele, 1910)	A	Guppy strawberry-cockle
Americardia media (Linnaeus, 1758)	A	Atlantic strawberry-cockle
Cerastoderma elegantulum Beck, 1842	A	elegant dwarf-cockle
Cerastoderma pinnulatum (Conrad, 1831)	A	northern dwarf-cockle
Clinocardium blandum (Gould, 1850)	P	smooth cockle
Clinocardium californiense (Deshayes, 1839)	P	California cockle
Clinocardium ciliatum (Fabricius, 1780)	A, P, Ac	hairy cockle
Clinocardium fucanum (Dall, 1907)	P	strait cockle
Clinocardium nuttallii (Conrad, 1837)	P	Nuttall cockle
Dinocardium robustum (Lightfoot, 1786)	A	Atlantic giant-cockle
Laevicardium elatum (Sowerby, 1833)	P	giant eggcockle
Laevicardium laevigatum (Linnaeus, 1758)	A	eggcockle
Laevicardium mortoni (Conrad, 1830)	A	Morton eggcockle
Laevicardium pictum (Ravenel, 1861)	A	painted eggcockle
Laevicardium substriatum (Conrad, 1837)	P	Pacific eggcockle
Laevicardium sybariticum (Dall, 1886)	A	delicate eggcockle
Nemocardium centifilosum (Carpenter, 1864)	P	hundred-line cockle
Nemocardium peramabile (Dall, 1881)	A	lovely micro-cockle
Nemocardium tinctum (Dall, 1881)	A	dyed micro-cockle
Nemocardium transversum Rehder and Abbott, 1951	A	transverse micro-cockle
Papyridea semisulcata (J. E. Gray, 1825)	A	frilled papercockle
Papyridea soleniformis (Bruguière, 1789)	A	spiny papercockle
Serripes groenlandicus (Bruguière, 1789)	A, P, Ac	Greenland cockle
Serripes laperousii (Deshayes, 1839)	P	broad cockle
Trachycardium egmontianum (Shuttleworth, 1856)	A	Florida pricklycockle
Trachycardium isocardia (Linnaeus, 1758)	A	even pricklycockle
Trachycardium magnum (Linnaeus, 1758)	A	magnum pricklycockle
Trachycardium muricatum (Linnaeus, 1758)	A	yellow pricklycockle

SCIENTIFIC NAME	OCCURRENCE	COMMON NAME
Trachycardium quadragenarium (Conrad, 1837) ...	P	spiny pricklycockle
Trigoniocardia antillarum (d'Orbigny, 1842)	A	Antillean cockle

Mactridae

Anatina anatina (Spengler, 1802)	A	smooth duckclam
Mactra californica Conrad, 1837................	P	California surfclam
Mactra dolabriformis (Conrad, 1867)	P	hatchet surfclam
Mactra fragilis Gmelin, 1791...................	A	fragile surfclam
Mactra nasuta Gould, 1851.....................	P	bent surfclam
Mactrellona alata (Spengler, 1802)..............	A	winged surfclam
Mactromeris polynyma (Stimpson, 1860)	A, P, Ac	Arctic surfclam
Mulinia cleryana (d'Orbigny, 1846)..............	A	Clery surfclam
Mulinia lateralis (Say, 1822)	A	dwarf surfclam
Raeta plicatella (Lamarck, 1818)	A	channeled duckclam
Raeta undulata (Gould, 1851)..................	P	Pacific duckclam
Rangia cuneata (Sowerby, 1831)	A	Atlantic rangia
Rangia flexuosa (Conrad, 1840)	A	brown rangia
Spisula catilliformis Conrad, 1867...............	P	dish surfclam
Spisula falcata (Gould, 1850)	P	hooked surfclam
Spisula hemphillii (Dall, 1894)..................	P	Hemphill surfclam
Spisula planulata (Conrad, 1837)	P	flat surfclam
Spisula solidissima (Dillwyn, 1817)..............	A	Atlantic surfclam
Tresus capax (Gould, 1850)....................	P	fat gaper
Tresus nuttallii (Conrad, 1837)	P	Pacific gaper
Tresus "*pajaroana* (Conrad, 1857)," auct.	P	lost gaper

Mesodesmatidae

Ervilia concentrica (Holmes, 1860)..............	A	concentric ervilia
Ervilia nitens (Montagu, 1808)..................	A	shining ervilia
Ervilia subcancellata E. A. Smith, 1885..........	A	subcancellate ervilia
Mesodesma arctatum (Conrad, 1830)	A	Arctic wedgeclam
Mesodesma deauratum (Turton, 1822)	A	gilded wedgeclam

Solenidae

Ensis directus Conrad, 1843	A	Atlantic jackknife
Ensis minor Dall, 1900.........................	A	minor jackknife
Ensis myrae S. S. Berry, 1953	P	California jackknife
Siliqua alta (Broderip and Sowerby, 1829)........	P	Alaska razor
Siliqua costata Say, 1822	A	Atlantic razor
Siliqua lucida (Conrad, 1837)	P	transparent razor
Siliqua patula (Dixon, 1789)	P	Pacific razor
Siliqua sloati Hertlein, 1961	P	Sloat razor
Siliqua squama de Blainville, 1824	A	rough razor
Solen obliquus Spengler, 1794...................	A	oblique jackknife
Solen rosaceus Carpenter, 1864	P	rosy jackknife
Solen sicarius Gould, 1850	P	sickle jackknife
Solen viridis Say, 1821.........................	A	green jackknife

Tellinidae

Cymatoica orientalis (Dall, 1890)	A	western tellin
Leporimetis intastriata (Say, 1827)..............	A	Atlantic fat-tellin
Leporimetis obesa (Deshayes, 1855).............	P	California fat-tellin

SCIENTIFIC NAME	OCCURRENCE	COMMON NAME
Macoma acolasta Dall, 1921	P	morsel macoma
Macoma balthica (Linnaeus, 1758)	A, P, Ac	Baltic macoma
Macoma brevifrons (Say, 1834)	A	short macoma
Macoma brota Dall, 1916	P, Ac	heavy macoma
Macoma calcarea (Gmelin, 1791)	A, P, Ac	chalky macoma
Macoma carlottensis Whiteaves, 1880	P, Ac	Charlotte macoma
Macoma cerina C. B. Adams, 1845	A	waxy macoma
Macoma constricta (Bruguière, 1792)	A	constricted macoma
Macoma crassula (Deshayes, 1855)	A, P, Ac	thick macoma
Macoma dexioptera Baxter, 1977	P	rightwing macoma
Macoma elimata Dunnill and Coan, 1968	P	file macoma
Macoma expansa Carpenter, 1864	P	expanded macoma
Macoma extenuata Dall, 1900	A	slender macoma
Macoma indentata Carpenter, 1864	P	indented macoma
Macoma inquinata (Deshayes, 1855)	P	stained macoma
Macoma lama Bartsch, 1929	P, Ac	Aleutian macoma
Macoma leptonoidea Dall, 1895	P	Lepton macoma
Macoma limula Dall, 1895	A	little-file macoma
Macoma lipara Dall, 1916	P	sleek macoma
Macoma loveni (A. S. Jensen, 1905)	A, P, Ac	Lovén macoma
Macoma middendorffi Dall, 1884	P	Middendorff macoma
Macoma mitchelli Dall, 1895	A	Mitchell macoma
Macoma moesta (Deshayes, 1855)	A, P, Ac	flat macoma
Macoma nasuta (Conrad, 1837)	P	bent-nose macoma
Macoma obliqua (Sowerby, 1817)	P	oblique macoma
Macoma phenax Dall, 1900	A	cheating macoma
Macoma pseudomera Dall and Simpson, 1901	A	Mera macoma
Macoma pulleyi Boyer, 1969	A	Pulley macoma
Macoma secta (Conrad, 1837)	P	white-sand macoma
Macoma tageliformis Dall, 1900	A	Tagelus macoma
Macoma tenta (Say, 1834)	A	elongate macoma
Macoma yoldiformis Carpenter, 1864	P	Yoldia macoma
Strigilla carnaria (Linnaeus, 1758)	A	large strigilla
Strigilla gabbi Olsson and McGinty, 1958	A	Gabb strigilla
Strigilla mirabilis (Philippi, 1841)	A	white strigilla
Strigilla pisiformis (Linnaeus, 1758)	A	pea strigilla
Strigilla producta Tryon, 1870	A	ovate strigilla
Strigilla pseudocarnaria Boss, 1969	A	false strigilla
Strigilla surinamensis Boss, 1972	A	Surinam strigilla
Tellidora cristata (Récluz, 1842)	A	white-crest tellin
Tellina aequistriata Say, 1824	A	striate tellin
Tellina agilis Stimpson, 1857	A	northern dwarf-tellin
Tellina alternata Say, 1822	A	alternate tellin
Tellina americana Dall, 1900	A	American tellin
Tellina angulosa Gmelin, 1791	A	angulate tellin
Tellina bodegensis Hinds, 1845	P	Bodega tellin
Tellina candeana d'Orbigny, 1842	A	Cande tellin
Tellina carpenteri Dall, 1900	P	Carpenter tellin
Tellina colorata Dall, 1900	A	tinted tellin
Tellina consobrina d'Orbigny, 1842	A	consobrine tellin
Tellina cristallina Spengler, 1798	A	crystal tellin
Tellina euvitrea Boss, 1964	A	shiny tellin

SCIENTIFIC NAME	OCCURRENCE	COMMON NAME
Tellina exerythra Boss, 1964	A	bloodless tellin
Tellina fausta Pulteney, 1799	A	favored tellin
Tellina gouldii Hanley, 1846	A	cuneate tellin
Tellina guildingii Hanley, 1844	A	Guilding tellin
Tellina idae Dall, 1891	P	Ida tellin
Tellina iris Say, 1822	A	rainbow tellin
Tellina juttingae Altena, 1965	A	Jutting tellin
Tellina laevigata Linnaeus, 1758	A	smooth tellin
Tellina lineata Turton, 1819	A	rose-petal tellin
Tellina listeri Röding, 1798	A	speckled tellin
Tellina lutea W. Wood, 1828	P, Ac	Alaska great-tellin
Tellina magna Spengler, 1798	A	great-tellin
Tellina martinicensis d'Orbigny, 1842	A	Martinique tellin
Tellina mera Say, 1834	A	pure tellin
Tellina meropsis Dall, 1900	P	oval tellin
Tellina modesta (Carpenter, 1864)	P	plain tellin
Tellina nitens C. B. Adams, 1845	A	shiny dwarf-tellin
Tellina nuculoides (Reeve, 1854)	P	salmon tellin
Tellina paramera Boss, 1964	A	perfect tellin
Tellina persica Dall and Simpson, 1901	A	apricot tellin
Tellina probina Boss, 1964	A	slandered tellin
Tellina punicea Born, 1778	A	watermelon tellin
Tellina radiata Linnaeus, 1758	A	sunrise tellin
Tellina similis Sowerby, 1806	A	candystick tellin
Tellina squamifera Deshayes, 1855	A	crenulate tellin
Tellina sybaritica Dall, 1881	A	sybaritic tellin
Tellina tampaensis Conrad, 1866	A	Tampa tellin
Tellina tenella A. E. Verrill, 1874	A	delicate tellin
Tellina texana Dall, 1900	A	Say tellin
Tellina versicolor DeKay, 1843	A	many-colored tellin
Tellina vespuciana (d'Orbigny, 1842)	A	Vespucci tellin

Donacidae

SCIENTIFIC NAME	OCCURRENCE	COMMON NAME
Donax californicus Conrad, 1837	P	California beanclam
Donax gouldii Dall, 1921	P	Gould beanclam
Donax texasianius Philippi, 1847	A	Texas coquina
Donax variabilis Say, 1822	A	variable coquina
Iphigenia brasiliana (Lamarck, 1818)	A	giant coquina

Psammobiidae

SCIENTIFIC NAME	OCCURRENCE	COMMON NAME
Asaphis deflorata (Linnaeus, 1758)	A	gaudy sanguin
Gari californica (Conrad, 1849)	P	California sunsetclam
Gari fucata (Hinds, 1845)	P	painted sunsetclam
Heterodonax bimaculatus (Linnaeus, 1758)	A	false-bean
Heterodonax pacificus (Conrad, 1837)	P	Pacific false-bean
Nuttallia nuttallii (Conrad, 1837)	P	mahogany clam
Sanguinolaria cruenta (Lightfoot, 1786)	A	operculate sanguin
Sanguinolaria sanguinolenta (Gmelin, 1791)	A	Atlantic sanguin

Semelidae

SCIENTIFIC NAME	OCCURRENCE	COMMON NAME
Abra aequalis (Say, 1822)	A	Atlantic abra

SCIENTIFIC NAME	OCCURRENCE	COMMON NAME
Abra lioica (Dall, 1881)	A	smooth abra
Abra longicallis americana A. E. Verrill and Bush, 1898	A, Ac	long abra
Cumingia californica Conrad, 1837	P	California semele
Cumingia coarctata Sowerby, 1833	A	contracted semele
Cumingia tellinoides (Conrad, 1831)	A	Tellin semele
Semele bellastriata (Conrad, 1837)	A	cancellate semele
Semele decisa (Conrad, 1837)	P	clipped semele
Semele incongrua Carpenter, 1864	P	incongruous semele
Semele proficua (Pulteney, 1799)	A	Atlantic semele
Semele pulchra (Sowerby, 1832)	P	beautiful semele
Semele purpurascens (Gmelin, 1791)	A	purplish semele
Semele rubropicta Dall, 1871	P	rose-painted semele
Semele rupicola Dall, 1915	P	rock semele
Semelina nuculoides (Conrad, 1841)	A	nut semele
Theora lubrica Gould, 1861	P[I]	Asian semele

Solecurtidae

Solecurtus cumingianus Dunker, 1861	A	corrugate solecurtus
Solecurtus sanctaemarthae d'Orbigny, 1842	A	St. Martha solecurtus
Tagelus affinis (C. B. Adams, 1852)	P	neighbor tagelus
Tagelus californianus (Conrad, 1837)	P	California tagelus
Tagelus divisus (Spengler, 1794)	A	purplish tagelus
Tagelus plebeius (Lightfoot, 1786)	A	stout tagelus
Tagelus subteres (Conrad, 1837)	P	lesser tagelus

Dreissenidae

Mytilopsis leucophaeata (Conrad, 1831)	A	dark falsemussel
Mytilopsis sallei (Récluz, 1849)	A	Santo Domingo falsemussel

Arcticidae

Arctica islandica (Linnaeus, 1767)	A	ocean quahog

Bernardinidae

Bernardina bakeri Dall, 1910	P	Baker bernardclam
Halodakra salmonea (Carpenter, 1864)	P	salmon bernardclam
Halodakra subtrigona (Carpenter, 1857)	P	subtrigonal bernardclam

Trapeziidae

Coralliophaga coralliophaga (Gmelin, 1791)	A	coralclam

Vesicomyidae

Vesicomya vesica (Dall, 1886)	A	fat vesicomya

Glossidae

Meiocardia agassizii Dall, 1889	A	ox-heart clam

SCIENTIFIC NAME	OCCURRENCE	COMMON NAME
Corbiculidae		
Corbicula fluminea (Müller, 1774)	F [I]	Asian clam
Polymesoda caroliniana (Bosc, 1801)	A	Carolina marshclam
Polymesoda maritima (d'Orbigny, 1842)	A	southern marshclam
Veneridae		
Agriopoma texasianum (Dall, 1892)	A	Texas venus
Amiantis callosa (Conrad, 1837)	P	white venus
Anomalocardia auberiana (d'Orbigny, 1842)	A	pointed-venus
Anomalocardia brasiliana (Gmelin, 1791)	A	Carib pointed-venus
Callista eucymata (Dall, 1890)	A	glory-of-the-seas venus
Chione californiensis (Broderip, 1835)	P	California venus
Chione cancellata (Linnaeus, 1767)	A	cross-barred venus
Chione clenchi Pulley, 1952	A	Clench venus
Chione fluctifraga (Sowerby, 1853)	P	smooth venus
Chione grus (Holmes, 1858)	A	gray pygmy-venus
Chione intapurpurea (Conrad, 1849)	A	lady-in-waiting venus
Chione latilirata (Conrad, 1841)	A	imperial venus
Chione paphia (Linnaeus, 1767)	A	king venus
Chione puber (Bory Saint-Vincent, 1827)	A	downy venus
Chione pygmaea (Lamarck, 1818)	A	pygmy-venus
Chione undatella (Sowerby, 1835)	P	frilled venus
Circomphalus strigillinus (Dall, 1902)	A	empress venus
Compsomyax subdiaphanus (Carpenter, 1864)	P	milky venus
Cyclinella tenuis (Récluz, 1852)	A	thin cyclinella
Dosinia discus (Reeve, 1850)	A	disk dosinia
Dosinia elegans Conrad, 1843	A	elegant dosinia
Gemma gemma (Totten, 1834)	A, P [I]	amethyst gemclam
Globivenus callimorpha (Dall, 1902)	A	lily venus
Globivenus fordi (Yates, 1890)	P	Ford venus
Gouldia cerina (C. B. Adams, 1845)	A	waxy gouldclam
Humilaria kennerleyi (Reeve, 1863)	P	Kennerley venus
Irusella lamellifera (Conrad, 1837)	P	lamellar venus
Liocyma fluctuosa (Gould, 1841)	A, P, Ac	wavy liocyma
Liocyma viride Dall, 1871	P, Ac	green liocyma
Macrocallista maculata (Linnaeus, 1758)	A	calico clam
Macrocallista nimbosa (Lightfoot, 1786)	A	sunray venus
Mercenaria campechiensis (Gmelin, 1791)	A	southern quahog
Mercenaria mercenaria (Linnaeus, 1758)	A, P [I]	northern quahog
Parastarte triquetra (Conrad, 1846)	A	brown gemclam
Periglypta listeri (J. E. Gray, 1838)	A	princess venus
Pitar albidus (Gmelin, 1791)	A	white pitar
Pitar arestus (Dall and Simpson, 1901)	A	pleasing pitar
Pitar cordatus (Schwengel, 1951)	A	Schwengel pitar
Pitar fulminatus (Menke, 1828)	A	lightning pitar
Pitar morrhuanus Linsley, 1848	A	false quahog
Pitar newcombianus (Gabb, 1865)	P	Newcomb pitar
Pitar pilula Rehder, 1943	A	little-pill pitar
Pitar simpsoni (Dall, 1895)	A	Simpson pitar
Pitar zonatus (Dall, 1902)	A	banded pitar
Protothaca granulata (Gmelin, 1791)	A	beaded venus
Protothaca laciniata (Carpenter, 1864)	P	fringed littleneck

SCIENTIFIC NAME	OCCURRENCE	COMMON NAME
Protothaca staminea (Conrad, 1837)	P	Pacific littleneck
Protothaca tenerrima (Carpenter, 1857)	P	thin-shell littleneck
Psephidia cymata Dall, 1913	P	wavy dwarf-venus
Psephidia lordi (Baird, 1863)	P	Lord dwarf-venus
Psephidia ovalis Dall, 1902	P	oval dwarf-venus
Psephidia stephensae Hertlein and Grant, 1972	P	Stephens dwarf-venus
Saxidomus giganteus (Deshayes, 1839)	P	butter clam
Saxidomus nuttalli Conrad, 1837	P	Washington clam
Tapes philippinarum (A. Adams and Reeve, 1850)	P[I]	Japanese littleneck
Tivela abaconis Dall, 1902	A	Abaco tivela
Tivela floridana Rehder, 1939	A	Florida tivela
Tivela mactroides (Born, 1778)	A	trigonal tivela
Tivela stultorum (Mawe, 1823)	P	pismo clam
Transennella confusa S. Gray, 1982	P	confusing transennella
Transennella conradina Dall, 1883	A	Conrad transennella
Transennella cubaniana (d'Orbigny, 1842)	A	Cuba transennella
Transennella stimpsoni Dall, 1902	A	Stimpson transennella
Transennella tantilla (Gould, 1853)	P	purple transennella
Ventricolaria rigida (Dillwyn, 1817)	A	rigid venus
Ventricolaria rugatina (Heilprin, 1887)	A	queen venus

Petricolidae

Petricola californiensis Pilsbry and Lowe, 1933	P	California petricola
Petricola lapicida (Gmelin, 1791)	A	boring petricola
Petricola pholadiformis Lamarck, 1818	A, P	false angelwing
Rupellaria cancellatum A. E. Verrill, 1885	A	cancellate rupellaria
Rupellaria carditoides (Conrad, 1837)	P	hearty rupellaria
Rupellaria tellimyalis (Carpenter, 1864)	P	thin rupellaria
Rupellaria typica (Jonas, 1844)	A	Atlantic rupellaria

Cooperellidae

Cooperella atlantica Rehder, 1943	A	Atlantic cooperclam
Cooperella subdiaphana (Carpenter, 1864)	P	shiny cooperclam

ORDER MYOIDA

Myidae

Cryptomya californica (Conrad, 1837)	P	California softshell
Mya arenaria Linnaeus, 1758	A, P	softshell
Mya profundior Grant and Gale, 1931	P	deep softshell
Mya pseudoarenaria Schlesch, 1931	P, Ac	false softshell
Mya truncata Linnaeus, 1758	A, P, Ac	truncate softshell
Mya uzenensis Nomura and Zinbo, 1937	P	Siberia softshell
Paramya subovata (Conrad, 1845)	A	subovate softshell
Platyodon cancellatus (Conrad, 1837)	P	boring softshell
Sphenia antillensis Dall and Simpson 1901	A	Antillean sphenia
Sphenia luticola (Valenciennes, 1846)	P	Pacific sphenia
Sphenia ovoidea Carpenter, 1864	P	ovoid sphenia
Sphenia tumida J. E. Lewis, 1968	A	tumid sphenia

SCIENTIFIC NAME	OCCURRENCE	COMMON NAME

Corbulidae

Corbula barrattiana C. B. Adams, 1852	A	Barratt corbula
Corbula chittyana C. B. Adams, 1852	A	snubnose corbula
Corbula contracta Say, 1822	A	contracted corbula
Corbula cubaniana d'Orbigny, 1842	A	Cuba corbula
Corbula cymella Dall, 1881	A	wavy corbula
Corbula dietziana C. B. Adams, 1852	A	Dietz corbula
Corbula kelseyi Dall, 1916	P	Kelsey corbula
Corbula krebsiana C. B. Adams, 1852	A	Krebs corbula
Corbula luteola Carpenter, 1864	P	western corbula
Corbula nasuta Sowerby, 1833	P	snubby corbula
Corbula nuciformis Sowerby, 1833	P	nut corbula
Corbula porcella Dall, 1916	P	ribbed corbula
Corbula swiftiana C. B. Adams, 1852	A	Swift corbula
Varicorbula operculata (Philippi, 1848)	A	oval varicorbula

Spheniopsidae

Grippina californica Dall, 1912	P	California grippclam

Gastrochaenidae

Gastrochaena hians (Gmelin, 1791)	A	Atlantic rocellaria
Gastrochaena ovata Sowerby, 1834	A	ovate rocellaria
Gastrochaena stimpsoni Tryon, 1861	A	Stimpson rocellaria
Spengleria rostrata (Spengler, 1783)	A	rostrate rocellaria

Hiatellidae

Cyrtodaria kurriana Dunker, 1861	A, P, Ac	Kurr propellerclam
Cyrtodaria siliqua (Spengler, 1793)	A	northern propellerclam
Hiatella arctica (Linnaeus, 1767)	A, P, Ac	Arctic hiatella
Hiatella azaria (Dall, 1881)	A	dirt hiatella
Hiatella pholadis (Linnaeus, 1771)	A, P, Ac	hole-hugger hiatella
Hiatella striata Fleuriau, 1802	Ac	striate hiatella
Panomya ampla Dall, 1898	P	ample roughmya
Panomya arctica (Lamarck, 1818)	A, P, Ac	Arctic roughmya
Panomya beringiana Dall, 1916	P	Bering roughmya
Panopea abrupta (Conrad, 1849)	P	Pacific geoduck
Panopea bitruncata Conrad, 1872	A	Atlantic geoduck
Saxicavella pacifica Dall, 1916	P	little saxicave

Pholadidae

Barnea subtruncata (Sowerby, 1834)	P	Pacific mud-piddock
Barnea truncata (Say, 1822)	A	Atlantic mud-piddock
Chaceia ovoidea (Gould, 1851)	P	wartneck piddock
Cyrtopleura costata (Linnaeus, 1758)	A	angelwing
Diplothyra smithii Tryon, 1862	A	oyster piddock
Jouannetia quillingi Turner, 1955	A	Quilling piddock
Martesia cuneiformis (Say, 1822)	A	wedge piddock
Martesia fragilis A. E. Verrill and Bush, 1890	A, P	fragile piddock

SCIENTIFIC NAME	OCCURRENCE	COMMON NAME
Martesia striata (Linnaeus, 1758)	A	striate piddock
Netastoma japonicum (Yokoyama, 1920)	P	northern rostrate piddock
Netastoma rostratum (Valenciennes, 1846)	P	rostrate piddock
Parapholas californica (Conrad, 1837)	P	scaleside piddock
Penitella conradi Valenciennes, 1846	P	abalone piddock
Penitella fitchi Turner, 1955	P	Fitch piddock
Penitella "*gabbii* (Tryon, 1863)," auct.	P	Gabb piddock
Penitella "*kamakurensis* (Yokoyama, 1922)," auct.	P	Alaska piddock
Penitella penita (Conrad, 1837)	P	flap-tip piddock
Penitella turnerae Evans and Fisher, 1966	P	Turner piddock
Pholas campechiensis Gmelin, 1791	A	Campeche angelwing
Xylophaga atlantica Richards, 1942	A	Atlantic woodeater
Xylophaga washingtona Bartsch, 1921	P	Washington woodeater
Zirfaea crispata (Linnaeus, 1758)	A	great piddock
Zirfaea pilsbryi Lowe, 1931	P	Pilsbry piddock

Teredinidae

Bankia carinata (J. E. Gray, 1827)	A	carinate shipworm
Bankia fimbriatula Moll and Roch, 1931	A	fimbriate shipworm
Bankia gouldi Bartsch, 1908	A, P	Gould shipworm
Bankia setacea (Tryon, 1863)	P	feathery shipworm
Lyrodus pedicellatus (Quatrefages, 1849)	A, P	blacktip shipworm
Nototeredo knoxi (Bartsch, 1917)	A	Knox shipworm
Nototeredo norvagicus (Spengler, 1792)	A	Norway shipworm
Psiloteredo megotara (Hanley, 1848)	A	big-ear shipworm
Teredo bartschi W. Clapp, 1923	A	Bartsch shipworm
Teredo navalis Linnaeus, 1758	A, P	naval shipworm
Teredora malleolus (Turton, 1822)	A	malleate shipworm

ORDER PHOLADOMYOIDA

Pholadomyidae

Panacca arata (A. E. Verrill and S. Smith, 1881)	A	panacca clam

Lyonsiidae

Entodesma beana (d'Orbigny, 1842)	A	pearly entodesma
Entodesma pictum (Sowerby, 1834)	P	painted entodesma
Entodesma saxicola (Baird, 1863)	P	rock entodesma
Lyonsia arenosa (Möller, 1842)	A, P, Ac	sandy lyonsia
Lyonsia bracteata (Gould, 1850)	P	scaly lyonsia
Lyonsia californica Conrad, 1837	P	California lyonsia
Lyonsia floridana Conrad, 1849	A	Florida lyonsia
Lyonsia gouldii Dall, 1915	P	Gould lyonsia
Lyonsia granulifera A. E. Verrill and Bush, 1898	A	granulate lyonsia
Lyonsia hyalina Conrad, 1831	A	glassy lyonsia
Lyonsia nesiotes Dall, 1915	P	island lyonsia
Lyonsia pugetensis Dall, 1913	P	giant lyonsia

SCIENTIFIC NAME	OCCURRENCE	COMMON NAME
Lyonsia striata Montagu, 1815	A	striate lyonsia
Mytilimeria nuttalli Conrad, 1837	P	bladderclam

Pandoridae

Pandora arenosa Conrad, 1834	A	sandy pandora
Pandora bilirata Conrad, 1855	P	bilirate pandora
Pandora bushiana Dall, 1886	A	Bush pandora
Pandora filosa (Carpenter, 1864)	P	threaded pandora
Pandora glacialis Leach, 1819	A, P, Ac	glacial pandora
Pandora gouldiana Dall, 1886	A	Gould pandora
Pandora inflata Boss and Merrill, 1965	A	inflated pandora
Pandora inornata A. E. Verrill and Bush, 1898	A	inornate pandora
Pandora punctata Conrad, 1837	P	punctate pandora
Pandora trilineata Say, 1822	A	threeline pandora
Pandora wardiana A. Adams, 1859	P	Ward pandora

Thraciidae

Asthenothaerus hemphilli Dall, 1886	A	Hemphill thracid
Asthenothaerus villosior Carpenter, 1864	P	hairy thracid
Bushia elegans (Dall, 1886)	A	elegant bushclam
Cyathodonta dubiosa Dall, 1915	P	doubtful thracid
Cyathodonta semirugosa (Reeve, 1859)	A	wavy thracid
Thracia adamsi MacGinitie, 1959	Ac	Adams thracia
Thracia beringi Dall, 1915	P	Bering thracia
Thracia challisiana Dall, 1915	P	Challis thracia
Thracia conradi Couthouy, 1838	A	Conrad thracia
Thracia corbuloides Blainville, 1824	A	Corbula thracia
Thracia curta Conrad, 1837	P	short thracia
Thracia devexa G. O. Sars, 1878	A, P, Ac	sloping thracia
Thracia distorta Montagu, 1808	A	distorted thracia
Thracia morrisoni R. E. Petit, 1964	A	Morrison thracia
Thracia myopsis Möller, 1842	A, P, Ac	Arctic thracia
Thracia phaseolina Lamarck, 1822	A	kidneybean thracia
Thracia rugosa Lamarck, 1818	A	rugose thracia
Thracia septentrionalis Jeffreys, 1872	A	northern thracia
Thracia stimpsoni Dall, 1886	A	Stimpson thracia
Thracia trapezoides Conrad, 1849	P	trapezoid thracia

Periplomatidae

Halistrepta sulcata (Dall, 1904)	P	sulcate spoonclam
Periploma affine A. E. Verrill and Bush, 1898	A	related spoonclam
Periploma aleuticum (Krause, 1885)	Ac	Aleutian spoonclam
Periploma anguliferum (Philippi, 1847)	A	angled spoonclam
Periploma discus Stearns, 1891	P	round spoonclam
Periploma fragile (Totten, 1835)	A	fragile spoonclam
Periploma leanum (Conrad, 1831)	A	Lea spoonclam
Periploma margaritaceum (Lamarck, 1801)	A	unequal spoonclam
Periploma papyratium (Say, 1822)	A	paper spoonclam
Periploma planiusculum Sowerby, 1834	P	flat spoonclam
Periploma tenerum P. Fischer, 1882	A	delicate spoonclam

SCIENTIFIC NAME	OCCURRENCE	COMMON NAME
Poromyidae		
Poromya albida Dall, 1886	A	white poromya
Poromya elongata Dall, 1886	A	elongate poromya
Poromya granulata (Nyst and Westendorp, 1839)	A, Ac	granular poromya
Poromya rostrata Rehder, 1943	A	rostrate poromya
Poromya trosti Strong and Hertlein, 1937	P	Trost poromya
Verticordiidae		
Halicardia flexuosa (A. E. Verrill and S. Smith, 1881)	A	flexed verticord
Haliris aequacostata (Howard, 1950)	P	even-ribbed verticord
Policordia insculpta (Jeffreys, 1881)	A	sculptureless verticord
Verticordia acuticostata Philippi, 1884	A	sharp-ribbed verticord
Verticordia fischeriana Dall, 1881	A	Fischer verticord
Verticordia ornata (d'Orbigny, 1842)	A, P	ornate verticord
Verticordia seguenzae Dall, 1886	A	Seguenza verticord
Verticordia trapezoides Seguenza, 1876	A	trapezoid verticord
Verticordia woodii E. A. Smith, 1885	A	Wood verticord
Cuspidariidae		
Cardiomya balboae (Dall, 1916)	P	Balboa cardiomya
Cardiomya californica (Dall, 1886)	P	California cardiomya
Cardiomya costata (Sowerby, 1834)	P	costate cardiomya
Cardiomya costellata (Deshayes, 1837)	A	little-ribbed cardiomya
Cardiomya glypta (Bush, 1885)	A	carved cardiomya
Cardiomya isolirata Bernard, 1969	P	lirate cardiomya
Cardiomya oldroydi (Dall, 1924)	P	Oldroyd cardiomya
Cardiomya ornatissima (d'Orbigny, 1842)	A	ornate cardiomya
Cardiomya pectinata (Carpenter, 1864)	P	pectinate cardiomya
Cardiomya perrostrata (Dall, 1881)	A	rostrate cardiomya
Cardiomya planetica (Dall, 1908)	P	planet cardiomya
Cardiomya striata (Jeffreys, 1876)	A, Ac	striate cardiomya
Cuspidaria alternata d'Orbigny, 1846	A	alternate dipperclam
Cuspidaria glacialis (G. O. Sars, 1878)	A, P, Ac	glacial dipperclam
Cuspidaria jeffreysi (Dall, 1881)	A	Jeffrey dipperclam
Cuspidaria media A. E. Verrill and Bush, 1898	A	median dipperclam
Cuspidaria microrhina Dall, 1886	A	little-snout dipperclam
Cuspidaria obesa (Lovén, 1846)	A, Ac	obese dipperclam
Cuspidaria parapodema Bernard, 1969	P	California dipperclam
Cuspidaria pellucida Stimpson, 1853	A	translucent dipperclam
Cuspidaria rostrata (Spengler, 1793)	A, Ac	rostrate dipperclam
Cuspidaria subglacialis Dall, 1913	A	subglacial dipperclam
Leiomya claviculata (Dall, 1881)	A	clavicle myoneria
Myonera lamellifera (Dall, 1881)	A	shingled myoneria
Plectodon granulatus (Dall, 1881)	A	grainy plectodon
Plectodon scaber Carpenter, 1864	P	rough plectodon

SCIENTIFIC NAME	OCCURRENCE	COMMON NAME

CLASS SCAPHOPODA—TUSKSHELLS AND TOOTHSHELLS

ORDER DENTALIIDA

Dentaliidae

Antalis antillarum (d'Orbigny, 1842)	A	Antillean tuskshell
Antalis berryi (A. G. Smith and Gordon, 1948)	P	Berry tuskshell
Antalis ceratum (Dall, 1881)	A	waxy tuskshell
Antalis circumcinctum (Watson, 1879)	A	banded tuskshell
Antalis entale occidentale (Stimpson, 1851)	A	occidental tuskshell
Antalis pilsbryi (Rehder, 1942)	A	Pilsbry tuskshell
Antalis pretiosum (Sowerby, 1860)	P	wampum tuskshell
Antalis taphrium (Dall, 1889)	A	green tuskshell
Dentalium agassizi Pilsbry and Sharp, 1897	P	stained tuskshell
Dentalium americanum Chenu, 1843	A	American tuskshell
Dentalium gouldii Dall, 1889	A	Gould tuskshell
Dentalium laqueatum A. E. Verrill, 1885	A	reticulate tuskshell
Dentalium neohexagonum Sharp and Pilsbry, 1897	P	hexagon tuskshell
Dentalium vallicolens Raymond, 1904	P	trench tuskshell
Fissidentalium carduus (Dall, 1889)	A	thistle tuskshell
Fissidentalium floridense (J. B. Henderson, 1920)	A	yellow tuskshell
Fissidentalium megathyris (Dall, 1890)	P	costate tuskshell
Fissidentalium meridionale (Pilsbry and Sharp, 1897)	A	meridian tuskshell
Graptacme calamus (Dall, 1889)	A	reed tuskshell
Graptacme eborea (Conrad, 1846)	A	ivory tuskshell
Graptacme inversa (Deshayes, 1826)	P	turned tuskshell
Graptacme perlonga (Dall, 1881)	A	slender tuskshell
Graptacme semipolita (Broderip and Sowerby, 1829)	P	half-smooth tuskshell
Graptacme semistriolata (Guilding, 1834)	A	scratched tuskshell
Pseudantalis stenoschizum (Pilsbry and Sharp, 1897)	A	narrow-slit tuskshell

Laevidentaliidae

Laevidentalium callipeplum (Dall, 1889)	A	gorgeous tuskshell
Rhabdus dalli (Pilsbry and Sharp, 1897)	P	Dall tuskshell
Rhabdus rectius (Carpenter, 1865)	P	straight tuskshell
Rhabdus watsoni (Sharp and Pilsbry, 1897)	P	Watson tuskshell

Gadilinidae

Episiphon didymus (Watson, 1879)	A	ovate tuskshell
Episiphon sowerbyi (Guilding, 1834)	A	annulate tuskshell

SCIENTIFIC NAME	OCCURRENCE	COMMON NAME
ORDER GADILIDA		
Entalinidae		
Heteroschismoides callithrix (Dall, 1889)	A	nineside toothshell
Pulsellidae		
Compressidens ophiodon (Dall, 1881)	A	flattened toothshell
Compressidens pressus (Pilsbry and Sharp, 1897)	A	compressed toothshell
Compressidens stearnsii Pilsbry and Sharp, 1897	P	Stearns toothshell
Pulsellum aberrans (Whiteaves, 1887)	P	aberrant toothshell
Pulsellum lobatum (Sowerby, 1860)	A, Ac	lobate toothshell
Pulsellum occidentale (J. B. Henderson, 1920)	A	westerly toothshell
Pulsellum salishorum E. Marshall, 1980	P	Salish toothshell
Pulsellum verrilli (J. B. Henderson, 1920)	A	Verrill toothshell
Siphonodentaliidae		
Polyschides agassizii (Dall, 1881)	A	Agassiz toothshell
Polyschides californicus (Pilsbry and Sharp, 1898)	P	California toothshell
Polyschides carolinensis (Bush, 1885)	A	Carolina toothshell
Polyschides elongatus (J. B. Henderson, 1920)	A	elongate toothshell
Polyschides grandis (A. E. Verrill, 1884)	A	grand toothshell
Polyschides greenlawi (Dall, 1889)	A	Greenlaw toothshell
Polyschides miamiensis J. B. Henderson, 1920	A	Miami toothshell
Polyschides pandionis (A. E. Verrill and S. Smith, 1880)	A	kingly toothshell
Polyschides parvus (J. B. Henderson, 1920)	A	little toothshell
Polyschides quadridentatus (Dall, 1881)	A	fourtooth toothshell
Polyschides rushii (Pilsbry and Sharp, 1898)	A	Rush toothshell
Polyschides spectabilis (A. E. Verrill, 1885)	A	notable toothshell
Polyschides tetrodon (Pilsbry and Sharp, 1898)	A	tetrodon toothshell
Polyschides tolmiei (Dall, 1897)	P	Tolmie toothshell
Polyschides watsoni (Dall, 1881)	A	Watson toothshell
Siphonodentalium quadrifissatum (Pilsbry and Sharp, 1898)	P	fourslit toothshell
Gadilidae		
Cadulus transitorius (J. B. Henderson, 1920)	A	inflated toothshell
Gadila fusiformis Pilsbry and Sharp, 1898	P	fusiform toothshell
Gadila hepburni (Dall, 1897)	P	Hepburn toothshell
Gadila mayori (J. B. Henderson, 1920)	A	Mayor toothshell

SCIENTIFIC NAME	OCCURRENCE	COMMON NAME

CLASS GASTROPODA—GASTROPODS

Order Archaeogastropoda

Pleurotomariidae

Perotrochus adansonianus (Crosse and P. Fischer, 1861)	A	
Perotrochus amabilis (Bayer, 1963)	A	lovely slitsnail
Perotrochus quoyanus (P. Fischer and Bernardi, 1856)	A	

Scissurellidae

Anatoma baxteri J. H. McLean, 1984	P	
Anatoma crispata (Fleming, 1828)	A, P, Ac	crispate scissurelle
Scissurella cingulata O. G. Costa, 1861	A	belt scissurelle
Scissurella lyra S. S. Berry, 1947	P	lyre scissurelle
Scissurella proxima Dall, 1927	A	Florida scissurelle
Scissurella soyoae (Habe, 1951)	P	
Sinezona rimuloides (Carpenter, 1865)	P	rim scissurelle

Haliotididae

Haliotis assimilis Dall, 1878	P	threaded abalone
Haliotis corrugata W. Wood, 1828	P	pink abalone
Haliotis cracherodii Leach, 1814	P	black abalone
Haliotis fulgens Philippi, 1845	P	green abalone
Haliotis kamtschatkana Jonas, 1845	P	pinto abalone
Haliotis pourtalesii Dall, 1881	A	
Haliotis rufescens Swainson, 1822	P	red abalone
Haliotis sorenseni Bartsch, 1940	P	white abalone
Haliotis walallensis Stearns, 1899	P	flat abalone

Fissurellidae

Diodora aguayoi Pérez Farfante, 1943	A	
Diodora arcuata Sowerby, 1862	A	arcuate keyhole limpet
Diodora arnoldi J. H. McLean, 1966	P	neat-rib keyhole limpet
Diodora aspera (Rathke, 1833)	P	rough keyhole limpet
Diodora bermudensis (Dall and Bartsch, 1911)	A	Bermuda keyhole limpet
Diodora cayenensis (Lamarck, 1822)	A	Cayenne keyhole limpet
Diodora dysoni (Reeve, 1850)	A	
Diodora fluviana (Dall, 1889)	A	Gulf Stream keyhole limpet
Diodora habanensis Christiaens, 1975	A	
Diodora jaumei Aguayo and Rehder, 1936	A	
Diodora meta (von Ihering, 1927)	A	
Diodora minuta (Lamarck, 1822)	A	dwarf keyhole limpet
Diodora sayi (Dall, 1899)	A	
Diodora tanneri A. E. Verrill, 1883	A	
Diodora variegata Sowerby, 1862	A	variegate keyhole limpet
Diodora viridula (Lamarck, 1822)	A	green keyhole limpet

SCIENTIFIC NAME	OCCURRENCE	COMMON NAME
Diodora wetmorei Pérez Farfante, 1945	A	
Emarginula dentigera Heilprin, 1889	A	toothed emarginula
Emarginula nordica Pérez Farfante, 1947	A	northern emarginula
Emarginula phrixodes Dall, 1927	A	ruffled emarginula
Emarginula pumila (A. Adams, 1851)	A	pygmy emarginula
Emarginula sicula J. E. Gray, 1825	A	dagger emarginula
Emarginula tuberculosa Libassi, 1859	A	tuberculate emarginula
Fissurella angusta (Gmelin, 1791)	A	narrow keyhole limpet
Fissurella barbadensis (Gmelin, 1791)	A	Barbados keyhole limpet
Fissurella fascicularis Lamarck, 1822	A	wobbly keyhole limpet
Fissurella nodosa (Born, 1778)	A	knobby keyhole limpet
Fissurella punctata P. Fischer, 1857	A	punctate keyhole limpet
Fissurella rosea (Gmelin, 1791)	A	rosy keyhole limpet
Fissurella volcano Reeve, 1849	P	volcano keyhole limpet
Fissurellidea bimaculata Dall, 1871	P	two-spot keyhole limpet
Hemitoma emarginata (Blainville, 1825)	A	emarginate emarginula
Hemitoma octoradiata (Gmelin, 1791)	A	eight-rib emarginula
Laevinesta atlantica (Pérez Farfante, 1947)	A	Atlantic nesta
Lucapina aegis (Reeve, 1850)	A	shield fleshy limpet
Lucapina eolis Pérez Farfante, 1945	A	
Lucapina philippiana (Finlay, 1930)	A	
Lucapina sowerbii (Sowerby, 1835)	A	
Lucapina suffusa (Reeve, 1850)	A	cancellate fleshy limpet
Lucapinella callomarginata (Dall, 1871)	P	rim fleshy limpet
Lucapinella limatula (Reeve, 1850)	A	file fleshy limpet
Megathura crenulata (Sowerby, 1825)	P	giant keyhole limpet
Puncturella asturiana (P. Fischer, 1882)	A	Atlantic puncturella
Puncturella billsae Pérez Farfante, 1947	A	
Puncturella caryophylla Dall, 1914	P	clove puncturella
Puncturella cooperi Carpenter, 1864	P	
Puncturella cucullata (Gould, 1846)	P	hood puncturella
Puncturella decorata I. M. Cowan and J. H. McLean, 1968	P	painted puncturella
Puncturella erecta Dall, 1889	A	erect puncturella
Puncturella galeata (Gould, 1846)	P	helmet puncturella
Puncturella granulata Seguenza, 1863	A	granulate puncturella
Puncturella longifissa Dall, 1914	P	long-slot puncturella
Puncturella major Dall, 1891	P	great puncturella
Puncturella multistriata Dall, 1914	P	many-rib puncturella
Puncturella noachina (Linnaeus, 1771)	A, P, Ac	diluvian puncturella
Puncturella punctocostata S. S. Berry, 1947	P	dot-rib puncturella
Rimula aequisculpta Dall, 1927	A	webbed rimula
Rimula californiana S. S. Berry, 1964	P	California rimula
Rimula dorriae Pérez Farfante, 1947	A	coarse rimula
Rimula frenulata Dall, 1889	A	bridle rimula
Rimula pycnonema Pilsbry, 1943	A	threaded rimula
Scelidotoma bella (Gabb, 1865)	P	elegant emarginula

Acmaeidae

Acmaea funiculata (Carpenter, 1864)	P	corded white limpet
Acmaea mitra Rathke, 1833	P	whitecap limpet
Collisella alveus (Conrad, 1831)	A, P	bowl limpet

SCIENTIFIC NAME	OCCURRENCE	COMMON NAME
Collisella asmi (Middendorff, 1847)	P	black limpet
Collisella borealis Lindberg, 1982	P	boreal limpet
Collisella conus (Test, 1945)	P	
Collisella digitalis (Rathke, 1833)	P	ribbed limpet
Collisella instabilis (Gould, 1846)	P	unstable limpet
Collisella jamaicensis (Gmelin, 1791)	A	Jamaica limpet
Collisella leucopleura (Gmelin, 1791)	A	black-rib limpet
Collisella limatula (Carpenter, 1864)	P	file limpet
Collisella ochracea (Dall, 1871)	P	yellow limpet
Collisella paradigitalis (Fritchman, 1960)	P	
Collisella pelta (Rathke, 1833)	P	shield limpet
Collisella scabra (Gould, 1846)	P	rough limpet
Collisella triangularis (Carpenter, 1864)	P	triangular limpet
Lottia gigantea (Sowerby, 1834)	P	owl limpet
Notoacmea depicta (Hinds, 1842)	P	painted limpet
Notoacmea fenestrata (Reeve, 1855)	P	fenestrate limpet
Notoacmea insessa (Hinds, 1842)	P	seaweed limpet
Notoacmea paleacea (Gould, 1853)	P	surfgrass limpet
Notoacmea persona (Rathke, 1833)	P	mask limpet
Notoacmea scutum (Rathke, 1833)	P	plate limpet
Notoacmea testudinalis (Müller, 1776)	A, P, Ac	plant limpet
Patelloida pustulata (Helbling, 1779)	A	spotted limpet
Problacmaea apicina (Dall, 1879)	P	cap limpet
Problacmaea moskalevi Golikov and Kussakin, 1972	P	
Problacmaea rubella (Fabricius, 1780)	Ac	reddish limpet
Problacmaea sybaritica (Dall, 1871)	P	lush limpet
Rhodopetala rosea (Dall, 1872)	P	pink limpet
Tectura rosacea (Carpenter, 1864)	P	Pacific rosy limpet

Lepetidae

SCIENTIFIC NAME	OCCURRENCE	COMMON NAME
Iothia lindbergi J. H. McLean, 1985	P	
Lepeta alba (Dall, 1869)	P	white blind limpet
Lepeta caeca (Müller, 1776)	A, P, Ac	northern blind limpet
Lepeta caecoides Carpenter, 1865	P, Ac	dead-end blind limpet
Lepeta concentrica Middendorff, 1851	P, Ac	ringed blind limpet

Cocculinidae

SCIENTIFIC NAME	OCCURRENCE	COMMON NAME
Cocculina beanii Dall, 1882	A	
Cocculina casanica Dall, 1919	P	Alaska cocculina
Cocculina rathbuni Dall, 1882	A	
Cocculina reticulata A. E. Verrill, 1885	A	reticulate cocculina

Addisoniidae

SCIENTIFIC NAME	OCCURRENCE	COMMON NAME
Addisonia brophyi J. H. McLean, 1985	P	
Addisonia paradoxa Dall, 1882	A	

Trochidae

SCIENTIFIC NAME	OCCURRENCE	COMMON NAME
Bathybembix bairdii (Dall, 1889)	P	
Calliostoma adelae Schwengel, 1951	A	Keys topsnail

SCIENTIFIC NAME	OCCURRENCE	COMMON NAME
Calliostoma annulatum (Lightfoot, 1786)	P	purple-ring topsnail
Calliostoma bairdii A. E. Verrill and S. Smith, 1880	A	
Calliostoma barbouri Clench and Aguayo, 1946	A	
Calliostoma bernardi J. H. McLean, 1984	P	
Calliostoma canaliculatum (Lightfoot, 1786)	P	channeled topsnail
Calliostoma euglyptum (A. Adams, 1854)	A	sculptured topsnail
Calliostoma fascinans Schwengel and McGinty, 1942	A	enchanting topsnail
Calliostoma gemmulatum Carpenter, 1864	P	gem topsnail
Calliostoma gloriosum Dall, 1871	P	glorious topsnail
Calliostoma javanicum (Lamarck, 1822)	A	chocolate-line topsnail
Calliostoma jujubinum (Gmelin, 1791)	A	mottled topsnail
Calliostoma keenae J. H. McLean, 1970	P	
Calliostoma ligatum (Gould, 1849)	P	blue topsnail
Calliostoma marionae Dall, 1906	A	mahogany topsnail
Calliostoma occidentale (Mighels and C. B. Adams, 1842)	A	boreal topsnail
Calliostoma platinum Dall, 1890	P	silvery topsnail
Calliostoma psyche Dall, 1889	A	psyche topsnail
Calliostoma pulchrum (C. B. Adams, 1850)	A	beautiful topsnail
Calliostoma roseolum Dall, 1881	A	rosy topsnail
Calliostoma sapidum Dall, 1881	A	
Calliostoma splendens Carpenter, 1864	P	splendid topsnail
Calliostoma supragranosum Carpenter, 1864	P	granulose topsnail
Calliostoma tricolor Gabb, 1865	P	tricolor topsnail
Calliostoma turbinum Dall, 1895	P	spindle topsnail
Calliostoma variegatum Carpenter, 1864	P	variegate topsnail
Calliostoma yucatecanum Dall, 1881	A	depressed topsnail
Calliotropis carlotta (Dall, 1889)	P	Charlotte Island spiny margarite
Calliotropis regalis (A. E. Verrill and S. Smith, 1880)	A	regal spiny margarite
Cidarina cidaris (Carpenter, 1864)	P	
Cittarium pica (Linnaeus, 1758)	A	West Indian topsnail
Dentistyla asperrima (Dall, 1881)	A	rough topsnail
Euchelus guttarosea Dall, 1889	A	red-spot topsnail
Gaza superba (Dall, 1881)	A	superb gaza
Halistylus pupoides (Carpenter, 1864)	P	pupa halistyle
Lirularia acuticostata (Carpenter, 1864)	P	sharp-rib lirularia
Lirularia bicostata (J. H. McLean, 1964)	P	two-rib margarite
Lirularia discors J. H. McLean, 1984	P	incongruous lirularia
Lirularia lirulata (Carpenter, 1864)	P	lirulate margarite
Lirularia optabilis (Carpenter, 1864)	P	choice margarite
Lirularia parcipicta (Carpenter, 1864)	P	few-spot lirularia
Lirularia succincta (Carpenter, 1864)	P	girdled lirularia
Lischkeia imperialis (Dall, 1881)	A	imperial spiny margarite
Margarites althorpensis Dall, 1919	P	Port Althorp margarite
Margarites costalis (Gould, 1841)	A, P	boreal rosy margarite
Margarites frigidus Dall, 1919	P, Ac	polar margarite
Margarites giganteus (Leche, 1878)	P, Ac	giant margarite
Margarites groenlandicus (Gmelin, 1791)	A, Ac	Greenland margarite
Margarites healyi Dall, 1919	Ac	
Margarites helicinus (Phipps, 1774)	A, P, Ac	spiral margarite
Margarites hickmanae J. H. McLean, 1984	P	

SCIENTIFIC NAME	OCCURRENCE	COMMON NAME
Margarites keepi A. G. Smith and Gordon, 1948	P	
Margarites multilineatus DeKay, 1843	A	many-line margarite
Margarites olivaceus (T. Brown, 1827)	A, P, Ac	olive margarite
Margarites pupillus (Gould, 1849)	P	little margarite
Margarites rhodia Dall, 1921	P	Pacific rosy margarite
Margarites salmoneus (Carpenter, 1864)	P	salmon margarite
Margarites simbla Dall, 1913	P	beehive margarite
Margarites smithi Bartsch, 1927	P	
Margarites vahlii (Möller, 1842)	A, P, Ac	
Margarites vorticifer (Dall, 1873)	P, Ac	vortex margarite
Microgaza rotella inornata Quinn, 1979	A	
Microgaza rotella rotella Dall, 1881	A	dwarf gaza
Mirachelus clinocnemus Quinn, 1979	A	stooped mirachelus
Mirachelus corbis (Dall, 1889)	A	basket mirachelus
Norrisia norrisi (Sowerby, 1838)	P	norrissnail
Pseudostomatella erythrocoma (Dall, 1889)	A	false stomatella
Solariella intermedia (Leche, 1878)	P	intermediate solarelle
Solariella lacunella (Dall, 1881)	A	channeled solarelle
Solariella laevis Friele, 1886	A	smooth solarelle
Solariella lamellosa A. E. Verrill and S. Smith, 1880	A	lamellose solarelle
Solariella lewisae Willett, 1946	P	
Solariella micraulax J. H. McLean, 1964	P	fine-groove solarelle
Solariella nuda Dall, 1896	P	naked solarelle
Solariella obscura (Couthouy, 1838)	A, P, Ac	obscure solarelle
Solariella peramabilis Carpenter, 1864	P	lovely solarelle
Solariella periscopia (Dall, 1927)	A	look-around solarelle
Solariella rhyssa Dall, 1919	P	wrinkled solarelle
Solariella varicosa (Mighels and C. B. Adams, 1842)	A, P	varicose solarelle
Synaptocochlea picta (d'Orbigny, 1842)	A	painted false stomatella
Tegula aureotincta Forbes, 1850	P	gilded tegula
Tegula brunnea Philippi, 1848	P	brown tegula
Tegula eiseni Jordan, 1936	P	banded tegula
Tegula excavata (Lamarck, 1822)	A	green-base tegula
Tegula fasciata (Born, 1778)	A	silky tegula
Tegula funebralis (A. Adams, 1855)	P	black tegula
Tegula gallina Forbes, 1850	P	speckled tegula
Tegula hotessieriana (d'Orbigny, 1842)	A	Caribbean tegula
Tegula lividomaculata (C. B. Adams, 1845)	A	West Indian tegula
Tegula montereyi (Kiener, 1850)	P	Monterey tegula
Tegula pulligo (Gmelin, 1791)	P	dusky tegula
Tegula regina Stearns, 1892	P	queen tegula
Turcica caffea (Gabb, 1865)	P	two-tooth topsnail

Seguenziidae

SCIENTIFIC NAME	OCCURRENCE	COMMON NAME
Ancistrobasis depressa Dall, 1889	A	depressed basilissa
Seguenzia giovia Dall, 1919	P	California seguenzia
Seguenzia monocingulata Seguenza, 1876	A	pygmy seguenzia

SCIENTIFIC NAME	OCCURRENCE	COMMON NAME
Cyclostrematidae		
Arene bairdii (Dall, 1889)	A	warty cyclostreme
Arene briareus (Dall, 1881)	A	briar cyclostreme
Arene cruentata (Mühlfeld, 1829)	A	star cyclostreme
Arene farallonensis (A. G. Smith, 1952)	P	Farallon cyclostreme
Arene tricarinata (Stearns, 1872)	A	gem cyclostreme
Arene variabilis (Dall, 1889)	A	variable cyclostreme
Arene venustula Aguayo and Rehder, 1936	A	graceful cyclostreme
Cyclostrema cancellatum Marryat, 1818	A	cancellate cyclostreme
Cyclostrema huesonicum Dall, 1927	A	Key West cyclostreme
Cyclostrema tortuganum (Dall, 1927)	A	Tortugas cyclostreme
Liotia fenestrata Carpenter, 1864	P	California cyclostreme
Macrarene cookeana (Dall, 1918)	P	
Parviturbo acuticostatus (Carpenter, 1864)	P	sharp-rib cyclostreme
Parviturbo calidimaris Pilsbry and McGinty, 1945	A	tropical cyclostreme
Parviturbo francesae Pilsbry and McGinty, 1945	A	
Parviturbo rehderi Pilsbry and McGinty, 1945	A	
Parviturbo weberi Pilsbry and McGinty, 1945	A	
Sansonia tuberculata (Watson, 1886)	A	tuberculate cyclostreme
Skeneidae		
Skenea californica (Bartsch, 1907)	P	California skenea
Skenea carmelensis A. G. Smith and Gordon, 1948	P	Carmel skenea
Skenea concordia (Bartsch, 1920)	P	beaded skenea
Skenea coronadoensis (Arnold, 1903)	P	Coronado skenea
Turbinidae		
Astralium phoebium (Röding, 1798)	A	longspine starsnail
Homalopoma albidum (Dall, 1881)	A	white dwarf-turban
Homalopoma baculum Carpenter, 1864	P	berry dwarf-turban
Homalopoma carpenteri (Pilsbry, 1888)	P	
Homalopoma draperi J. H. McLean, 1984	P	
Homalopoma engbergi (Willett, 1929)	P	
Homalopoma grippi (Dall, 1911)	P	
Homalopoma indutum (Watson, 1879)	A	two-faced dwarf-turban
Homalopoma juanense (Dall, 1919)	P	northwest dwarf-turban
Homalopoma luridum (Dall, 1885)	P	dark dwarf-turban
Homalopoma paucicostatum (Dall, 1871)	P	few-rib dwarf-turban
Homalopoma radiatum (Dall, 1918)	P	rayed dwarf-turban
Lithopoma americanum (Gmelin, 1791)	A	American starsnail
Lithopoma caelatum (Gmelin, 1791)	A	carved starsnail
Lithopoma gibberosum (Dillwyn, 1817)	P	red turban
Lithopoma tectum (Lightfoot, 1786)	A	West Indian starsnail
Lithopoma tuber (Linnaeus, 1767)	A	green starsnail
Lithopoma undosum (W. Wood, 1828)	P	wavy turban
Moelleria costulata (Möller, 1842)	A, Ac	ribbed moelleria
Spiromoelleria kachemakensis Baxter and J. H. McLean, 1984	P	
Spiromoelleria quadrae (Dall, 1897)	P	

SCIENTIFIC NAME	OCCURRENCE	COMMON NAME
Turbo cailletii P. Fischer and Bernardi, 1856	A	filose turban
Turbo canaliculatus Hermann, 1781	A	channeled turban
Turbo castanea Gmelin, 1791	A	chestnut turban

Phasianellidae

Tricolia affinis (C. B. Adams, 1850)	A	checkered pheasant
Tricolia bella (M. Smith, 1937)	A	shouldered pheasant
Tricolia compta (Gould, 1855)	P	banded pheasant
Tricolia cruenta Robertson, 1958	A	stained pheasant
Tricolia pterocladica Robertson, 1958	A	rhodophyte pheasant
Tricolia pulloides (Carpenter, 1865)	P	sullied pheasant
Tricolia rubrilineata (Strong, 1928)	P	red-line pheasant
Tricolia substriata (Carpenter, 1864)	P	low-line pheasant
Tricolia thalassicola Robertson, 1958	A	turtlegrass pheasant

Neritidae

Nerita fulgurans Gmelin, 1791	A	Antillean nerite
Nerita peloronta Linnaeus, 1758	A	bleeding tooth
Nerita tessellata Gmelin, 1791	A	checkered nerite
Nerita versicolor Gmelin, 1791	A	four-tooth nerite
Neritina clenchi Russell, 1940	A (E), F	
Neritina reclivata (Say, 1822)	A (E), F	olive nerite
Neritina virginea (Linnaeus, 1758)	A (E)	virgin nerite
Puperita pupa (Linnaeus, 1767)	A	zebra nerite
Smaragdia viridis (Linnaeus, 1758)	A	emerald nerite

Phenacolepadidae

Phenacolepas hamillei (P. Fischer, 1857)	A	

Helicinidae

Alcadia striata (Lamarck, 1822)	T[I]	striate drop
Helicina clappi Pilsbry, 1909	T	rainbow drop
Hendersonia occulta (Say, 1831)	T	cherrystone drop
Lucidella tantilla Pilsbry, 1902	T	ochre drop
Oligyra orbiculata Say, 1818	T	globular drop

ORDER MESOGASTROPODA

Valvatidae

Valvata bicarinata I. Lea, 1841	F	two-ridge valvata
Valvata humeralis Say, 1829	F	glossy valvata
Valvata lewisi Currier, 1868	F	fringed valvata
Valvata mergella Westerlund, 1883	F	rams-horn valvata
Valvata perdepressa Walker, 1906	F	purplecap valvata
Valvata piscinalis (Müller, 1774)	F	European stream valvata
Valvata sincera Say, 1824	F	mossy valvata
Valvata tricarinata (Say, 1817)	F	threeridge valvata
Valvata utahensis Call, 1884	F	desert valvata

SCIENTIFIC NAME	OCCURRENCE	COMMON NAME
Valvata virens Tryon, 1863	F	emerald valvata
Valvata winnebagoensis F. C. Baker, 1928	F	flanged valvata

Viviparidae

Campeloma brevispirum F. C. Baker, 1928	F	*classification uncertain*
Campeloma crassulum Rafinesque, 1819	F	ponderous campeloma
Campeloma decampi (W. G. Binney, 1865)	F	slender campeloma
Campeloma decisum (Say, 1817)	F	pointed campeloma
Campeloma exile (Anthony, 1860)	F	*classification uncertain*
Campeloma floridense Call, 1886	F	purple-throat campeloma
Campeloma geniculum (Conrad, 1834)	F	ovate campeloma
Campeloma gibbum (Currier, 1867)	F	*classification uncertain*
Campeloma leptum Mattox, 1940	F	*classification uncertain*
Campeloma lewisi Walker, 1915	F	*classification uncertain*
Campeloma limum (Anthony, 1860)	F	file campeloma
Campeloma milesi (I. Lea, 1863)	F	*classification uncertain*
Campeloma parthenum Vail, 1979	F	maiden campeloma
Campeloma regulare (I. Lea, 1841)	F	cylinder campeloma
Campeloma rufum (Haldeman, 1841)	F	*classification uncertain*
Campeloma tannum Mattox, 1940	F	*classification uncertain*
Cipangopaludina chinensis malleata (Reeve, 1863)	F [I]	Chinese mysterysnail
Cipangopaludina japonica (von Martens, 1861)	F [I]	Japanese mysterysnail
Lioplax cyclostomaformis (I. Lea, 1841)	F	cylindrical lioplax
Lioplax pilsbryi Walker, 1905	F	Choctaw lioplax
Lioplax subcarinata (Say, 1816)	F	ridged lioplax
Lioplax sulculosa (Menke, 1827)	F	furrowed lioplax
Lioplax talquinensis Vail, 1979	F	*classification uncertain*
Tulotoma magnifica (Conrad, 1834)	F [X]	tulotoma
Viviparus georgianus (I. Lea, 1834)	F	banded mysterysnail
Viviparus intertextus (Say, 1829)	F	rotund mysterysnail
Viviparus subpurpureus (Say, 1829)	F	olive mysterysnail

Pilidae

Marisa cornuarietis (Linnaeus, 1758)	F [I]	giant rams-horn
Pomacea bridgesi (Reeve, 1856)	F	spiketop applesnail
Pomacea paludosa (Say, 1829)	F	Florida applesnail

Bithyniidae

Bithynia tentaculata (Linnaeus, 1758)	F	mud bithynia

Hydrobiidae

Amnicola aldrichi (Call and Beecher, 1886)	F	Hoosier amnicola
Amnicola bakerianus Pilsbry, 1917	F	*classification uncertain*
Amnicola browni Carpenter, 1872	F	slender duskysnail
Amnicola clarki Pilsbry, 1917	F	*classification uncertain*
Amnicola cora Hubricht, 1979	F	Foushee cavesnail
Amnicola dalli (Pilsbry and Beecher, 1892)	F	peninsula amnicola
Amnicola decisus Haldeman, 1845	F	*classification uncertain*
Amnicola granum (Say, 1822)	F	squat duskysnail

SCIENTIFIC NAME	OCCURRENCE	COMMON NAME
Amnicola greggi Pilsbry, 1935	F	Rocky Mountain duskysnail
Amnicola limosus (Say, 1817)	F	mud amnicola
Amnicola missouriensis Pilsbry, 1898	F	Missouri amnicola
Amnicola pilsbry Walker, 1906	F	lake duskysnail
Amnicola proserpina Hubricht, 1940	F	Proserpine cavesnail
Amnicola pupoideus (Gould, 1841)	F	pupa duskysnail
Amnicola retromargo F. G. Thompson, 1968	F	indented duskysnail
Amnicola rhombostoma F. G. Thompson, 1968	F	squaremouth amnicola
Amnicola stygius Hubricht, 1971	F	stygian amnicola
Amnicola walkeri Pilsbry, 1898	F	Canadian duskysnail
Antrobia culveri Hubricht, 1972	F	Tumbling Creek cavesnail
Antroselatus spiralis Hubricht, 1963	F	shaggy cavesnail
Aphaostracon asthenes F. G. Thompson, 1968	F	Blue Spring hydrobe
Aphaostracon chalarogyrus F. G. Thompson, 1968	F	freemouth hydrobe
Aphaostracon hypohyalinum F. G. Thompson, 1968	F	Suwanee hydrobe
Aphaostracon monas (Pilsbry, 1899)	F	Wekiwa hydrobe
Aphaostracon pachynotum F. G. Thompson, 1968	F	thick--shell hydrobe
Aphaostracon pycnum F. G. Thompson, 1968	F	dense hydrobe
Aphaostracon rhadinum F. G. Thompson, 1968	F	slough hydrobe
Aphaostracon theiocrenetum F. G. Thompson, 1968	F	Clifton Spring hydrobe
Aphaostracon xynoelictum F. G. Thompson, 1968	F	Fenney Spring hydrobe
Birgella subglobosus (Say, 1825)	F	globe siltsnail
Bythinella hemphilli Pilsbry, 1890	F	*classification uncertain*
Cincinnatia comalensis (Pilsbry and Ferriss, 1906)	F	Comal siltsnail
Cincinnatia floridana (Frauenfeld, 1863)	F	hyacinth siltsnail
Cincinnatia fraterna F. G. Thompson, 1968	F	creek siltsnail
Cincinnatia helicogyra F. G. Thompson, 1968	F	crystal siltsnail
Cincinnatia integra (Say, 1829)	F	midland siltsnail
Cincinnatia mica F. G. Thompson, 1968	F	Ichetucknee siltsnail
Cincinnatia monroensis (Dall, 1885)	F	Enterprise siltsnail
Cincinnatia parva F. G. Thompson, 1968	F	pygmy siltsnail
Cincinnatia peracuta Pilsbry and Walker, 1889	F	*classification uncertain*
Cincinnatia petrifons F. G. Thompson, 1968	F	Rock Springs siltsnail
Cincinnatia ponderosa F. G. Thompson, 1968	F	ponderous siltsnail
Cincinnatia vanhyningi (Vanatta, 1934)	F	Seminole siltsnail
Cincinnatia wekiwae F. G. Thompson, 1968	F	Wekiwa siltsnail
Clappia cahabensis Clench, 1965	F	Cahaba pebblesnail
Clappia umbilicata (Walker, 1904)	F [X]	umbilicate pebblesnail
Cochliopa texana Pilsbry, 1935	F	*classification uncertain*
Cochliopina riograndensis (Pilsbry and Ferriss, 1906)	F	spiral pebblesnail
Floridiscrobs dysbatus (Pilsbry and McGinty, 1949)	A	
Fluminicola avernalis Pilsbry, 1935	F	Moapa pebblesnail
Fluminicola columbiana Hemphill, 1899	F	Columbia pebblesnail
Fluminicola erythopoma Pilsbry, 1899	F	Ash Meadows pebblesnail
Fluminicola fuscus (Haldeman, 1847)	F	ashy pebblesnail
Fluminicola hindsi (Baird, 1863)	F	vagrant pebblesnail
Fluminicola merriami (Pilsbry and Beecher, 1892)	F	Pahranagat pebblesnail
Fluminicola minutissimus Pilsbry, 1907	F	pixie pebblesnail
Fluminicola modoci Hannibal, 1912	F	Modoc pebblesnail
Fluminicola nevadensis Walker, 1916	F	Cortez Hills pebblesnail

SCIENTIFIC NAME	OCCURRENCE	COMMON NAME
Fluminicola nuttallianus (I. Lea, 1838)..........	F	.... dusky pebblesnail
Fluminicola seminalis (Hinds, 1842).............	F	............nugget pebblesnail
Fluminicola turbiniformis (Tryon, 1865)..........	F	............turban pebblesnail
Fluminicola virens (I. Lea, 1838)	F	.... Olympia pebblesnail
Fontelicella californiensis Gregg and Taylor, 1965 ...	F	.... Laguna Mountain springsnail
Fontelicella deserta (Pilsbry, 1916)..............	F	.... desert springsnail
Fontelicella hendersoni (Pilsbry, 1933)...........	F	........Harney Lake springsnail
Fontelicella idahoensis (Pilsbry, 1933)	F	............Idaho springsnail
Fontelicella intermedia (Tryon, 1865)............	F	..... Crooked Creek springsnail
Fontelicella micrococcus (Pilsbry, 1893).........	F	.... Oasis Valley springsnail
Fontelicella neomexicana (Pilsbry, 1916)	F	.... Socorro springsnail
Fontelicella pilsbryana (J. L. Baily and R. I. Baily, 1952)..............	F	.... Bear Lake springsnail
Fontelicella robusta (Walker, 1908)	F	.... Jackson Lake springsnail
Fontelicella stearnsiana (Pilsbry, 1899)	F	............Yaqui springsnail
Fontigens antroecetes Hubricht, 1972	F	.........*classification uncertain*
Fontigens binneyana (Hannibal, 1912)	F	.........*classification uncertain*
Fontigens cryptica Hubricht, 1963	F	.... hidden springsnail
Fontigens holsingeri Hubricht, 1976.............	F	............tapered cavesnail
Fontigens nickliniana (I. Lea, 1838)	F	............watercress snail
Fontigens orolibas Hubricht, 1957	F	.... Blue Ridge springsnail
Fontigens tartarea Hubricht, 1963	F	.... organ cavesnail
Fontigens turritella Hubricht, 1976...............	F	........Greenbrier cavesnail
Fontigens weberi Pilsbry, 1950	F	.........*classification uncertain*
Gillia altilis (I. Lea, 1841)......................	F	..... buffalo pebblesnail
Hadoceras taylori Hershler, 1986................	F	.... nymph trumpet
Heleobops docima F. G. Thompson, 1968........	E	.... oolite hydrobe
Hoyia sheldoni (Pilsbry, 1890)...................	F	............storm hydrobe
Hyalopyrgus aequicostatus (Pilsbry, 1890)........	F	.......smooth-rib hydrobe
Hyalopyrgus brevissimus (Pilsbry, 1890)	F	.... regal hydrobe
Hydrobia booneae Morrison, 1973	A (E)	.... Boone hydrobe
Hydrobia totteni Morrison, 1954................	A (E)	.... minute hydrobe
Lepyrium showalteri (I. Lea, 1861)...............	F	.... flat pebblesnail
Littoridinops monroensis (Frauenfeld, 1863)	F, E	.... cockscomb hydrobe
Littoridinops palustris F. G. Thompson, 1968	E	.... bantum hydrobe
Littoridinops tenuipes Couper, 1844	F, E	.... henscomb hydrobe
Marstonia agarhecta F. G. Thompson, 1969	F	.... Ocmulgee marstonia
Marstonia arga F. G. Thompson, 1977	F	...........ghost marstonia
Marstonia castor F. G. Thompson, 1977	F	.........beaverpond marstonia
Marstonia halcyon F. G. Thompson, 1977........	F	........halcyon marstonia
Marstonia lustrica (Pilsbry, 1890)..............	F	.... boreal marstonia
Marstonia ogmorhaphe F. G. Thompson, 1977	F	.... royal marstonia
Marstonia olivacea (Pilsbry, 1895)	F	.... olive marstonia
Marstonia pachyta F. G. Thompson, 1977........	F	.... armored marstonia
Notogillia sathon F. G. Thompson, 1969.........	F	.... satyr siltsnail
Notogillia wetherbyi (Dall, 1885)................	F	.... alligator siltsnail
Onobops crassus F. G. Thompson, 1968	E	.... spartina hydrobe
Onobops jacksoni (Bartsch, 1953)................	E	.... fine-lined hydrobe
Paludestrina bottimeri Walker, 1925.............	F	.........*classification uncertain*
Phreatodrobia conica Hershler and Longley, 1986 .	F	.... Hueco cavesnail
Phreatodrobia imitata Hershler and Longley, 1986................	F	............mimic cavesnail

SCIENTIFIC NAME	OCCURRENCE	COMMON NAME
Phreatodrobia micra (Pilsbry and Ferriss, 1906) ...	F	flattened cavesnail
Phreatodrobia nugax (Pilsbry and Ferriss, 1906) ...	F	domed cavesnail
Phreatodrobia plana Hershler and Longley, 1986 ..	F	disc cavesnail
Phreatodrobia punctata Hershler and Longley, 1986.................	F	high-hat cavesnail
Phreatodrobia rotunda Hershler and Longley, 1986..	F	beaked cavesnail
Probythinella lacustris (F. C. Baker, 1928)	F	delta hydrobe
Pyrgophorus platyrachis F. G. Thompson, 1968 ...	F	serrate crownsnail
Pyrgophorus spinosus (Call and Pilsbry, 1886).....	F	spiny crownsnail
Pyrgulopsis archimedes S. S. Berry, 1947	F	Archimedes pyrg
Pyrgulopsis letsoni (Walker, 1901)	F	gravel pyrg
Pyrgulopsis nevadensis (Stearns, 1883)...........	F	corded pyrg
Pyrgulopsis ozarkensis Hinkley, 1915.............	F	Ozark pyrg
Pyrgulopsis scalariformis (Wolf, 1869)	F	moss pyrg
Rhapinema dacryon F. G. Thompson, 1969.......	F	teardrop snail
Somatogyrus alcoviensis Krieger, 1972	F	reverse pebblesnail
Somatogyrus amnicoloides Walker, 1915	F	Oachita pebblesnail
Somatogyrus aureus Tryon, 1865	F	golden pebblesnail
Somatogyrus biangulatus Walker, 1906	F	angular pebblesnail
Somatogyrus constrictus Walker, 1904............	F	knotty pebblesnail
Somatogyrus coosaensis Walker, 1904............	F	Coosa pebblesnail
Somatogyrus crassilabris Walker, 1915	F	thick-lip pebblesnail
Somatogyrus crassus Walker, 1904...............	F	stocky pebblesnail
Somatogyrus currierianus (I. Lea, 1863)	F	Tennessee pebblesnail
Somatogyrus decipiens Walker, 1909	F	hidden pebblesnail
Somatogyrus depressus (Tryon, 1862)	F	sandbar pebblesnail
Somatogyrus excavatus Walker, 1906............	F	ovate pebblesnail
Somatogyrus georgianus Walker, 1904............	F	Cherokee pebblesnail
Somatogyrus hendersoni Walker, 1909............	F	fluted pebblesnail
Somatogyrus hinkleyi Walker, 1904	F	granite pebblesnail
Somatogyrus humerosus Walker, 1906............	F	Atlas pebblesnail
Somatogyrus integra (Say, 1829)	F	Ohio pebblesnail
Somatogyrus nanus Walker, 1904................	F	dwarf pebblesnail
Somatogyrus obtusus Walker, 1904	F	moon pebblesnail
Somatogyrus parvulus Tryon, 1865...............	F	sparrow pebblesnail
Somatogyrus pennsylvanicus Walker, 1904	F	shale pebblesnail
Somatogyrus pilsbryanus Walker, 1904	F	Tallapoosa pebblesnail
Somatogyrus pumilus (Conrad, 1834).............	F	compact pebblesnail
Somatogyrus pygmaeus Walker, 1909	F	pygmy pebblesnail
Somatogyrus quadratus Walker, 1906	F	quadrate pebblesnail
Somatogyrus rheophilas F. G. Thompson, 1984 ...	F	flint pebblesnail
Somatogyrus sargenti Pilsbry, 1895	F	mud pebblesnail
Somatogyrus strengi Pilsbry and Walker, 1906	F	rolling pebblesnail
Somatogyrus substriatus Walker, 1906............	F	Choctaw pebblesnail
Somatogyrus tenax F. G. Thompson, 1969	F	Savannah pebblesnail
Somatogyrus tennesseensis Walker, 1906..........	F	opaque pebblesnail
Somatogyrus trothis Doherty, 1878...............	F	*classification uncertain*
Somatogyrus tryoni Pilsbry and F. C. Baker, 1927.................	F	coldwater pebblesnail
Somatogyrus virginicus Walker, 1904.............	F	panhandle pebblesnail
Somatogyrus walkerianus Aldrich, 1905...........	F	Gulf coast pebblesnail
Somatogyrus wheeleri Walker, 1915..............	F	channelled pebblesnail

SCIENTIFIC NAME	OCCURRENCE	COMMON NAME
Spilochlamys conica F. G. Thompson, 1968	F	conical siltsnail
Spilochlamys gravis F. G. Thompson, 1968	F	armored siltsnail
Spilochlamys turgida F. G. Thompson, 1969	F	pumpkin siltsnail
Spurwinkia salsa (Pilsbry, 1905)	F	saltmarsh hydrobe
Stiobia nana F. G. Thompson, 1978	F	sculpin snail
Stygopyrgus bartonensis Hershler and Longley, 1986	F	Barton cavesnail
Texadina spinctostoma (Abbott and Ladd, 1953)	E	narrowmouth hydrobe
Tryonia cheatumi (Pilsbry, 1935)	F	Phantom tryonia
Tryonia clathrata Stimpson, 1865	F	grated tryonia
Tryonia diaboli (Pilsbry and Ferriss, 1906)	F	devil tryonia
Tryonia imitator (Pilsbry, 1899)	F	mimic tryonia
Tryonia protea (Gould, 1855)	F	desert tryonia

Pomatiopsidae

Pomatiopsis binneyi Tryon, 1863	F	robust walker
Pomatiopsis californica Pilsbry, 1899	F	Pacific walker
Pomatiopsis chacei Pilsbry, 1937	F	marsh walker
Pomatiopsis cincinnatiensis (I. Lea, 1840)	F	brown walker
Pomatiopsis hinkleyi Pilsbry, 1896	F	classification uncertain
Pomatiopsis lapidaria (Say, 1817)	F	slender walker

Thiaridae

Melanoides tuberculatus (Müller, 1774)	F	red-rim melania
Melanoides turriculus (I. Lea, 1850)	F	fawn melania
Tarebia granifera (Lamarck, 1822)	F	quilted melania

Pleuroceridae

Elimia acuta (I. Lea, 1831)	F	acute elimia
Elimia alabamensis (I. Lea, 1861)	F	mud elimia
Elimia albanyensis (I. Lea, 1864)	F	black-crest elimia
Elimia ampla (Anthony, 1854)	F	ample elimia
Elimia annettae (Goodrich, 1841)	F	Lilyshoals elimia
Elimia arachnoidea (Anthony, 1854)	F	spider elimia
Elimia aterina (I. Lea, 1863)	F	coal elimia
Elimia athearni (Clench and Turner, 1956)	F	knobby elimia
Elimia bellacrenata (Haldeman, 1841)	F	princess elimia
Elimia bellula (I. Lea, 1861)	F	walnut elimia
Elimia bentoniensis (I. Lea, 1862)	F	rusty elimia
Elimia boykiniana (I. Lea, 1840)	F	flaxen elimia
Elimia brevis (Reeve, 1860)	F	short-spire elimia
Elimia caelatura (Reeve, 1860)	F	rippled elimia
Elimia cahawbensis (I. Lea, 1841)	F	Cahaba elimia
Elimia capillaris (I. Lea, 1861)	F	spindle elimia
Elimia carinicostata (I. Lea, 1854)	F	fluted elimia
Elimia carinifera (Lamarck, 1822)	F	sharp-crest elimia
Elimia catenaria (Say, 1822)	F	gravel elimia
Elimia chiltonensis (Goodrich, 1941)	F	prune elimia
Elimia clara (Anthony, 1854)	F	riffle elimia
Elimia clausa (I. Lea, 1861)	F [X]	closed elimia
Elimia clavaeformis (I. Lea, 1841)	F	club elimia

SCIENTIFIC NAME	OCCURRENCE	COMMON NAME
Elimia clenchi (Goodrich, 1924)	F	slackwater elimia
Elimia cochilaris (I. Lea, 1868)	F	cockle elimia
Elimia comalensis (Pilsbry, 1890)	F	Balcones elimia
Elimia comma (Conrad, 1834)	F	hispid elimia
Elimia costifera (Reeve, 1861)	F	corded elimia
Elimia crenatella (I. Lea, 1860)	F	lacey elimia
Elimia curreyana (I. Lea, 1841)	F	amber elimia
Elimia curvicostata (Reeve, 1861)	F	graphite elimia
Elimia cylindracea (Conrad, 1834)	F	cylinder elimia
Elimia dickinsoni (Clench and Turner, 1956)	F	stately elimia
Elimia dislocata (Reeve, 1861)	F	lapped elimia
Elimia ebenum (I. Lea, 1841)	F	ebony elimia
Elimia edgariana (I. Lea, 1841)	F	Cumberland elimia
Elimia fascinans (I. Lea, 1861)	F	banded elimia
Elimia flava (I. Lea, 1862)	F	yellow elimia
Elimia floridensis (Reeve, 1860)	F	rasp elimia
Elimia fusiformis (I. Lea, 1861)	F [X]	fusiform elimia
Elimia gerhardti (I. Lea, 1862)	F	coldwater elimia
Elimia hartmaniana (I. Lea, 1861)	F [X]	high-spired elimia
Elimia haysiana (I. Lea, 1843)	F	silt elimia
Elimia hydei (Conrad, 1834)	F	gladiator elimia
Elimia impressa (I. Lea, 1841)	F [X]	constricted elimia
Elimia inclinans (I. Lea, 1862)	F	slanted elimia
Elimia interrupta (Haldeman, 1840)	F	knotty elimia
Elimia interveniens (I. Lea, 1862)	F	slowwater elimia
Elimia jonesi (Goodrich, 1936)	F [X]	hearty elimia
Elimia laeta (Jay, 1839)	F [X]	ribbed elimia
Elimia laqueata (Say, 1829)	F	panel elimia
Elimia livescens (Menke, 1830)	F	liver elimia
Elimia mutabilis (I. Lea, 1862)	F	oak elimia
Elimia nassula (Conrad, 1834)	F	round-rib elimia
Elimia olivula (Conrad, 1834)	F	caper elimia
Elimia paupercula (I. Lea, 1862)	F	sooty elimia
Elimia perstriata (I. Lea, 1852)	F	engraved elimia
Elimia pilsbryi (Goodrich, 1927)	F [X]	rough-lined elimia
Elimia plicatostriata (Wetherby, 1876)	F	carved elimia
Elimia porrecta (I. Lea, 1863)	F	nymph elimia
Elimia potosiensis (I. Lea, 1841)	F	pyramid elimia
Elimia proxima (Say, 1825)	F	sprite elimia
Elimia pupaeformis (I. Lea, 1864)	F [X]	pupa elimia
Elimia pupoidea (Anthony, 1854)	F	bot elimia
Elimia pybasi (I. Lea, 1862)	F	spring elimia
Elimia pygmaea (H. H. Smith, 1936)	F [X]	pygmy elimia
Elimia semicarinata (Say, 1829)	F	fine-ridged elimia
Elimia showalteri (I. Lea, 1860)	F	compact elimia
Elimia simplex (Say, 1825)	F	smooth elimia
Elimia striatula (I. Lea 1842)	F	file elimia
Elimia strigosa (I. Lea, 1841)	F	brook elimia
Elimia symmetrica (Haldeman, 1841)	F	symmetrical elimia
Elimia taitiana (I. Lea, 1841)	F	dented elimia
Elimia teres (I. Lea, 1841)	F	elegant elimia
Elimia troostiana (I. Lea, 1838)	F	mossy elimia
Elimia vanhyningiana (Goodrich, 1921)	F	goblin elimia
Elimia vanuxemiana (I. Lea, 1843)	F	cobble elimia

SCIENTIFIC NAME	OCCURRENCE	COMMON NAME
Elimia varians (I. Lea, 1861)	F [X]	puzzle elimia
Elimia variata (I. Lea, 1861)	F	squat elimia
Elimia viennaensis (I. Lea, 1862)	F	slough elimia
Elimia virginica (Say, 1817)	F	Piedmont elimia
Gyrotoma excisa (I. Lea, 1843)	F [X]	excised slitshell
Gyrotoma lewisii (I. Lea, 1869	F [X]	striate slitshell
Gyrotoma pagoda (I. Lea, 1845)	F [X]	pagoda slitshell
Gyrotoma pumila (I. Lea, 1860)	F [X]	ribbed slitshell
Gyrotoma pyramidata (Shuttleworth, 1845)	F [X]	pyramid slitshell
Gyrotoma walkeri (H. H. Smith, 1924)	F [X]	round slitshell
Io fluvialis (Say, 1825)	F	spiny riversnail
Juga acutifilosa (Stearns, 1890)	F	topaz juga
Juga bulbosa (Gould, 1847)	F	bulb juga
Juga hemphilli (J. Henderson, 1935)	F	barren juga
Juga interioris (Goodrich, 1944)	F	smooth juga
Juga laurae (Goodrich, 1944)	F	oasis juga
Juga nigrina (I. Lea, 1856)	F	black juga
Juga occata (Hinds, 1844)	F	scalloped juga
Juga plicifera (I. Lea, 1838)	F	pleated juga
Juga silicula (Gould, 1847)	F	glassy juga
Leptoxis ampla (Anthony, 1855)	F	round rocksnail
Leptoxis arkansensis (Hinkley, 1915)	F	Arkansas mudalia
Leptoxis carinata (Bruguière, 1792)	F	crested mudalia
Leptoxis clipeata (H. H. Smith, 1922)	F [X]	agate rocksnail
Leptoxis compacta (Anthony, 1854)	F	oblong rocksnail
Leptoxis crassa (Haldeman, 1841)	F	boulder snail
Leptoxis dilatata (Conrad, 1835)	F	seep mudalia
Leptoxis formanii (I. Lea, 1843)	F [X]	interrupted rocksnail
Leptoxis formosa (I. Lea, 1860)	F	maiden rocksnail
Leptoxis ligata (Anthony, 1860)	F [X]	rotund rocksnail
Leptoxis lirata (H. H. Smith, 1922)	F [X]	lirate rocksnail
Leptoxis melanoidus (Conrad, 1834)	F	black mudalia
Leptoxis minor (Hinkley, 1912)	F	knob mudalia
Leptoxis occultata (H. H. Smith, 1922)	F [X]	bigmouth rocksnail
Leptoxis picta (Conrad, 1834)	F	spotted rocksnail
Leptoxis plicata (Conrad, 1834)	F	plicate rocksnail
Leptoxis praerosa (Say, 1821)	F	onyx rocksnail
Leptoxis showalterii (I. Lea, 1860)	F [X]	Coosa rocksnail
Leptoxis taeniata (Conrad, 1834)	F	painted rocksnail
Leptoxis trilineata (Say, 1829)	F	broad mudalia
Leptoxis umbilicata (Wetherby, 1876)	F	umbilicate rocksnail
Leptoxis virgata (I. Lea, 1841)	F	smooth mudalia
Leptoxis vittata (I. Lea, 1860)	F [X]	striped rocksnail
Lithasia armigera (Say, 1821)	F	armored rocksnail
Lithasia curta (I. Lea, 1868)	F	knobby rocksnail
Lithasia duttoniana (I. Lea, 1841)	F	helmet rocksnail
Lithasia geniculata (Haldeman, 1840)	F	ornate rocksnail
Lithasia hubrichti Clench, 1956	F	Big Black rocksnail
Lithasia jayana (I. Lea, 1841)	F	rugose rocksnail
Lithasia lima (Conrad, 1834)	F	warty rocksnail
Lithasia obovata (Say, 1829)	F	Shawnee rocksnail
Lithasia salebrosa (Conrad, 1834)	F	muddy rocksnail
Lithasia verrucosa (Rafinesque, 1820)	F	varicose rocksnail
Pleurocera acuta Rafinesque, 1831	F	sharp hornsnail

SCIENTIFIC NAME	OCCURRENCE	COMMON NAME
Pleurocera alveare (Conrad, 1834)	F	rugged hornsnail
Pleurocera annulifera (Conrad, 1834)	F	ringed hornsnail
Pleurocera brumbyi (I. Lea, 1852)	F	spiral hornsnail
Pleurocera canaliculata (Say, 1821)	F	silty hornsnail
Pleurocera corpulenta (Anthony, 1854)	F	corpulent hornsnail
Pleurocera curta (Haldeman, 1841)	F	shortspire hornsnail
Pleurocera foremani (I. Lea, 1843)	F	rough hornsnail
Pleurocera gradata (Anthony, 1854)	F	bottle hornsnail
Pleurocera nobilis (I. Lea, 1845)	F	noble hornsnail
Pleurocera parva (I. Lea, 1862)	F	dainty hornsnail
Pleurocera postelli (I. Lea, 1862)	F	broken hornsnail
Pleurocera prasinata (Conrad, 1834)	F	smooth hornsnail
Pleurocera pyrenella (Conrad, 1834)	F	skirted hornsnail
Pleurocera showalteri (I. Lea, 1862)	F	upland hornsnail
Pleurocera trochiformis (Conrad, 1834)	F	sulcate hornsnail
Pleurocera uncialis (Reeve, 1861)	F	pagoda hornsnail
Pleurocera vestita (Conrad, 1834)	F	brook hornsnail
Pleurocera walkeri Goodrich, 1928	F	telescope hornsnail

Annulariidae

Chondropoma dentatum (Say, 1825)	T	crenulate horn

Lacunidae

Haloconcha minor Dall, 1919	P	lesser lacuna
Haloconcha reflexa (Dall, 1884)	P	reflexed lacuna
Lacuna carinata Gould, 1848	P	carinate lacuna
Lacuna crassior (Montagu, 1803)	A, P	thick lacuna
Lacuna marmorata Dall, 1919	P	chink snail
Lacuna pallidula (E. M. da Costa, 1778)	A	pale lacuna
Lacuna parva (E. M. da Costa, 1778)	A	tiny lacuna
Lacuna succinea S. S. Berry, 1953	P	amber lacuna
Lacuna unifasciata Carpenter, 1856	P	one-band lacuna
Lacuna vaginata Dall, 1918	P	delightful lacuna
Lacuna variegata Carpenter, 1864	P	variegate lacuna
Lacuna vincta (Montagu, 1803)	A, P, Ac	northern lacuna

Littorinidae

Algamorda newcombiana (Hemphill, 1876)	P	
Littorina angulifera (Lamarck, 1822)	A	mangrove periwinkle
Littorina angustior (Mörch, 1876)	A	slender periwinkle
Littorina irrorata (Say, 1822)	A	marsh periwinkle
Littorina keenae Rosewater, 1978	P	eroded periwinkle
Littorina lineolata d'Orbigny, 1840	A	lineolate periwinkle
Littorina littorea (Linnaeus, 1758)	A	common periwinkle
Littorina meleagris (Potiez and Michaud, 1838)	A	white-spot periwinkle
Littorina mespillum (Mühlfeld, 1824)	A	brown periwinkle
Littorina nebulosa (Lamarck, 1822)	A	cloudy periwinkle
Littorina neglecta Bean, 1844	A	obscure periwinkle
Littorina obtusata (Linnaeus, 1758)	A	yellow periwinkle
Littorina saxatilis (Olivi, 1792)	A, P, Ac	rough periwinkle
Littorina scutulata Gould, 1849	P	checkered periwinkle

SCIENTIFIC NAME	OCCURRENCE	COMMON NAME
Littorina sitkana Philippi, 1846	P	Sitka periwinkle
Littorina ziczac (Gmelin, 1791)	A	zebra periwinkle
Nodilittorina tuberculata (Menke, 1828)	A	pricklywinkle
Tectarius muricatus (Linnaeus, 1758)	A	beaded periwinkle
Tectininus nodulosus (Pfeiffer, 1839)	A	false pricklywinkle

Rissoidae

SCIENTIFIC NAME	OCCURRENCE	COMMON NAME
Alvania acutelirata (Carpenter, 1864)	P	
Alvania areolata Stimpson, 1851	A	
Alvania auberiana (d'Orbigny, 1842)	A	West Indian alvania
Alvania castanea Möller, 1842	A	
Alvania exarata Stimpson, 1851	A	
Alvania filosa Carpenter, 1864	P	
Alvania latior (Mighels and C. B. Adams, 1842)	P	
Alvania microglypta Haas, 1943	P	fine-cut alvania
Alvania moerchi (Collins, 1886)	A, P, Ac	
Alvania multilineata (Stimpson, 1851)	A	
Alvania oldroydae Bartsch, 1911	P	
Alvania precipitata (Dall, 1889)	A	
Alvania rosana Bartsch, 1911	P	Santa Rosa alvania
Alvania trachisma Bartsch, 1911	P	
Alvania xanthias (Watson, 1886)	A	sharp-rib alvania
Benthonella gaza Dall, 1889	A	
Benthonella nisonis Dall, 1889	A	regal benthonella
Boreocingula castanea (Möller, 1842)	A, Ac	
Boreocingula globula (Möller, 1842)	A	
Boreocingula martyni (Dall, 1887)	P	
Cingula alaskana Bartsch, 1912	P	
Cingula globuloides Warén, 1972	A	
Cingula jacksoni Bartsch, 1953	A	
Cingula montereyensis Bartsch, 1912	P	
Cingula robusta scipio Dall, 1887	P	
Elachisina floridana (Rehder, 1943)	A	
Folinia bermudezi (Aguayo and Rehder, 1936)	A	
Frigidoalvania brychia (A. E. Verrill, 1884)	A, Ac	
Frigidoalvania janmayeni (Friele, 1878)	Ac	Jan Mayen alvania
Manzonia aequisculpa (Keep, 1887)	P	
Manzonia almo (Bartsch, 1911)	P	
Manzonia cosmia (Bartsch, 1911)	P	cosmic alvania
Manzonia purpurea (Dall, 1871)	P	purple alvania
Microstelma vestale (Rehder, 1943)	A	
Onoba aculeus (Gould, 1841)	A	pointed cingula
Onoba alaskana (Dall, 1886)	P	
Onoba aurivillii (Dall, 1886)	P	
Onoba bakeri (Bartsch, 1910)	P	
Onoba carpenteri (Weinkauff, 1885)	P	
Onoba castanella (Dall, 1886)	P	
Onoba cerinella Dall, 1887	P	
Onoba dalli (Bartsch, 1927)	P	
Onoba dinora (Bartsch, 1917)	P	
Onoba forresterensis (Willett, 1934)	P	
Onoba kyskensis (Bartsch, 1911)	P	

SCIENTIFIC NAME	OCCURRENCE	COMMON NAME
Onoba palmeri (Dall, 1919)	P	
Onoba pelagica (Stimpson, 1851)	A	carinate alvania
Onoba muriei (Bartsch and Rehder, 1939)	P	
Pusillina pseudoareolata (Warén, 1974)	A	
Rissoa toroensis Olsson and McGinty, 1958	A	
Rissoina californica Bartsch, 1915	P	
Rissoina cancellata Philippi, 1847	A	
Rissoina decussata (Montagu, 1803)	A	
Rissoina keenae A. G. Smith and Gordon, 1948	P	
Rissoina mayori Dall, 1927	A	
Rissoina multicostata (C. B. Adams, 1850)	A	
Rissoina newcombei Dall, 1897	P	
Rissoina sagraiana d'Orbigny, 1842	A	
Rissoina striosa (C. B. Adams, 1850)	A	
Schwartziella bakeri (Bartsch, 1902)	P	
Schwartziella bryerea (Montagu, 1803)	A	
Schwartziella catesbyana (d'Orbigny, 1842)	A	
Schwartziella cleo (Bartsch, 1915)	P	
Schwartziella coronadoensis (Bartsch, 1915)	P	
Schwartziella dalli (Bartsch, 1915)	P	
Schwartziella hannai (A. G. Smith and Gordon, 1948)	P	
Stosicia aberrans (C. B. Adams, 1850)	A	
Zebina browniana (d'Orbigny, 1842)	A	smooth risso

Barleeiidae

SCIENTIFIC NAME	OCCURRENCE	COMMON NAME
Amphithalamus inclusus Carpenter, 1864	P	
Amphithalamus lacunatus Carpenter, 1864	P	
Amphithalamus vallei Aguayo and Jaume, 1947	A	
Barleeia acuta (Carpenter, 1864)	P	acute barleysnail
Barleeia alderi (Carpenter, 1856)	P	
Barleeia bentleyi Bartsch, 1920	P	
Barleeia californica Bartsch, 1920	P	California barleysnail
Barleeia carpenteri Bartsch, 1920	P	
Barleeia haliotiphila Carpenter, 1864	P	abalone barleysnail
Barleeia oldroydi Bartsch, 1920	P	
Barleeia sanjuanensis Bartsch, 1920	P	
Barleeia subtenuis Carpenter, 1864	P	fragile barleysnail
Barleeia tincta Guppy, 1895	A	Caribbean barleysnail
Caelatura rustica (Watson, 1886)	A	
Lirobarleeia kelseyi (Dall and Bartsch, 1902)	P	

Assimineidae

SCIENTIFIC NAME	OCCURRENCE	COMMON NAME
Assiminea californica (Tryon, 1865)	P(E)	California assiminea
Assiminea infima S. S. Berry, 1947	F[3]	badwater snail
Assiminea succinea (Pfeiffer, 1840)	A(E)	Atlantic assiminea

---

[3]Occurs in association with desert saline pools.

SCIENTIFIC NAME	OCCURRENCE	COMMON NAME
Falsicingulidae		
Falsicingula aleutica (Dall, 1887)	P	
Pelycidiidae		
Pelycidion kelseyi (Bartsch, 1911)	P	
Elachisinidae		
Elachisina grippi Dall, 1918	P	
Truncatellidae		
Truncatella californica Pfeiffer, 1857	P	California truncatella
Truncatella caribaeensis Reeve, 1842	A	Caribbean truncatella
Truncatella floridana Hubricht, 1983	A	*classification uncertain*
Truncatella pulchella Pfeiffer, 1839	A	beautiful truncatella
Truncatella regina Hubricht, 1983	A	*classification uncertain*
Truncatella scalaris (Michaud, 1830)	A	ladder truncatella
Truncatella stimpsonii Stearns, 1872	P	
Rissoellidae		
Rissoella caribaea Rehder, 1943	A	Caribbean risso
Rissoella hertleini A.G. Smith and Gordon, 1948	P	Monterey risso
Skeneopsidae		
Skeneopsis alaskana Dall, 1919	P	Alaska skenea
Skeneopsis planorbis (Fabricius, 1780)	A	flat skenea
Omalogyridae		
Omalogyra atomus (Philippi, 1841)	A	atom snail
Vitrinellidae		
Anticlimax athleenae Pilsbry and McGinty, 1946	A	
Anticlimax pilsbryi McGinty, 1945	A	cupola vitrinella
Aorotrema cistronium (Dall, 1889)	A	
Aorotrema erraticum Pilsbry and McGinty, 1945	A	
Aorotrema pontogenes (Schwengel and McGinty, 1942)	A	
Circulus cosmius Bartsch, 1907	P	
Circulus dalli Bush, 1897	A	
Circulus multistriatus (A. E. Verrill, 1884)	A	threaded vitrinella
Circulus rossellinus Dall, 1919	P	
Circulus semisculptus (Olsson and McGinty, 1958)	A	
Circulus suppressus (Dall, 1889)	A	suppressed vitrinella
Cyclostremiscus beauii (P. Fischer, 1857)	A	
Cyclostremiscus jeannae (Pilsbry and McGinty, 1945)	A	

SCIENTIFIC NAME	OCCURRENCE	COMMON NAME
Cyclostremiscus ornatus Olsson and McGinty, 1958	A	
Cyclostremiscus pentagonus (Gabb, 1873)		
Didianema pauli Pilsbry and McGinty, 1945	A	
Epicynia devexa Keen, 1946	P	
Epicynia inornata (d'Orbigny, 1842)	A	hairy vitrinella
Epicynia multicarinata (Dall, 1889)	A	fringed vitrinella
Parviturboides interruptus (C. B. Adams, 1850)	A	interrupted vitrinella
Pleuromalaxis balesi Pilsbry and McGinty, 1945	A	
Scissilabra dalli Bartsch, 1907	P	splitlip vitrinella
Solariorbis arnoldi Bartsch, 1927	P	
Solariorbis blakei Rehder, 1944	A	
Solariorbis infracarinatus Gabb, 1881	A	
Solariorbis mooreanus Vanatta, 1904	A	
Teinostoma bibbianum Dall, 1919	P	
Teinostoma biscaynense Pilsbry and McGinty, 1945	A	Biscayne vitrinella
Teinostoma carinicallus Pilsbry and McGinty, 1946	A	
Teinostoma clavium Pilsbry and McGinty, 1945	A	
Teinostoma cocolitoris Pilsbry and McGinty, 1945	A	
Teinostoma cryptospira A. E. Verrill, 1884	A	
Teinostoma goniogyrus Pilsbry and McGinty, 1945	A	
Teinostoma lerema Pilsbry and McGinty, 1945	A	
Teinostoma lituspalmarum Pilsbry and McGinty, 1945	A	
Teinostoma megastoma (C.B. Adams, 1850)	A	
Teinostoma minusculum (Bush, 1897)	A	
Teinostoma multistriatum A. E. Verrill, 1884	A	
Teinostoma nesaeum Pilsbry and McGinty, 1945	A	
Teinostoma obtectum Pilsbry and McGinty, 1945	A	
Teinostoma parvicallum Pilsbry and McGinty, l945	A	
Teinostoma pilsbryi McGinty, 1945	A	
Teinostoma reclusum (Dall, 1889)	A	
Teinostoma salvanium Dall, 1919	P	
Teinostoma sapiella Dall, 1919	P	
Teinostoma semistriatum (d'Orbigny, 1842)	A	
Teinostoma supravallatum (Carpenter, 1864)	P	upright vitrinella
Vitrinella alaskensis Bartsch, 1907	P	
Vitrinella berryi (Bartsch, 1907)	P	
Vitrinella bicaudata Pilsbry and McGinty, 1946	A	two-tail vitrinella
Vitrinella carinata (d'Orbigny, 1842)	A	
Vitrinella columbiana Bartsch, 1921	P	
Vitrinella diaphana (d'Orbigny, 1842)	A	
Vitrinella eshnaurae Bartsch, 1907	P	
Vitrinella filifera Pilsbry and McGinty, 1946	A	
Vitrinella floridana Pilsbry and McGinty, 1946	A	Florida vitrinella
Vitrinella helicoidea C. B. Adams, 1850	A	helix vitrinella
Vitrinella hemphilli Vanatta, 1913	A	
Vitrinella oldroydi Bartsch, 1907	P	
Vitrinella praecox Pilsbry and McGinty, 1946	A	premature vitrinella
Vitrinella smithi Bartsch, 1927	P	

SCIENTIFIC NAME	OCCURRENCE	COMMON NAME
Vitrinella stearnsi Bartsch, 1907	P	
Vitrinella terminalis Pilsbry and McGinty, 1946	A	terminal vitrinella
Vitrinella texana Moore, 1965	A	Texas vitrinella
Vitrinella thomasi (Pilsbry, 1945)	A	
Vitrinella tryoni Bush, 1897	A	
Vitrinella williamsoni Dall, 1892	P	
Vitrinorbis diegensis (Bartsch, 1907)	P	San Diego vitrinella

Tornidae

Cochliolepis parasitica Stimpson, 1858	A	parasitic scalesnail
Cochliolepis striata Dall, 1889	A	striate scalesnail
Tornus calianus (Dall, 1919)	P	

Caecidae

Caecum antillarum Carpenter, 1858	A	Antillean caecum
Caecum bipartitum de Folin, 1870	A	
Caecum californicum Dall, 1885	P	California caecum
Caecum carolinianum Dall, 1892	A	Carolina caecum
Caecum carpenteri Bartsch, 1920	P	
Caecum clava de Folin, 1867	A	club caecum
Caecum condylum Moore, 1969	A	bone caecum
Caecum cooperi S. Smith, 1860	A	
Caecum cornucopiae Carpenter, 1858	A	horn-of-plenty caecum
Caecum crebricinctum Carpenter, 1864	P	many-named caecum
Caecum cubitatum de Folin, 1868	A	smooth caecum
Caecum cycloferum de Folin, 1867	A	fatlip caecum
Caecum dalli Bartsch, 1920	P	
Caecum floridanum Stimpson, 1851	A	Florida caecum
Caecum gurgulio Carpenter, 1858	A	windpipe caecum
Caecum heladum Olsson and Harbison, 1953	A	fine-line caecum
Caecum imbricatum Carpenter, 1858	A	imbricate caecum
Caecum johnsoni Winkley, 1908	A	
Caecum nitidum Stimpson, 1851	A	little horn caecum
Caecum plicatum Carpenter, 1858	A	plicate caecum
Caecum profundicola Bartsch, 1920	P	deepwater caecum
Caecum pulchellum Stimpson, 1851	A	beautiful caecum
Caecum ryssotitum de Folin, 1867	A	minute caecum
Caecum textile de Folin, 1867	A	textile caecum
Caecum tortile Dall, 1892	A	twisted caecum
Caecum vestitum de Folin, 1870	A	Vera Cruz caecum
Fartulum occidentale (Bartsch, 1920)	P	western caecum
Fartulum orcutti (Dall, 1885)	P	

Turritellidae

Tachyrhynchus erosus (Couthouy, 1838)	A, P, Ac	eroded turretsnail
Tachyrhynchus lacteolus (Carpenter, 1865)	P	milky turretsnail
Tachyrhynchus pratomus Dall, 1919	P	
Tachyrhynchus reticulatus (Mighels and C.B. Adams, 1842)	A, P, Ac	
Tachyrhynchus stearnsii Dall, 1919	P	
Turritella acropora Dall, 1889	A	boring turretsnail

SCIENTIFIC NAME	OCCURRENCE	COMMON NAME
Turritella cooperi Carpenter, 1864	P	
Turritella exoleta (Linnaeus, 1758)	A	eastern turretsnail
Turritella mariana Dall, 1908	P	
Turritella orthosymmetrica S. S. Berry, 1953	P	symmetrical turretsnail
Turritella variegata (Linnaeus, 1758)	A	variegate turretsnail
Turritellopsis acicula (Stimpson, 1851)	A, P, Ac	needle turretsnail
Vermicularia fargoi Olsson, 1951	A	
Vermicularia fewkesi Yates, 1890	P	
Vermicularia knorrii (Deshayes, 1843)	A	Florida wormsnail
Vermicularia radicula Stimpson, 1851	A	northern wormsnail
Vermicularia spirata (Philippi, 1836)	A	West Indian wormsnail

Siliquariidae

Siliquaria squamata Blainville, 1827	A	slit wormsnail

Vermetidae

Dendropoma lituella (Mörch, 1861)	P	flat wormsnail
Petaloconchus compactus (Carpenter, 1864)	P	compact wormsnail
Petaloconchus erectus (Dall, 1888)	A	erect wormsnail
Petaloconchus montereyensis Dall, 1919	P	Monterey wormsnail
Petaloconchus varians (d'Orbigny, 1841)	A	variable wormsnail
Serpulorbis decussatus (Gmelin, 1791)	A	decussate wormsnail
Serpulorbis squamiger (Carpenter, 1856)	P	scaled wormsnail
Spiroglyptus annulatus Daudin, 1800	A	ringed wormsnail
Spiroglyptus irregularis (d'Orbigny, 1842)	A	irregular wormsnail
Spiroglyptus rastrus (Mörch, 1861)	P	California wormsnail

Planaxidae

Planaxis lineatus (E. M. da Costa, 1778)	A	dwarf planaxis
Planaxis nucleus (Bruguière, 1789)	A	black planaxis

Modulidae

Modulus modulus (Linnaeus, 1758)	A	buttonsnail

Potamididae

Batillaria minima (Gmelin, 1791)	A	West Indian false cerith
Batillaria zonalis (Bruguière, 1792)	P [I]	Japanese false cerith
Cerithidea californica (Haldeman, 1840)	P (E)	California hornsnail
Cerithidea costata (E. M. da Costa, 1778)	A (E)	costate hornsnail
Cerithidea pliculosa (Menke, 1829)	A (E)	plicate hornsnail
Cerithidea scalariformis (Say, 1825)	A (E)	ladder hornsnail

Cerithiidae

Alaba catalinensis Bartsch, 1920	P	
Alaba incerta (d'Orbigny, 1842)	A	varicose cerith
Alaba jeanettae Bartsch, 1910	P	
Alaba serrana A. G. Smith and Gordon, 1948	P	
Bittium alternatum (Say, 1822)	A	
Bittium armillatum (Carpenter, 1864)	P	

SCIENTIFIC NAME	OCCURRENCE	COMMON NAME
Bittium asperum Gabb, 1861	P	
Bittium attenuatum Carpenter, 1864	P	slender cerith
Bittium challisae Bartsch, 1917	P	
Bittium eschrichtii (Middendorff, 1849)	P	threaded cerith
Bittium fastigiatum Carpenter, 1864	P	
Bittium fetellum Bartsch, 1911	P	
Bittium interfossum (Carpenter, 1864)	P	
Bittium johnstonae Bartsch, 1911	P	
Bittium larum Bartsch, 1911	P	
Bittium munitum (Carpenter, 1864)	P	
Bittium oldroydae Bartsch, 1911	P	
Bittium purpureum (Carpenter, 1864)	P	
Bittium quadrifilatum Carpenter, 1864	P	four-thread cerith
Bittium rugatum (Carpenter, 1864)	P	
Bittium sanjuanense Bartsch, 1917	P	
Bittium serra Bartsch, 1917	P	
Bittium tumidum Bartsch, 1907	P	
Bittium vancouverense Dall and Bartsch, 1910	P	
Bittium varium (Pfeiffer, 1840)	A	grass cerith
Cerithium atratum (Born, 1778)	A	dark cerith
Cerithium eburneum Bruguière, 1792	A	ivory cerith
Cerithium guinaicum Philippi, 1849	A	Guinea cerith
Cerithium litteratum (Born, 1778)	A	stocky cerith
Cerithium lutosum Menke, 1828	A	variable cerith
Cerithium muscarum Say, 1822	A	flyspeck cerith
Finella adamsi (Dall, 1889)	A	
Finella barbarensis Bartsch, 1911	P	Santa Barbara cerith
Finella californica (Dall and Bartsch, 1901)	P	California cerith
Finella dubia (d'Orbigny, 1842)	A	
Finella hamlini Bartsch, 1911	P	
Finella io Bartsch, 1911	P	
Finella phanea Bartsch, 1911	P	
Finella tenuisculpta (Carpenter, 1864)	P	fine-sculpted cerith
Litiopa melanostoma Rang, 1829	A	sargassum snail

Cerithiopsidae

Cerithiopsis alcima Bartsch, 1911	P	
Cerithiopsis antefilosa Bartsch, 1911	P	
Cerithiopsis antemunda Bartsch, 1911	P	
Cerithiopsis arnoldi Bartsch, 1911	P	
Cerithiopsis berryi Bartsch, 1911	P	
Cerithiopsis bicolor (C. B. Adams, 1850)	A	
Cerithiopsis carpenteri Bartsch, 1911	P	
Cerithiopsis cesta Bartsch, 1911	P	
Cerithiopsis charlottensis Bartsch, 1917	P	
Cerithiopsis columna Carpenter, 1864	P	
Cerithiopsis cosmia Bartsch, 1907	P	
Cerithiopsis costulata Möller, 1842	A	
Cerithiopsis crystallina Dall, 1881	A	
Cerithiopsis diegensis Bartsch, 1911	P	
Cerithiopsis diomedea Bartsch, 1911	P	
Cerithiopsis emersonii (C. B. Adams, 1838)	A	

SCIENTIFIC NAME	OCCURRENCE	COMMON NAME
Cerithiopsis fraseri Bartsch, 1921	P	
Cerithiopsis fusiformis (C. B. Adams, 1850)	A	
Cerithiopsis gloriosa Bartsch, 1911	P	
Cerithiopsis greeni (C. B. Adams, 1839)	A	
Cerithiopsis grippi Bartsch, 1917	P	
Cerithiopsis ingens Bartsch, 1907	P	
Cerithiopsis io Dall and Bartsch, 1911	A	
Cerithiopsis montereyensis Bartsch, 1911	P	
Cerithiopsis onealensis Bartsch, 1921	P	
Cerithiopsis paramoea Bartsch, 1911	P	
Cerithiopsis pedroana Bartsch, 1907	P	
Cerithiopsis pulchellum Jeffreys, 1858 (non C. B. Adams, 1850)	A	
Cerithiopsis rowelli Bartsch, 1911	P	
Cerithiopsis signa Bartsch, 1921	P	
Cerithiopsis stejnegeri Dall, 1884	P	
Cerithiopsis stephensae Bartsch, 1909	P	
Cerithiopsis truncata Dall, 1886	P	
Cerithiopsis tubercularis floridana Dall, 1892	A	
Cerithiopsis tumida Bartsch, 1907	P	
Cerithiopsis vanhyningi Bartsch, 1918	A	
Cerithiopsis virginica J. B. Henderson and Bartsch, 1914	A	
Cerithiopsis willetti Bartsch, 1921	P	
Seila adamsi (H. C. Lea, 1845)	A	
Seila montereyensis Bartsch, 1907	P	

Mathildidae

SCIENTIFIC NAME	OCCURRENCE	COMMON NAME
Mathilda barbadensis Dall, 1889	A	Barbados mathilda
Mathilda hendersoni Dall, 1927	A	
Mathilda yucatecana (Dall, 1881)	A	Yucatan mathilda

Architectonicidae

SCIENTIFIC NAME	OCCURRENCE	COMMON NAME
Architectonica nobilis Röding, 1798	A	common sundial
Discotectonica discus (Philippi, 1844)	A	keeled sundial
Heliacus alleryi (di Monterosato, 1873)	A	
Heliacus architae (O. G. Costa, 1830)	A	noduled sundial
Heliacus bisulcatus (d'Orbigny, 1842)	A	beaded sundial
Heliacus borealis (A. E. Verrill and S. Smith, 1880)	A	boreal sundial
Heliacus cylindricus (Gmelin, 1791)	A	Atlantic cylinder sundial
Heliacus perrieri (Rochebrune, 1881)	A	channeled sundial
Philippia krebsii (Mörch, 1875)	A	smooth sundial
Pseudomalaxis lamellifera Rehder, 1935	A	lamellate false-dial
Pseudomalaxis nobilis A. E. Verrill, 1885	A	noble false-dial
Spirolaxis centrifuga (di Monterosato, 1890)	A	exquisite false-dial

Triphoridae

SCIENTIFIC NAME	OCCURRENCE	COMMON NAME
Metaxia convexa (Carpenter, 1857)	P	
Metaxia metaxae (delle Chiaje, 1829)	A	
Metaxia rugulosa (C. B. Adams, 1850)	A	

SCIENTIFIC NAME	OCCURRENCE	COMMON NAME
Metaxia taeniolata (Dall, 1889)	A	
Triphora callipyrga (Bartsch, 1907)	P	
Triphora carpenteri (Bartsch, 1907)	P	
Triphora catalinensis (Bartsch, 1907)	P	
Triphora decorata (C. B. Adams, 1850)	A	mottled triphora
Triphora intermedia (Dall, 1881)	A	
Triphora lilacina (Dall, 1889)	A	
Triphora longissima (Dall, 1881)	A	
Triphora melanura (C. B. Adams, 1850)	A	white Atlantic triphora
Triphora montereyensis (Bartsch, 1907)	P	
Triphora nigrocincta (C. B. Adams, 1839)	A	black-line triphora
Triphora ornata (Deshayes, 1832)	A	
Triphora pedroana (Bartsch, 1907)	P	San Pedro triphora
Triphora peninsularis (Bartsch, 1907)	P	
Triphora pulchella (C. B. Adams, 1850)	A	beautiful triphora
Triphora pyrrha J. B. Henderson and Bartsch, 1914	A	
Triphora stearnsi (Bartsch, 1907)	P	
Triphora turristhomae (Holten, 1802)	A	St. Thomas triphora

Janthinidae

SCIENTIFIC NAME	OCCURRENCE	COMMON NAME
Janthina exigua Lamarck, 1816	A	dwarf janthina
Janthina globosa Swainson, 1822	A, P	elongate janthina
Janthina janthina (Linnaeus, 1758)	A, P	janthina
Janthina pallida W. Thompson, 1840	A	pale janthina
Récluzia rollandiana Petit de la Saussaye, 1853	A	brown janthina

Epitoniidae

SCIENTIFIC NAME	OCCURRENCE	COMMON NAME
Acirsa borealis (Lyell, 1841)	A, P	chalky wentletrap
Alexania floridana (Pilsbry, 1945)	A	smooth wentletrap
Amaea mitchelli (Dall, 1896)	A	
Amaea retifera (Dall, 1889)	A	reticulate wentletrap
Asperiscala bellastriata (Carpenter, 1864)	P	well-threaded wentletrap
Asperiscala cookeana (Dall, 1917)	P	
Asperiscala lowei (Dall, 1906)	P	
Boreoscala greenlandica (G. Perry, 1811)	A, P	Greenland wentletrap
Cirsotrema dalli Rehder, 1945	A	
Cirsotrema pilsbryi (McGinty, 1940)	A	
Couthouyella striatula (Couthouy, 1839)	A	rough wentletrap
Depressiscala nautlae (Mörch, 1874)	A	slender wentletrap
Depressiscala nitidella (Dall, 1889)	A	mottled wentletrap
Depressiscala polita (Sowerby, 1844)	P	polished wentletrap
Epitonium albidum (d'Orbigny, 1842)	A	bladed wentletrap
Epitonium angulatum (Say, 1830)	A	angulate wentletrap
Epitonium apiculatum (Dall, 1889)	A	semismooth wentletrap
Epitonium babylonium (Dall, 1889)	A	tower wentletrap
Epitonium blainei Clench and Turner, 1953	A	
Epitonium candeanum (d'Orbigny, 1842)	A	
Epitonium championi Clench and Turner, 1952	A	
Epitonium denticulatum (Sowerby, 1844)	A	tooth-rib wentletrap
Epitonium echinaticostum (d'Orbigny, 1842)	A	wide-coil wentletrap
Epitonium foliaceicostum (d'Orbigny, 1842)	A	wrinkle-rib wentletrap

SCIENTIFIC NAME	OCCURRENCE	COMMON NAME
Epitonium fractum Dall, 1927	A	humble wentletrap
Epitonium frielei (Dall, 1889)	A	
Epitonium humphreysii (Kiener, 1838)	A	
Epitonium krebsii (Mörch, 1874)	A	
Epitonium lamellosum (Lamarck, 1822)	A	lamellose wentletrap
Epitonium matthewsae Clench and Turner, 1952	A	
Epitonium multistriatum (Say, 1826)	A	many-rib wentletrap
Epitonium novangliae (Couthouy, 1838)	A	New England wentletrap
Epitonium occidentale (Nyst, 1871)	A	fine-ribbed wentletrap
Epitonium pourtalesii (A. E. Verrill and S. Smith, 1880)	A	
Epitonium rupicola (Kurtz, 1860)	A	brown-band wentletrap
Epitonium rushii (Dall, 1889)	A	frosted wentletrap
Epitonium sericifilum (Dall, 1889)	A	silky wentletrap
Epitonium tollini Bartsch, 1938	A	
Epitonium unifasciatum (Sowerby, 1844)	A	one-band wentletrap
Nitidiscala caamanoi (Dall and Bartsch, 1910)	P	tabulate wentletrap
Nitidiscala californica (Dall, 1917)	P	California wentletrap
Nitidiscala catalinae (Dall, 1908)	P	
Nitidiscala catalinensis (Dall, 1917)	P	
Nitidiscala hindsii (Carpenter, 1856)	P	
Nitidiscala indianorum (Carpenter, 1864)	P	money wentletrap
Nitidiscala sawinae (Dall, 1903)	P	
Nitidiscala tincta (Carpenter, 1864)	P	tinted wentletrap
Nystiella atlantis Clench and Turner, 1952	A	Atlantic wentletrap
Nystiella concava (Dall, 1889)	A	concave wentletrap
Opalia abbotti Clench and Turner, 1952	A	
Opalia andrewsii (A. E. Verrill, 1882)	A	
Opalia aurifilia (Dall, 1889)	A	fine-mesh wentletrap
Opalia borealis Carpenter, 1865	P	boreal wentletrap
Opalia burryi Clench and Turner, 1950	A	
Opalia crenata (Linnaeus, 1758)	A	coarse wentletrap
Opalia eolis Clench and Turner, 1950	A	cancellate wentletrap
Opalia funiculata (Carpenter, 1857)	P	scalloped wentletrap
Opalia hotessieriana (d'Orbigny, 1842)	A	pitted wentletrap
Opalia infrequens (C. B. Adams, 1852)	P	sparse wentletrap
Opalia montereyensis (Dall, 1907)	P	Monterey wentletrap
Opalia pumilio (Mörch, 1874)	A	dwarf wentletrap
Opalia spongiosa (Carpenter, 1866)	P	spongy wentletrap
Sthenorytis pernobilis (P. Fischer and Bernardi, 1857)	A	noble wentletrap

Aclididae

SCIENTIFIC NAME	OCCURRENCE	COMMON NAME
Aclis carolinensis Bartsch, 1911	A	Carolina aclis
Aclis eolis Bartsch, 1947	A	
Aclis hypergonia Schwengel and McGinty, 1942	A	angular aclis
Aclis lata Dall, 1889	A	wide aclis
Aclis occidentalis Hemphill, 1894	P	Pacific aclis
Aclis shepardiana Dall, 1919	P	
Aclis striata A. E. Verrill, 1880	A	striate aclis
Aclis tenuis A. E. Verrill, 1882	A	thin aclis
Aclis turrita Carpenter, 1864	P	turreted aclis
Aclis underwoodae (Bartsch, 1947)	A	

SCIENTIFIC NAME	OCCURRENCE	COMMON NAME
Bermudaclis tampaensis Bartsch, 1947	A	Tampa aclis
Henrya morrisoni Bartsch, 1947	A	
Schwengelia hendersoni (Dall, 1927)	A	
Eulimidae		
Asterophila japonica Randall and Heath, 1912	P	
Balcis thersites (Carpenter, 1864)	P	
Cythnia albida Carpenter, 1864	P	
Ersilia stancyki Warén, 1980	A	brittlestar snail
Eulima fulvocincta C. B. Adams, 1850	A	brown-varice eulima
Eulima schwengelae (Bartsch, 1938)	P	
Eulimostraca hemphilli (Dall, 1884)	A	brown eulima
Eulimostraca subcarinata (d'Orbigny, 1842)	A	brown-line eulima
Melanella arcuata (C. B. Adams, 1850)	A	twisted eulima
Melanella bakeri Bartsch, 1917	P	
Melanella berryi Bartsch, 1917	P	
Melanella californica Bartsch, 1917	P	
Melanella catalinensis Bartsch, 1917	P	Catalina eulima
Melanella columbiana Bartsch, 1917	P	
Melanella comoxensis Bartsch, 1917	P	
Melanella compacta Carpenter, 1864	P	
Melanella conoidea Kurtz and Stimpson, 1851	A	conoidal eulima
Melanella delmontensis (A. G. Smith and Gordon, 1948)	P	
Melanella elongata Bucquoy, Dollfus and Dautzenberg, 1883	A	
Melanella eulimoides (C. B. Adams, 1850)	A	grooved eulima
Melanella gibba de Folin, 1867	A	
Melanella gracilis (C. B. Adams, 1850)	A	
Melanella grippi Bartsch, 1917	P	
Melanella jamaicensis (C. B. Adams, 1845)	A	Jamaica eulima
Melanella lastra Bartsch, 1917	P	
Melanella macra Bartsch, 1917	P	
Melanella micans (Carpenter, 1864)	P	
Melanella montereyensis Bartsch, 1917	P	Monterey eulima
Melanella oldroydi Bartsch, 1917	P	
Melanella peninsularis Bartsch, 1917	P	Baja eulima
Melanella randolphi Vanatta, 1899	P	
Melanella rutila (Carpenter, 1864)	P	auburn eulima
Melanella tacomaensis Bartsch, 1917	P	Tacoma eulima
Melanella titubans (S. S. Berry, 1956)	P	
Niso aeglees Bush, 1885	A	brown-line niso
Niso hendersoni Bartsch, 1953	A	
Niso hipolitensis Bartsch, 1917	P	Hipolito niso
Niso lomana Bartsch, 1917	P	
Oceanida graduata de Folin, 1871	A	shouldered eulima
Oceanida inglei Lyons, 1978	A	
Pelseneeria stimpsoni (A. E. Verrill, 1872)	A	
Sabinella troglodytes (Thiele, 1925)	A	pencil-spine eulima
"*Scalenostoma*" *babylonia* Bartsch, 1912	P	keeled eulima
Strombiformis alaskensis Bartsch, 1917	P	Alaska eulima
Strombiformis almo Bartsch, 1917	P	
Strombiformis auricinctus Abbott, 1958	A	gold-stripe eulima

SCIENTIFIC NAME	OCCURRENCE	COMMON NAME
Strombiformis bifasciatus d'Orbigny, 1842	A	two-band eulima
Strombiformis bilineatus Alder, 1848	A	two-line eulima
Strombiformis californicus Bartsch, 1917	P	California eulima
Strombiformis patula (Dall and Simpson, 1901)	A	largemouth eulima

Entoconchidae

Enteroxenos parasitichopoli (Tikasingh, 1961)	P	
Entocolax ludwigii Voigt, 1888	P	
Thyonicola americanus Tikasingh, 1961	P	

Aporrhaididae

Aporrhais occidentalis Beck, 1836	A, Ac	American pelicanfoot

Strombidae

Strombus alatus Gmelin, 1791	A	Florida fighting conch
Strombus costatus Gmelin, 1791	A	milk conch
Strombus gallus Linnaeus, 1758	A	roostertail conch
Strombus gigas Linnaeus, 1758	A	queen conch[4]
Strombus pugilis Linnaeus, 1758	A	West Indian fighting conch
Strombus raninus Gmelin, 1791	A	hawkwing conch

Hipponicidae

Hipponix antiquatus (Linnaeus, 1767)	A, P	white hoofsnail
Hipponix subrufus (Lamarck, 1819)	A	orange hoofsnail
Hipponix tumens Carpenter, 1864	P	ribbed hoofsnail
Sabia conica (Schumacher, 1817)	P [I]	

Fossaridae

Fossarus bellus (Dall, 1889)	A	beautiful fossarus
Fossarus compactus (Dall, 1889)	A	compact fossarus
Fossarus elegans A. E. Verrill and S. Smith, 1882	A	elegant fossarus
Macromphalina adamsii (P. Fischer, 1857)	A	
Macromphalina californica Dall, 1903	P	California macromphaline
Macromphalina floridana Moore, 1965	A	Florida macromphaline
Macromphalina palmalitoris Pilsbry and McGinty, 1950	A	Palm Beach macromphaline
Macromphalina pierrot Gardner, 1948	A	

Vanikoroidae

Vanikoro oxychone Mörch, 1877	A	West Indian vanikoro

Capulidae

Capulus californicus Dall, 1900	P	California capsnail
Capulus incurvatus (Gmelin, 1791)	A	incurved capsnail
Capulus ungaricus (Linnaeus, 1767)	A, Ac	fools capsnail

---

[4]Also commonly referred to as pink conch.

SCIENTIFIC NAME	OCCURRENCE	COMMON NAME
Trichotropidae		
Torellia ammonia Dall, 1919	P	rams-horn hairysnail
Torellia fimbriata A. E. Verrill and S. Smith, 1882	A	fringed hairysnail
Torellia vallonia Dall, 1919	P	acorn hairysnail
Torellia vestita Jeffreys, 1867	A	veiled hairysnail
Trichotropis bicarinata (Sowerby, 1825)	A, P, Ac	two-keel hairysnail
Trichotropis borealis Broderip and Sowerby, 1829	A, P, Ac	boreal hairysnail
Trichotropis cancellata Hinds, 1843	P	cancellate hairysnail
Trichotropis coronata Gould, 1860	P, Ac	crowned hairysnail
Trichotropis insignis Middendorff, 1849	P	gray hairysnail
Trichotropis kroyeri Philippi, 1848	P, Ac	
Trichotropis migrans Dall, 1881	A	wandering hairysnail
Calyptraeidae		
Calyptraea burchi A. G. Smith and Gordon, 1948	P	
Calyptraea centralis (Conrad, 1841)	A	circular Chinese-hat
Calyptraea fastigiata Gould, 1856	P	Pacific Chinese-hat
Cheilea equestris (Linnaeus, 1758)	A	false cup-and-saucer
Crepidula aculeata (Gmelin, 1791)	A, P	spiny slippersnail
Crepidula adunca Sowerby, 1825	P	hooked slippersnail
Crepidula convexa Say, 1822	A, P [I]	convex slippersnail
Crepidula fornicata (Linnaeus, 1758)	A, P [I]	common Atlantic slippersnail
Crepidula grandis Middendorff, 1849	P	great slippersnail
Crepidula maculosa Conrad, 1846	A	spotted slippersnail
Crepidula naticarum Williamson, 1905	P	cheeky slippersnail
Crepidula norrisiarum Williamson, 1905	P	
Crepidula nummaria Gould, 1846	P	western white slippersnail
Crepidula onyx Sowerby, 1824	P	onyx slippersnail
Crepidula perforans (Valenciennes, 1846)	P	white slippersnail
Crepidula plana Say, 1822	A	eastern white slippersnail
Crepidula striolata Menke, 1851	P	ridged slippersnail
Crepipatella charybdis (S. S. Berry, 1940)	P	greedy slippersnail
Crepipatella lingulata (Gould, 1846)	P	Pacific half-slippersnail
Crepipatella orbiculata (Dall, 1919)	P	round slippersnail
Crucibulum auricula (Gmelin, 1791)	A	West Indian cup-and-saucer
Crucibulum spinosum (Sowerby, 1824)	P	spiny cup-and-saucer
Crucibulum striatum Say, 1824	A	striate cup-and-saucer
Xenophoridae		
Xenophora caribaea Petit de la Saussaye, 1856	A	Caribbean carriersnail
Xenophora conchyliophora (Born, 1780)	A	American carriersnail
Xenophora longleyi Bartsch, 1931	A	shingled carriersnail
Lamellariidae		
Capulacmaea commoda (Middendorff, 1851)	P, Ac	widemouth lamellaria
Lamellaria diegoensis Dall, 1885	P	San Diego lamellaria
Lamellaria digueti Rochebrune, 1895	P	
Lamellaria koto Schwengel, 1944	A	

SCIENTIFIC NAME	OCCURRENCE	COMMON NAME
Lamellaria leucosphaera Schwengel, 1942	A	white-ball lamellaria
Lamellaria pellucida A. E. Verrill, 1880	A	transluscent lamellaria
Lamellaria perspicua (Linnaeus, 1758)	A	transparent lamellaria
Marsenina ampla A. E. Verrill, 1880	A	great lamellaria
Marsenina glabra (Couthouy, 1832)	A, Ac	bald lamellaria
Marsenina globosa L. M. Perry, 1939	A	rotund lamellaria
Marsenina rhombica (Dall, 1871)	P	rhombic lamellaria
Marsenina stearnsii (Dall, 1871)	P	
Marseniopsis sharonae (Willett, 1939)	P	
Onchidiopsis corys Balch, 1910	A	
Onchidiopsis glacialis (M. Sars, 1851)	A, P, Ac	icy lamellaria
Onchidiopsis hannai Dall, 1916	P	
Onchidiopsis kingmaruensis Russell, 1942	A	Baffin lamellaria
Velutina conica Dall, 1887	P	conical lamellaria
Velutina granulata Dall, 1919	P	granular lamellaria
Velutina lanigera Möller, 1842	A, P	wooly lamellaria
Velutina plicatilis (Müller, 1776)	A, P, Ac	oblique lamellaria
Velutina prolongata Carpenter, 1865	P	elongate lamellaria
Velutina rubra Willett, 1919	P	red lamellaria
Velutina undata (T. Brown, 1839)	A, P, Ac	wavy lamellaria
Velutina velutina (Müller, 1776)	A, P, Ac	smooth lamellaria

Triviidae

Erato albescens Dall, 1905	P	whitish erato
Erato columbella Menke, 1847	P	pigeon erato
Erato maugeriae J. E. Gray, 1832	A	green erato
Erato vitellina (Hinds, 1844)	P	appleseed erato
Trivia antillarum Schilder, 1922	A	Antilles trivia
Trivia californiana (J. E. Gray, 1827)	P	California trivia
Trivia candidula Gaskoin, 1835	A	white trivia
Trivia maltbiana Schwengel and McGinty, 1942	A	
Trivia nix (Schilder, 1922)	A	snowy trivia
Trivia pediculus (Linnaeus, 1758)	A	coffeebean trivia
Trivia quadripunctata (J. E. Gray, 1827)	A	fourspot trivia
Trivia ritteri Raymond, 1903	P	
Trivia solandri (Sowerby, 1832)	P	
Trivia suffusa (J. E. Gray, 1832)	A	pink trivia

Cypraeidae

Cypraea cervus Linnaeus, 1771	A	Atlantic deer cowrie
Cypraea cinerea Gmelin, 1791	A	Atlantic gray cowrie
Cypraea spadicea Swainson, 1823	P	chestnut cowrie
Cypraea spurca acicularis Gmelin, 1791	A	Atlantic yellow cowrie
Cypraea surinamensis G. Perry, 1811	A	Surinam cowrie
Cypraea zebra Linnaeus, 1758	A	measled cowrie

Ovulidae

Aperiovula abbotti Cate, 1973	A	
Cymbovula acicularis (Lamarck, 1810)	A	West Indian simnia
Cyphoma alleneae Cate, 1973	A	
Cyphoma aureocinctum (Dall, 1899)	A	gold-line cyphoma

SCIENTIFIC NAME	OCCURRENCE	COMMON NAME
Cyphoma gibbosum (Linnaeus, 1758)	A	flamingo tongue
Cyphoma macgintyi Pilsbry, 1939	A	
Cyphoma rhomba Cate, 1978	A	bullroarer cyphoma
Cyphoma sedlaki Cate, 1976	A	
Cyphoma signatum Pilsbry and McGinty, 1939	A	fingerprint cyphoma
Delonovolva aequalis vidleri (Sowerby, 1881)	P	
Neosimnia avena ruthturnerae Cate, 1973	A	
Neosimnia spelta capitia Cate, 1973	A	Keys simnia
Pedicularia californica Newcomb, 1864	P	California pedicularia
Pedicularia decussata (Gould, 1855)	A	hatched pedicularia
Phenacovolva piragua (Dall, 1889)	A	slender simnia
Primovula solemi Cate, 1973	A	robust miniovula
Pseudocyphoma gibbulum Cate, 1978	A	plump cyphoma
Pseudocyphoma intermedium (Sowerby, 1828)	A	intermediate cyphoma
Pseudosimnia pyrifera Cate, 1973	A	pear simnia
Pseudosimnia sphoni Cate, 1973	A	
Pseudosimnia vanhyningi (M. Smith, 1940)	A	
Simnialena marferula Cate, 1973	A	sea-whip simnia
Simnialena uniplicata (Sowerby, 1848)	A	one-tooth simnia
Spiculata advena Cate, 1978	A	
Spiculata barbarensis (Dall, 1892)	P	Santa Barbara simnia
Spiculata loebbeckeana (Weinkauff, 1881)	P	
Subsimnia bellamaris (S. S. Berry, 1946)	P	Pacific simnia
Volva volva striata (Lamarck, 1810)	A	lined egg spindle
Atlantidae		
Atlanta fusca Souleyet, 1852	A	
Atlanta helicinoides Souleyet, 1852	A	
Atlanta inclinata Souleyet, 1852	A, P	
Atlanta inflata Souleyet, 1852	P	
Atlanta lesueuri Souleyet, 1852	A, P	
Atlanta peronii Lesueur, 1817	A, P	
Atlanta turriculata d'Orbigny, 1836	P	
Oxygyrus keraudrenii (Lesueur, 1817)	A	
Protatlanta souleyeti (E. A. Smith, 1888)	A	
Carinariidae		
Cardiapoda placenta (Lesson, 1830)	A, P	flat cardiapod
Carinaria cithara Benson, 1835	P	harp carinaria
Carinaria cristata (Linnaeus, 1767)	P	
Carinaria galea Benson, 1835	P	helmet carinaria
Carinaria lamarcki Péron and Lesueur, 1810	A, P	
Pterotracheidae		
Firoloida demarestia Lesueur, 1817	A	
Pterotrachea coronata Niebuhr, 1775	A, P	
Pterotrachea keraudrenii Eydoux and Souleyet, 1832	A	
Pterotrachea scutata Gegenbaur, 1855	A, P	

SCIENTIFIC NAME	OCCURRENCE	COMMON NAME
Naticidae		
Amauropsis islandica (Gmelin, 1791)	A, P, Ac	Iceland moonsnail
Bulbus fragilis (Leach, 1819)	A, P, Ac	fragile moonsnail
Calinaticina oldroydii (Dall, 1897)	P	
Euspira heros (Say, 1822)	A	northern moonsnail
Euspira immaculata (Totten, 1835)	A	immaculate moonsnail
Euspira levicula (A. E. Verrill, 1880)	A	lightweight moonsnail
Euspira tenuis (Récluz, 1850)	A	thin moonsnail
Euspira triseriata (Say, 1826)	A	spotted moonsnail
Gyrodes depressus Seguenza, 1874	A	
Haliotinella patinaria (Guppy, 1876)	A	fingernail moonsnail
Natica affinis Gmelin, 1791	A, P, Ac	
Natica canrena (Linnaeus, 1758)	A	colorful moonsnail
Natica castrensis Dall, 1889	A	netted moonsnail
Natica clausa Broderip and Sowerby, 1829	A, P, Ac	Arctic moonsnail
Natica janthostoma Deshayes, 1839	P	purplemouth moonsnail
Natica livida Pfeiffer, 1840	A	livid moonsnail
Natica marochiensis (Gmelin, 1791)	A	Morocco moonsnail
Natica sagraiana d'Orbigny, 1842	A	lined moonsnail
Natica tedbayeri Rehder, 1986	A	
Naticarius verae Rehder, 1947	A	
Neverita duplicata (Say, 1822)	A	shark eye
Neverita nana (Möller, 1842)	A, P, Ac	tiny moonsnail
Neverita politiana (Dall, 1919)	P	polished moonsnail
Neverita reclusiana (Deshayes, 1839)	P	
Polinices altus (Pilsbry, 1929)	P	tall moonsnail
Polinices draconis (Dall, 1903)	P	
Polinices hepaticus (Röding, 1798)	A	brown moonsnail
Polinices lacteus (Guilding, 1834)	A	milk moonsnail
Polinices lewisii (Gould, 1847)	P	
Polinices pallidus (Broderip and Sowerby, 1829)	A, P, Ac	pale moonsnail
Polinices uberinus (d'Orbigny, 1842)	A	white moonsnail
Sigatica carolinensis (Dall, 1889)	A	Carolina moonsnail
Sigatica semisulcata (J. E. Gray, 1839)	A	semisulcate moonsnail
Sinum debile Gould, 1853	P	slight baby-ear
Sinum maculatum (Say, 1831)	A	brown baby-ear
Sinum minus (Dall, 1889)	A	dwarf baby-ear
Sinum perspectivum (Say, 1831)	A	white baby-ear
Sinum scopulosum (Conrad, 1849)	P	fat baby-ear
Stigmaulax sulcata (Born, 1778)	A	grooved moonsnail
Tectonatica pusilla (Say, 1822)	A	miniature moonsnail
Cassidae		
Casmaria ponderosa atlantica Clench, 1944	A	Atlantic casmaria
Cassis flammea (Linnaeus, 1758)	A	flame helmet
Cassis madagascariensis Lamarck, 1822	A	cameo helmet
Cassis tuberosa (Linnaeus, 1758)	A	Caribbean helmet
Cypraecassis testiculus (Linnaeus, 1758)	A	reticulate cowrie-helmet
Phalium coronadoi (Crosse, 1867)	A	Coronado bonnet
Phalium granulatum (Born, 1778)	A	Scotch bonnet
Sconsia striata (Lamarck, 1816)	A	royal bonnet

SCIENTIFIC NAME	OCCURRENCE	COMMON NAME

Ranellidae

Charonia tritonis variegata (Lamarck, 1816)..........................	A	.... Atlantic trumpet triton
Cymatium aquatile (Reeve, 1844)...............	A	....
Cymatium comptum (A. Adams, 1854)..........	A	.... dwarf triton
Cymatium corrugatum amictum (Reeve, 1844)	P	....
Cymatium corrugatum krebsii (Mörch, 1877)......	A	....
Cymatium cynocephalum (Lamarck, 1816)........	A	.... doghead triton
Cymatium femorale (Linnaeus, 1758)............	A	...............angular triton
Cymatium labiosum (W. Wood, 1828)..........	A	....................lip triton
Cymatium martinianum (d'Orbigny, 1846)........	A	.... hairy triton
Cymatium muricinum (Röding, 1798)............	A	...............knobbed triton
Cymatium nicobaricum (Röding, 1798)...........	A	.... goldmouth triton
Cymatium occidentale (Mörch, 1877)............	A	....
Cymatium parthenopeum (von Salis, 1793)	A	...................giant triton
Cymatium rehderi A. H. Verrill, 1950	A	...............twisted triton
Cymatium tenuiliratum (Lischke, 1873)	A	...............slender triton
Distorsio clathrata (Lamarck, 1816).............	A	.............Atlantic distorsio
Distorsio constricta macgintyi Emerson and Puffer, 1953	A	....
Distorsio perdistorta Fulton, 1938...............	A	.... hunchback distorsio
Fusitriton oregonensis (Redfield, 1848)...........	P	................Oregon triton
Linatella caudata (Gmelin, 1791)	A	...............ringed triton

Bursidae

Bufonaria bufo (Bruguière, 1792)	A	.... chestnut frogsnail
Bursa corrugata ponderosa (Reeve, 1844)	A	.... gaudy frogsnail
Bursa granularis cubaniana (d'Orbigny, 1842).....	A	.... Cuba frogsnail
Bursa grayana (Dunker, 1862)	A	.... elegant frogsnail
Bursa ranelloides tenuisculpta (Dautzenberg and P. Fischer, 1906)............	A	.........fine-sculpted frogsnail
Bursa rhodostoma thomae (d'Orbigny, 1842)......	A	..........St. Thomas frogsnail
Crossata californica (Hinds, 1843)	P	.... California frogsnail

Tonnidae

Eudolium crosseanum (di Monterosato, 1869)	A	.... straw tun
Eudolium thompsoni McGinty, 1955.............	A	.... mottled tun
Tonna galea (Linnaeus, 1758)..................	A	.... giant tun
Tonna maculosa (Dillwyn, 1817)................	A	.... Atlantic partridge tun

Oocorythidae

Oocorys bartschi Rehder, 1943	A	.... giant false-tun
Oocorys sulcata P. Fischer, 1883................	A	.... sulcate false-tun

Ficidae

Ficus carolae Clench, 1945.....................	A	................slender figsnail
Ficus communis Röding, 1798...................	A	.... Atlantic figsnail

SCIENTIFIC NAME	OCCURRENCE	COMMON NAME

ORDER NEOGASTROPODA

Muricidae

SCIENTIFIC NAME	OCCURRENCE	COMMON NAME
Acanthina lugubris (Sowerby, 1821)	P	dark unicorn
Acanthina paucilirata (Stearns, 1871)	P	checkered unicorn
Acanthina punctulata (Sowerby, 1825)	P	spotted unicorn
Acanthina spirata (de Blainville, 1832)	P	angular unicorn
Acanthotrophon striatoides Vokes, 1980	A	knobbed trophon
Aspella senex Dall, 1903	A	graybeard aspella
Attiliosa philippiana (Dall, 1889)	A	
Austrotrophon cerrosensis catalinensis I. S. Oldroyd, 1927	P	Catalina forreria
Boreotrophon albospinosus (Willett, 1931)	P	white-spine trophon
Boreotrophon avalonensis (Dall, 1902)	P	Avalon trophon
Boreotrophon bentleyi (Dall, 1908)	P	
Boreotrophon beringi (Dall, 1902)	P, Ac	tabulate trophon
Boreotrophon cepulus (Sowerby, 1880)	P	scallion trophon
Boreotrophon clathratus (Linnaeus, 1758)	A, Ac	clathrate trophon
Boreotrophon cymatus (Dall, 1902)	P	wavy trophon
Boreotrophon disparilis (Dall, 1891)	P	girdled trophon
Boreotrophon eucymatus (Dall, 1902)	P	grooved trophon
Boreotrophon macouni (Dall, 1910)	P	corded trophon
Boreotrophon multicostatus (Eschscholtz, 1829)	P	ribbed trophon
Boreotrophon orpheus (Gould, 1849)	P	threaded trophon
Boreotrophon pacificus (Dall, 1902)	P, Ac	elegant trophon
Boreotrophon rotundatus (Dall, 1902)	P	rotund trophon
Boreotrophon staphylinus (Dall, 1919)	P	carrot trophon
Boreotrophon stuarti (E. A. Smith, 1880)	P	winged trophon
Boreotrophon triangulatus (Carpenter, 1864)	P	triangular trophon
Boreotrophon truncatus (Strøm, 1768)	A, P	bobtail trophon
Calotrophon andrewsi Vokes, 1976	A	
Calotrophon ostrearum (Conrad, 1846)	A	mauve-mouth drill
Ceratostoma foliatum (Gmelin, 1791)	P	foliate thornmouth
Ceratostoma nuttalli (Conrad, 1837)	P	
Chicoreus florifer dilectus (A. Adams, 1855)	A	lace murex
Chicoreus mergus Vokes, 1974	A	
Dermomurex elizabethae (McGinty, 1940)	A	
Dermomurex pauperculus (C. B. Adams, 1850)	A	beggar aspella
Eupleura caudata (Say, 1822)	A	thick-lip drill
Eupleura sulcidentata Dall, 1890	A	sharp-rib drill
Favartia alveata (Kiener, 1842)	A	frilly dwarf triton
Favartia cellulosa (Conrad, 1846)	A	pitted murex
Favartia minirosea (Abbott, 1954)	A	rosy drill
Forreria belcheri (Hinds, 1843)	P	
Maxwellia gemma (Sowerby, 1879)	P	gem murex
Maxwellia santarosana (Dall, 1905)	P	Santa Rosa murex
Murex anniae M. Smith, 1940	A	
Murex belgladeensis Vokes, 1963	A	Belleglade murex
Murex cabritii Bernardi, 1859	A	
Murex rubidus F. C. Baker, 1897	A	rose murex
Murex tryoni Hidalgo in Tryon, 1880	A	
Murexiella glypta (M. Smith, 1938)	A	carved murex

SCIENTIFIC NAME	OCCURRENCE	COMMON NAME
Murexiella hidalgoi (Crosse, 1869)	A	
Murexiella levicula (Dall, 1889)	A	lightweight murex
Murexiella macgintyi (M. Smith, 1938)	A	
Muricanthus fulvescens (Sowerby, 1834)	A	giant eastern murex
Muricopsis oxytata (M. Smith, 1938)	A	hexagonal murex
Nipponotrophon fabricii (Müller, 1842)	A	
Nipponotrophon lasius (Dall, 1919)	P	sandpaper trophon
Nipponotrophon scitulus (Dall, 1891)	P	spiny trophon
Nodulotrophon dalli (Kobelt, 1878)	P, Ac	crown trophon
Nucella canaliculata (Duclos, 1832)	P	channeled dogwinkle
Nucella emarginata (Deshayes, 1839)	P	emarginate dogwinkle
Nucella lamellosa (Gmelin, 1791)	P	frilled dogwinkle
Nucella lapillus (Linnaeus, 1758)	A	Atlantic dogwinkle
Nucella lima (Gmelin, 1791)	P	file dogwinkle
Ocenebra atropurpurea Carpenter, 1865	P	purple rocksnail
Ocenebra barbarensis (Gabb, 1865)	P	Santa Barbara rocksnail
Ocenebra beta (Dall, 1919)	P	beta rocksnail
Ocenebra circumtexta (Stearns, 1871)	P	circled rocksnail
Ocenebra crispatissima S. S. Berry, 1953	P	curly rocksnail
Ocenebra foveolata (Hinds, 1844)	P	dim rocksnail
Ocenebra gracillima Stearns, 1871	P	graceful rocksnail
Ocenebra grippi (Dall, 1911)	P	
Ocenebra inornata (Récluz, 1851)	P	Japanese rocksnail
Ocenebra interfossa Carpenter, 1864	P	
Ocenebra lurida (Middendorff, 1848)	P	lurid rocksnail
Ocenebra minor Dall, 1919	P	minor rocksnail
Ocenebra painei (Dall, 1903)	P	ribbed rocksnail
Ocenebra squamulifera (Carpenter in Gabb, 1869)	P	scaly rocksnail
Ocenebra tracheia Dall, 1919	P	
Ocinebrina emipowlusi (Abbott, 1954)	A	
Paziella nuttingi (Dall, 1896)	A	
Paziella pazi (Crosse, 1869)	A	
Pazionotus stimpsonii (Dall, 1889)	A	
Phyllonotus pomum (Gmelin, 1791)	A	apple murex
Pterochelus ariomus (Clench and Pérez Farfante, 1945)	A	
Pteropurpura bequaerti (Clench and Pérez Farfante, 1945)	A	
Pteropurpura festiva (Hinds, 1844)	P	festive murex
Pteropurpura macroptera (Deshayes, 1839)	P	frill-wing murex
Pteropurpura trialata (Sowerby, 1834)	P	three-wing murex
Pteropurpura vokesae Emerson, 1964	P	wrinkle-wing murex
Pterotyphis triangularis (A. Adams, 1856)	A	
Pterynotus phaneus (Dall, 1889)	A	shining murex
Purpura patula (Linnaeus, 1758)	A	widemouth rocksnail
Roperia poulsoni (Carpenter, 1864)	P	
Siratus beauii (P. Fischer and Bernardi, 1857)	A	
Siratus cailleti (Petit de la Saussaye, 1856)	A	
Siratus consuela (A. H. Verrill, 1950)	A	
Siratus formosus (Sowerby, 1841)	A	Antilles murex
Thais deltoidea (Lamarck, 1822)	A	deltoid rocksnail
Thais haemastoma canaliculata (J. E. Gray, 1839)	A	
Thais haemastoma floridana (Conrad, 1837)	A	Florida rocksnail

SCIENTIFIC NAME	OCCURRENCE	COMMON NAME
Thais rustica (Lamarck, 1822)	A	rustic rocksnail
Trachypollia didyma (Schwengel, 1943)	A	twin drupe
Trachypollia lugubris (C. B. Adams, 1852)	P	dark drupe
Trachypollia nodulosa (C. B. Adams, 1845)	A	blackberry drupe
Trachypollia sclera Woodring, 1928	A	
Trophonopsis kamchatkana (Dall, 1902)	P	Kamchatka trophon
Trophonopsis keepi (Strong and Hertlein, 1937)	P	
Typhis sowerbii Broderip, 1833	A	frilly typhis
Urosalpinx cinerea (Say, 1822)	A, P [I]	Atlantic oyster drill
Urosalpinx macra A. E. Verrill, 1887	A	waxy drill
Urosalpinx perrugata (Conrad, 1846)	A	Gulf oyster drill
Urosalpinx sclera Dall, 1919	P	
Urosalpinx subangulata (Stearns, 1873)	P	
Urosalpinx tampaensis (Conrad, 1846)	A	Tampa drill

Coralliophilidae

SCIENTIFIC NAME	OCCURRENCE	COMMON NAME
Babelomurex dalli (Emerson and D'Attilio, 1963)	A	
Babelomurex mansfieldi (McGinty, 1940)	A	
Babelomurex oldroydi (I. S. Oldroyd, 1929)	P	
Coralliophila aberrans (C. B. Adams, 1850)	A	globose coralsnail
Coralliophila caribaea Abbott, 1958	A	Caribbean coralsnail
Coralliophila costata (de Blainville, 1832)	P	California coralsnail
Coralliophila galea (Reeve, 1846)	A	helmet coralsnail
Coralliophila scalariformis (Lamarck, 1822)	A	staircase coralsnail

Columbellidae

SCIENTIFIC NAME	OCCURRENCE	COMMON NAME
Aesopus chrysalloideus (Carpenter, 1864)	P	cocoon dovesnail
Aesopus eurytoideus (Carpenter, 1864)	P	
Aesopus goforthi Dall, 1912	P	
Aesopus myrmecoon Dall, 1916	P	ant-egg dovesnail
Aesopus sanctus Dall, 1919	P	Santa Monica dovesnail
Aesopus stearnsii (Tryon, 1883)	A	
Aesopus subturritus (Carpenter, 1864)	P	graceful dovesnail
Alia carinata (Hinds, 1844)	P	carinate dovesnail
Amphissa bicolor Dall, 1892	P	two-tone amphissa
Amphissa columbiana Dall, 1916	P	wrinkled amphissa
Amphissa cymata Dall, 1916	P	wavy amphissa
Amphissa haliaeeti (Jeffreys, 1867)	A	Atlantic amphissa
Amphissa reticulata Dall, 1916	P	reticulate amphissa
Amphissa undata (Carpenter, 1864)	P	
Amphissa versicolor Dall, 1871	P	variegate amphissa
Anachis avara (Say, 1822)	A	greedy dovesnail
Anachis catenata (Sowerby, 1844)	A	chain dovesnail
Anachis floridana Rehder, 1939	A	Florida dovesnail
Anachis hotessieriana (d'Orbigny, 1842)	A	
Anachis iontha (Ravenel, 1861)	A	lineate dovesnail
Anachis lafresnayi (P. Fischer and Bernardi, 1856)	A	well-ribbed dovesnail
Anachis obesa (C. B. Adams, 1845)	A	fat dovesnail
Anachis pulchella (de Blainville, 1829)	A	beautiful dovesnail
Anachis semiplicata Stearns, 1873	A	Gulf dovesnail
Anachis sparsa (Reeve, 1859)	A	sparse dovesnail

SCIENTIFIC NAME	OCCURRENCE	COMMON NAME
Anachis subturrita Carpenter, 1866.............	P	....
Columbella mercatoria (Linnaeus, 1758)	A	..........West Indian dovesnail
Columbella rusticoides Heilprin, 1887	A	.... rusty dovesnail
Cosmioconcha calliglypta (Dall and Simpson, 1901)..	A	.... flame dovesnail
Mitrella amiantis (Dall, 1919)	P	....
Mitrella argus d'Orbigny, 1842	A	.... argus dovesnail
Mitrella aurantiaca (Dall, 1871)	P	.... golden dovesnail
Mitrella callimorpha (Dall, 1919)	P	....
Mitrella clementensis Bartsch, 1927	P	.... San Clemente dovesnail
Mitrella diaphana (A. E. Verrill, 1882)...........	A	.... translucent dovesnail
Mitrella hypodra (Dall, 1916)	P	....
Mitrella idalina (Duclos, 1840)	A	....
Mitrella lunata (Say, 1826)	A	.... lunar dovesnail
Mitrella lutulenta (Dall, 1919)...................	P	.... muddy dovesnail
Mitrella multilineata (Dall, 1889)	A	..........brown-band dovesnail
Mitrella nycteis (Duclos, 1846)	A	.... fenestrate dovesnail
Mitrella ocellata (Gmelin, 1791)	A	.... white-spot dovesnail
Mitrella permodesta (Dall, 1890)................	P	.............shy dovesnail
Mitrella profundi (Dall, 1889)	A	.... deepwater dovesnail
Mitrella pura (A. E. Verrill, 1882)...............	A	.... simple dovesnail
Mitrella raveneli (Dall, 1889)...................	A	....
Mitrella rosacea (Gould, 1841)	A, P	.... rosy northern dovesnail
Mitrella tuberosa (Carpenter, 1864)	P	.... variegate dovesnail
Nassarina bushiae (Dall, 1889)	A	....
Nassarina glypta (Bush, 1885)	A	.... engraved dovesnail
Nassarina grayi Dall, 1889	A	....
Nassarina minor (C. B. Adams, 1845)	A	.............banded dovesnail
Nassarina monilifera (Sowerby, 1844)	A	.... many-spot dovesnail
Nassarina penicillata (Carpenter, 1864)	P	..........penciled dovesnail
Nitidella gausapata Gould, 1850.................	P	..........shaggy dovesnail
Nitidella gouldi (Carpenter, 1857)................	P	....
Nitidella laevigata (Linnaeus, 1758)	A	.... smooth dovesnail
Nitidella nitida (Lamarck, 1822)	A	.... glossy dovesnail
Nitidella parva Dunker, 1847	A	....
Buccinidae		
Antillophos candei (d'Orbigny, 1842)	A	.... beaded phos
Bailya intricata (Dall, 1884).....................	A	.............intricate phos
Bartschia significans Rehder, 1943	A	....
Beringius beringii (Middendorff, 1848)	P	....
Beringius crebricostatus (Dall, 1877)	P	.............thick-cord whelk
Beringius eyerdami A. G. Smith, 1959	P	....
Beringius frielei (Dall, 1894)	P	....
Beringius indentatus (Dall, 1919)	P	....
Beringius kennicottii (Dall, 1907)	P	....
Beringius malleatus (Dall, 1884)	P, Ac	.............hammered whelk
Beringius marshalli (Dall, 1919)	P	....
Beringius stimpsoni (Gould, 1860)................	P, Ac	....
Beringius turtoni (Bean, 1834)..................	A, Ac	....
Buccinum abyssorum A. E. Verrill, 1884	A	.............shingled whelk
Buccinum aleuticum Dall, 1894..................	P	.............Aleut whelk
Buccinum angulosum angulosum J. E. Gray, 1839 .	Ac	.... angular whelk

SCIENTIFIC NAME	OCCURRENCE	COMMON NAME
Buccinum angulosum subcostatum Dall, 1885	Ac	
Buccinum angulosum transliratum Dall, 1919	P, Ac	
Buccinum baerii (Middendorff, 1848)	P	
Buccinum castaneum castaneum Dall, 1877	P	chestnut whelk
Buccinum castaneum fluctuatum Dall, 1919	P	
Buccinum castaneum triplostephanum Dall, 1919	P	
Buccinum chishimanum Pilsbry, 1904	P	
Buccinum ciliatum Fabricius, 1780	A, P	
Buccinum cyaneum cyaneum Bruguière, 1792	A, P	bluish whelk
Buccinum cyaneum patulum G. O. Sars, 1878	A	
Buccinum cyaneum perdix Mörch, 1868	A	
Buccinum eugrammatum Dall, 1907	P	lirate whelk
Buccinum fischerianum Dall, 1871	P	
Buccinum fringillum Dall, 1877	P	finch whelk
Buccinum glaciale Linnaeus, 1761	A, P, Ac	glacial whelk
Buccinum gouldii A. E. Verrill, 1882	A	
Buccinum hertzensteini Verkruzen, 1882	P	
Buccinum humphreysianum Bennett, 1825	A, P	
Buccinum hydrophanum Hancock, 1846	A, Ac	
Buccinum inexhaustum Verkruzen, 1878	A	
Buccinum kodiakense Dall, 1907	P	Kodiak whelk
Buccinum micropoma Thorson, 1944	A, Ac	berry whelk
Buccinum ochotense (Middendorff, 1848)	P, Ac	Okhotsk whelk
Buccinum oedematum Dall, 1907	P	swollen whelk
Buccinum onismatopleura Dall, 1919	P, Ac	
Buccinum pemphigus major Dall, 1919	P	
Buccinum percrassum Dall, 1881	P, Ac	crude whelk
Buccinum physematum Dall, 1919	P, Ac	
Buccinum picturatum Dall, 1877	P	painted whelk
Buccinum planeticum Dall, 1919	P	wandering whelk
Buccinum plectrum Stimpson, 1865	A, P, Ac	sinuous whelk
Buccinum polare J. E. Gray, 1839	P, Ac	polar whelk
Buccinum scalariforme Möller, 1842	A, P, Ac	ladder whelk
Buccinum sericatum Hancock, 1846	A, P, Ac	silky whelk
Buccinum simulatum Dall, 1907	P	
Buccinum solenum Dall, 1919	P	
Buccinum striatissimum Sowerby, 1899	P	
Buccinum strigillatum fucanum Dall, 1907	P	
Buccinum tenellum Dall in Kobelt, 1883	P, Ac	
Buccinum totteni Stimpson, 1865	A, Ac	thin whelk
Buccinum undatum Linnaeus, 1758	A, Ac	waved whelk
Buccinum viridum Dall, 1889	P	turban whelk
Caducifer weberi Watters, 1983	A	banded phos
Cantharus cancellarius (Conrad, 1846)	A	cancellate cantharus
Cantharus multangulus (Philippi, 1848)	A	ribbed cantharus
Colus barbarinus (Dall, 1919)	P	Santa Barbara whelk
Colus bristolensis (Dall, 1919)	P	
Colus caelatus (A. E. Verrill and S. Smith, 1880)	A	carved whelk
Colus capponius (Dall, 1919)	P	
Colus errones (Dall, 1919)	P	wayward whelk
Colus esychus (Dall, 1907)	P, Ac	
Colus georgianus (Dall, 1920)	P	
Colus halidonus (Dall, 1919)	P	

SCIENTIFIC NAME	OCCURRENCE	COMMON NAME
Colus halimeris (Dall, 1919)	P	
Colus halli (Dall, 1873)	P	
Colus herendeenii (Dall, 1902)	P	thin-ribbed whelk
Colus hypolispus (Dall, 1891)	P, Ac	oblique whelk
Colus islandicus (Gmelin, 1791)	A, Ac	Iceland whelk
Colus jordani (Dall, 1913)	P	
Colus lividus (Mörch, 1862)	A	bruised whelk
Colus martensi (Krause, 1885)	P, Ac	
Colus morditus (Dall, 1919)	P	shrew whelk
Colus nobilis (Dall, 1919)	P	noble whelk
Colus obesus (A. E. Verrill, 1884)	A	plump whelk
Colus ombronius (Dall, 1919)	P	shady whelk
Colus periscelidus (Dall, 1891)	P	garter whelk
Colus pubescens (A. E. Verrill, 1882)	A, Ac	hairy whelk
Colus pulcius (Dall, 1919)	P, Ac	
Colus pygmaeus (Gould, 1841)	A	pygmy whelk
Colus roseus (Dall, 1877)	P, Ac	rosy whelk
Colus sabinii (J. E. Gray, 1824)	A	
Colus spitzbergensis (Reeve, 1855)	A, P, Ac	thick-ribbed whelk
Colus stimpsoni (Mörch, 1867)	A	
Colus timetus (Dall, 1919)	P	
Colus trombinus (Dall, 1919)	P	
Colus trophius (Dall, 1919)	P	
Colus ventricosus (J. E. Gray, 1839)	A	ventricose whelk
Engina caribbaea Bartsch and Rehder, 1939	A	Caribbean engina
Engina corinnae Crovo, 1971	A	
Engina strongi Pilsbry and Lowe, 1932	P	
Engina turbinella (Kiener, 1835)	A	white-spot engina
Exilioidea kelseyi (Dall, 1908)	P	
Exilioidea rectirostris (Carpenter, 1865)	P	
Kelletia kelleti (Forbes, 1850)	P	
Liomesus nassula Dall, 1901	P	basket whelk
Liomesus nux Dall, 1877	P	nut whelk
Liomesus ooides (Middendorff, 1848)	P, Ac	egg whelk
Macron lividus (A. Adams, 1855)	P	livid macron
Mohnia carolinensis (A. E. Verrill, 1884)	A	Carolina whelk
Mohnia simplex (A. E. Verrill, 1884)	A	
Neptunea amiantus (Dall, 1890)	P	
Neptunea beringiana (Dall, 1919)	P	
Neptunea communis (Middendorff, 1849)	P, Ac	
Neptunea despecta (Linnaeus, 1758)	A	
Neptunea eucosmia (Dall, 1891)	P	corded whelk
Neptunea heros J. E. Gray, 1850	P	
Neptunea insularis (Dall, 1895)	P	
Neptunea lyrata decemcostata (Say, 1826)	A	wrinkle whelk
Neptunea lyrata lyrata (Gmelin, 1791)	P, Ac	lyre whelk
Neptunea lyrata turnerae Clarke, 1956	A	
Neptunea magna (Dall, 1895)	P	helmet whelk
Neptunea middendorffiana MacGinitie, 1959	P	
Neptunea pribiloffensis (Dall, 1919)	P	Pribilof whelk
Neptunea smirnia (Dall, 1919)		
Neptunea stilesi A. G. Smith, 1968	P	inflated whelk
Neptunea tabulata (Baird, 1863)	P	tabled whelk
Neptunea ventricosa (Gmelin, 1791)	P, Ac	fat whelk

SCIENTIFIC NAME	OCCURRENCE	COMMON NAME
Neptunea vinosa (Dall, 1919)	P	wine whelk
Pisania auritula (Link, 1807)	A	gaudy cantharus
Pisania pusio (Linnaeus, 1758)	A	miniature trumpet triton
Pisania tincta (Conrad, 1846)	A	tinted cantharus
Plicifusus arcticus (Philippi, 1850)	A, P, Ac	arctic whelk
Plicifusus brunneus (Dall, 1877)	P	brown whelk
Plicifusus cretaceus (Reeve, 1847)	A	chalky whelk
Plicifusus griseus (Dall, 1890)	P	gray whelk
Plicifusus incisus Dall, 1919	P	
Plicifusus johanseni Dall, 1919	Ac	
Plicifusus kroyeri (Möller, 1842)	A, P, Ac	
Plicifusus laticordatus (Dall, 1907)	P	broad-cord whelk
Plicifusus oceanodromae Dall, 1919	P	seahorse whelk
Plicifusus syrtensis (Packard, 1867)	A	
Plicifusus verkruzeni (Kobelt, 1876)	P, Ac	
Plicifusus virens (Dall, 1877)	P	green whelk
Ptychosalpinx globulus (Dall, 1889)	A	globose whelk
Searlesia dira (Reeve, 1846)	P	dire whelk
Volutharpa ampullacea (Middendorff, 1848)	P	paper whelk
Volutopsius attenuatus (Dall, 1874)	P, Ac	elongate whelk
Volutopsius behringi (Middendorff, 1849)	P, Ac	
Volutopsius callorhinus callorhinus (Dall, 1877)	P	strombiform whelk
Volutopsius callorhinus stejnegeri (Dall, 1884)	P	
Volutopsius castaneus (Mörch, 1858)	P	volute whelk
Volutopsius deformis (Reeve, 1847)	P, Ac	warped whelk
Volutopsius filosus Dall, 1919	P	threaded whelk
Volutopsius fragilis (Dall, 1891)	P	fragile whelk
Volutopsius harpa (Mörch, 1858)	P	left-hand whelk
Volutopsius middendorffii (Dall, 1891)	P	tulip whelk
Volutopsius norvegicus (Gmelin, 1791)	A	Norway whelk
Volutopsius regularis (Dall, 1873)	P	regular whelk
Volutopsius rotundus Dall, 1919	P	rotund whelk
Volutopsius simplex Dall, 1907	P	simple whelk
Volutopsius stefanssoni Dall, 1919	P, Ac	shouldered whelk
Volutopsius trophonius Dall, 1902	P	frilled whelk

Colubrariidae

SCIENTIFIC NAME	OCCURRENCE	COMMON NAME
Colubraria lanceolata (Menke, 1828)	A	arrow dwarf triton
Colubraria obscura (Reeve, 1844)	A	obscure dwarf triton

Melongenidae

SCIENTIFIC NAME	OCCURRENCE	COMMON NAME
Busycon candelabrum (Lamarck, 1816)	A	splendid whelk
Busycon carica (Gmelin, 1791)	A	knobbed whelk
Busycon laeostomum Kent, 1982	A	snow whelk
Busycon pulleyi Hollister, 1958	A	prickly whelk
Busycon sinistrum Hollister, 1958	A	lightning whelk
Busycotypus canaliculatus (Linnaeus, 1758)	A, P [I]	channeled whelk
Busycotypus spiratus (Lamarck, 1816)	A	pearwhelk
Busycotypus plagosus (Conrad, 1863)	A	shouldered pearwhelk
Melongena corona (Gmelin, 1791)	A	crown conch

SCIENTIFIC NAME	OCCURRENCE	COMMON NAME
Nassariidae		
Ilyanassa obsoleta (Say, 1822)	A, P [I]	eastern mudsnail
Ilyanassa trivittata (Say, 1822)	A	threeline mudsnail
Nassarius acutus (Say, 1822)	A	sharp nassa
Nassarius albus (Say, 1826)	A	white nassa
Nassarius antillarum (d'Orbigny, 1842)	A	Antilles nassa
Nassarius consensus (Ravenel, 1861)	A	striate nassa
Nassarius fossatus (Gould, 1849)	P	channeled nassa
Nassarius fraterculus (Dunker, 1860)	P [I]	Japanese nassa
Nassarius hotessieri (d'Orbigny, 1845)	A	miniature nassa
Nassarius insculptus (Carpenter, 1864)	P	smooth western nassa
Nassarius mendicus cooperi (Forbes, 1850)	P	lean nassa
Nassarius mendicus mendicus (Gould, 1849)	P	lean western nassa
Nassarius perpinguis (Hinds, 1844)	P	fat western nassa
Nassarius polygonatus cinisculus (Reeve, 1853)	A	black-spot nassa
Nassarius rhinetes S. S. Berry, 1953	P	California nassa
Nassarius scissuratus (Dall, 1889)	A	carved nassa
Nassarius tegula (Reeve, 1853)	P	western mud nassa
Nassarius vibex (Say, 1822)	A	bruised nassa
Fasciolariidae		
Dolicholatirus cayohuesonicus (Sowerby, 1878)	A	Key West latirus
Dolicholatirus pauli (McGinty, 1955)	A	slender latirus
Fasciolaria bullisi Lyons, 1972	A	yellow tulip
Fasciolaria lilium branhamae Rehder and Abbott, 1951	A	
Fasciolaria lilium hunteria (G. Perry, 1811)	A	
Fasciolaria lilium lilium G. Fischer, 1807	A	banded tulip
Fasciolaria lilium tortugana Hollister, 1957	A	
Fasciolaria tulipa (Linnaeus, 1758)	A	true tulip
Fusinus aepynotus (Dall, 1889)	A	graceful spindle
Fusinus alcimus (Dall, 1889)	A	stout spindle
Fusinus amphiurgus (Dall, 1889)	A	slender spindle
Fusinus barbarensis (Trask, 1855)	P	Santa Barbara spindle
Fusinus benthalis (Dall, 1889)	A	modest spindle
Fusinus couei (Petit de la Saussaye, 1853)	A	Yucatan spindle
Fusinus eucosmius (Dall, 1889)	A	apricot spindle
Fusinus harfordii (Stearns, 1871)	P	
Fusinus helenae Bartsch, 1939	A	brown spindle
Fusinus kobelti (Dall, 1877)	P	
Fusinus luteopictus (Dall, 1877)	P	painted spindle
Fusinus monksae Dall, 1915	P	
Fusinus stegeri Lyons, 1978	A	ornamented spindle
Heilprinia timessus (Dall, 1889)	A	turnip spindle
Latirus angulatus (Röding, 1798)	A	short-tail latirus
Latirus carinifer Lamarck, 1822	A	yellow latirus
Latirus infundibulum (Gmelin, 1791)	A	brown-line latirus
Latirus nematus Woodring, 1928	A	threaded latirus
Leucozonia nassa (Gmelin, 1791)	A	chestnut latirus
Leucozonia ocellata (Gmelin, 1791)	A	white-spot latirus
Pleuroploca gigantea (Kiener, 1840)	A	horse conch

SCIENTIFIC NAME	OCCURRENCE	COMMON NAME
Olividae		
Jaspidella blanesi (Ford, 1898)	A	
Jaspidella jaspidea (Gmelin, 1791)	A	jasper dwarf olive
Jaspidella miris Olsson, 1956	A	
Oliva reticularis Lamarck, 1810	A	netted olive
Oliva sayana Ravenel, 1834	A	lettered olive
Olivella adelae Olsson, 1956	A	
Olivella baetica Carpenter, 1864	P	beatic dwarf olive
Olivella biplicata (Sowerby, 1825)	P	purple dwarf olive
Olivella bullula (Reeve, 1850)	A	bubble dwarf olive
Olivella dealbata (Reeve, 1850)	A	whitened dwarf olive
Olivella floralia (Duclos, 1853)	A	rice olive
Olivella fuscocincta Dall, 1889	A	
Olivella macgintyi Olsson, 1956	A	
Olivella minuta (Link, 1807)	A	minute dwarf olive
Olivella mutica (Say, 1822)	A	variable dwarf olive
Olivella nivea (Gmelin, 1791)	A	snowy dwarf olive
Olivella parva T. S. Oldroyd, 1921	P	
Olivella pedroana (Conrad, 1856)	P	San Pedro dwarf olive
Olivella perplexa Olsson, 1956	A	
Olivella pusilla (Marrat, 1871)	A	tiny dwarf olive
Olivella rotunda Dall, 1888	A	
Olivella stegeri Olsson, 1956	A	
Olivella thompsoni Olsson, 1956	A	
Olivella watermani McGinty, 1940	A	
Harpidae		
Morum dennisoni (Reeve, 1842)	A	Dennison morum
Morum lamarcki (Deshayes, 1844)	A	rose-mouth morum
Morum oniscus (Linnaeus, 1767)	A	Atlantic morum
Mitridae		
Mitra barbadensis (Gmelin, 1791)	A	Barbados miter
Mitra florida Gould, 1856	A	Florida miter
Mitra fultoni E. A. Smith, 1892	P	pitted miter
Mitra idae Melvill, 1893	P	half-pitted miter
Mitra nodulosa (Gmelin, 1791)	A	beaded miter
Mitra straminea A. Adams, 1853	A	Gulf Stream miter
Mitra swainsonii antillensis Dall, 1889	A	Antillean miter
Costellariidae		
Thala floridana (Dall, 1883)	A	
Thala foveata (Sowerby, 1874)	A	beaded thala
Thala gratiosa (Reeve, 1845)	P	esteemed miter
Vexillum dermestinum (Lamarck, 1811)	A	mottled miter
Vexillum epiphaneum (Rehder, 1943)	A	half-brown miter
Vexillum exiguum (C. B. Adams, 1845)	A	
Vexillum gemmatum (Sowerby, 1874)	A	gem miter
Vexillum hendersoni (Dall, 1927)	A	
Vexillum histrio (Reeve, 1844)	A	harlequin miter

SCIENTIFIC NAME	OCCURRENCE	COMMON NAME
Vexillum laterculatum (Sowerby, 1874)	A	honeycomb miter
Vexillum puella (Reeve, 1845)	A	white-spot miter
Vexillum pulchellum (Reeve, 1844)	A	beautiful miter
Vexillum styria (Dall, 1889)	A	dwarf deepsea miter
Vexillum sykesi (Melvill, 1925)	A	white-band miter
Vexillum trophonium (Dall, 1889)	A	
Vexillum wandoense (Holmes, 1860)	A	waxy miter

Volutomitridae

Microvoluta blakeana (Dall, 1889)	A	
Volutomitra groenlandica (Möller, 1842)	A	Greenland miter

Turbinellidae

Metzgeria californica Dall, 1903	P	California false spindle
Metzgeria montereyana A. G. Smith and Gordon, 1948	P	Monterey false spindle
Ptychatractus ligatus (Mighels and C. B. Adams, 1842)	A	ligate false spindle
Ptychatractus occidentalis Stearns, 1873	P	
Vasum muricatum (Born, 1778)	A	Caribbean vase

Volutidae

Arctomelon stearnsii Dall, 1872	P	Alaska volute
Enaeta cylleniformis (Sowerby, 1844)	A	sand lyria
Scaphella dubia (Broderip, 1827)	A	
Scaphella gouldiana (Dall, 1887)	A	banded volute
Scaphella junonia (Shaw, 1808)	A	junonia

Marginellidae

Cystiscus jewetti (Carpenter, 1857)	P	
Cystiscus politus (Carpenter, 1857)	P	polished marginella
Cystiscus politulus (Dall, 1919)	P	polite marginella
Cystiscus subtrigonus (Carpenter, 1864)	P	triangular marginella
Dentimargo aureocinctus (Stearns, 1872)	A	gold-line marginella
Dentimargo eburneolus (Conrad, 1834)	A	tan marginella
Granulina hadria (Dall, 1889)	A	
Granulina margaritula (Carpenter, 1857)	P	pear marginella
Granulina ovuliformis (d'Orbigny, 1841)	A	teardrop marginella
Hyalina pallida (Linnaeus, 1758)	A	pallid marginella
Marginella amabilis Redfield, 1852	A	queen marginella
Marginella apicina Menke, 1828	A	common Atlantic marginella
Marginella bella Conrad, 1868	A	la belle marginella
Marginella borealis (A. E. Verrill, 1884)	A	boreal marginella
Marginella carnea (Storer, 1837)	A	orange marginella
Marginella cassis Dall, 1889	A	helmet marginella
Marginella cineracea Dall, 1889	A	gray marginella
Marginella evelynae Bayer, 1943	A	
Marginella guttata (Dillwyn, 1817)	A	white-spot marginella
Marginella hartleyana Schwengel, 1941	A	
Marginella hematita Kiener, 1834	A	carmine marginella

SCIENTIFIC NAME	OCCURRENCE	COMMON NAME
Marginella idiochila Schwengel, 1943	A	
Marginella lavalleeana d'Orbigny, 1841	A	snowflake marginella
Marginella nobiliana Bayer, 1943	A	courtly marginella
Marginella perexilis Bavay, 1922	A	gaunt marginella
Marginella roosevelti Bartsch and Rehder, 1939	A	
Marginella roscida Redfield, 1860	A	seaboard marginella
Marginella virginiana Conrad, 1868	A	Virginia marginella
Marginellopsis serrei Bavay, 1911	A	
Persicula catenata (Montagu, 1803)	A	princess marginella
Persicula pulcherrima (Gaskoin, 1849)	A	decorated marginella
Volvarina albolineata (d'Orbigny, 1842)	A	white-line marginella
Volvarina avena (Kiener, 1834)	A	orange-band marginella
Volvarina avenacea (Deshayes, 1844)	A	little oat marginella
Volvarina subtriplicata (d'Orbigny, 1842)	A	threerib marginella
Volvarina taeniolata (Mörch, 1860)	P	California marginella
Volvarina torticula (Dall, 1881)	A	knave marginella
Volvarina veliei (Pilsbry, 1896)	A	

Cancellariidae

Admete californica Dall, 1908	P	California admete
Admete circumcincta (Dall, 1873)	P	corded admete
Admete couthouyi (Jay, 1839)	A, P, Ac	northern admete
Admete gracilior (Carpenter in Gabb, 1869)	P	slender admete
Admete modesta (Carpenter, 1865)	P	modest admete
Admete regina Dall, 1911	P, Ac	noble admete
Admete rhyssa (Dall, 1919)	P	wrinkled admete
Admete seftoni S. S. Berry, 1956	P	stubby admete
Admete unalashkensis (Dall, 1873)	P	Aleutian admete
Admete woodworthi (Dall, 1905)	P	graceful admete
Agatrix agassizii (Dall, 1889)	A	
Cancellaria cooperi Gabb, 1865	P	
Cancellaria corbicula Dall, 1908	P	basket nutmeg
Cancellaria crawfordiana (Dall, 1891)	P	
Cancellaria reticulata adelae Pilsbry, 1940	A	
Cancellaria reticulata reticulata (Linnaeus, 1767)	A	
Olssonella smithii (Dall, 1888)	A	
Trigonostoma rugosum (Lamarck, 1822)	A	rugose nutmeg
Trigonostoma tenerum (Philippi, 1848)	A	

Conidae

Conus amphiurgus Dall, 1889	A	
Conus armiger Crosse, 1858	A	mace cone
Conus attenuatus Reeve, 1844	A	slender cone
Conus californicus Hinds, 1844	P	California cone
Conus cancellatus Hwass, 1792	A	cancellate cone
Conus daucus Hwass, 1792	A	carrot cone
Conus delessertii Récluz, 1843	A	
Conus ermineus Born, 1778	A	agate cone
Conus flamingo Petuch, 1980	A	flamingo cone
Conus flavescens Sowerby, 1834	A	flame cone
Conus floridanus Gabb, 1868	A	Florida cone

SCIENTIFIC NAME	OCCURRENCE	COMMON NAME
Conus floridensis Sowerby, 1870	A	
Conus granulatus Linnaeus, 1758	A	glory-of-the-Atlantic cone
Conus jaspideus Gmelin, 1791	A	jasper cone
Conus macgintyi Pilsbry, 1955	A	
Conus mindanus Hwass, 1792	A	Bermuda cone
Conus mus Hwass, 1792	A	mouse cone
Conus patae Abbott, 1971	A	sunrise cone
Conus perryae Clench, 1942	A	
Conus rainesae McGinty, 1953	A	
Conus regius Gmelin, 1791	A	crown cone
Conus sennottorum Rehder and Abbott, 1951	A	speckled cone
Conus spurius Gmelin, 1791	A	alphabet cone
Conus stearnsi Conrad, 1869	A	dusky cone
Conus stimpsoni Dall, 1902	A	yellow cone
Conus villepini P. Fischer and Bernardi, 1857	A	

Terebridae

Hastula cinerea (Born, 1778)	A	gray auger
Hastula hastata (Gmelin, 1791)	A	shiny auger
Hastula maryleeae (R. L. Burch, 1965)	A	
Hastula salleana (Deshayes, 1859)	A	
Terebra acrior Dall, 1889	A	
Terebra arcas Abbott, 1954	A	Arcas auger
Terebra benthalis Dall, 1889	A	
Terebra concava Say, 1827	A	concave auger
Terebra crenifera Deshayes, 1859	P	western crenate auger
Terebra danai S. S. Berry, 1958	P	
Terebra dislocata (Say, 1822)	A	eastern auger
Terebra floridana Dall, 1889	A	yellow auger
Terebra glossema Schwengel, 1940	A	tongue auger
Terebra hemphilli Vanatta, 1924	P	
Terebra nassula Dall, 1889	A	woven auger
Terebra onslowensis Petuch, 1972	A	Onslow auger
Terebra pedroana Dall, 1908	P	San Pedro auger
Terebra protexta (Conrad, 1845)	A	fine-ribbed auger
Terebra rushii Dall, 1889	A	porcelain auger
Terebra taurina Lightfoot, 1786	A	flame auger
Terebra texana Dall, 1898	A	Texas auger
Terebra vinosa Dall, 1889	A	lilac auger

Turridae

Aforia circinata (Dall, 1873)	P	keeled aforia
Agathotoma stellata (Mörch, 1860)	P	
Antiplanes abarbareus Dall, 1919	P	
Antiplanes briseis Dall, 1919	P	
Antiplanes bulimoides Dall, 1919	P	
Antiplanes catalinae (Raymond, 1904)	P	
Antiplanes diaulax (Dall, 1908)	P	
Antiplanes hyperia Dall, 1919	P	
Antiplanes litus Dall, 1919	P	
Antiplanes major Bartsch, 1944	P	
Antiplanes perversus (Gabb, 1865)	P	

SCIENTIFIC NAME	OCCURRENCE	COMMON NAME
Antiplanes santarosanus (Dall, 1902)	P	
Antiplanes voyi (Gabb, 1866)	P	
Antiplanes willetti S. S. Berry, 1953	P	
Bactrocythara asarca (Dall and Simpson, 1901)	A	
Bathytoma viabrunnea (Dall, 1889)	A	
Bellaspira grippi (Dall, 1908)	P	
Bellaspira pentagonalis (Dall, 1889)	A	orange-rib drillia
Borsonella bartschi (Arnold, 1903)	P	
Borsonella civitella Dall, 1919	P	
Borsonella coronadoi (Dall, 1908)	P	
Borsonella diegensis (Dall, 1908)	P	
Borsonella nicoli Dall, 1919	P	
Borsonella nychia Dall, 1919	P	
Borsonella omphale Dall, 1919	P	
Borsonella pinosensis Bartsch, 1944	P	
Brachycythara barbarae Lyons, 1972	A	
Brachycythara biconica (C. B. Adams, 1850)	A	
Cerodrillia bealiana Schwengel and McGinty, 1942	A	
Cerodrillia clappi Bartsch and Rehder, 1939	A	
Cerodrillia girardi Lyons, 1972	A	
Cerodrillia perryae Bartsch and Rehder, 1939	A	
Cerodrillia schroederi Bartsch and Rehder, 1939	A	
Cerodrillia simpsoni (Dall in Simpson, 1887)	A	
Cerodrillia thea (Dall, 1883)	A	tea drillia
Cerodrillia verrilli (Dall, 1881)	A	
Clathromangelia fuscoligata (Dall, 1871)	P	
Clathromangelia interfossa (Carpenter, 1864)	P	
Clathurella canfieldi (Dall, 1871)	P	
Clathurella capaniola Dall, 1919	P	
Clathurella castianira Dall, 1919	P	
Clathurella conradiana Gabb, 1869	P	
Clathurella crystallina Gabb, 1865	P	
Clathurella rava (Hinds, 1843)	P	
Clathurella rigida (Hinds, 1843)	P	
Cochlespira elegans (Dall, 1881)	A	elegant star-turris
Cochlespira radiata (Dall, 1889)	A	lesser star-turris
Compsodrillia eucosmia (Dall, 1889)	A	
Compsodrillia haliostrephis (Dall, 1889)	A	
Crassispira cubana Melvill, 1923	A	
Crassispira montereyensis Stearns, 1871	P	
Crassispira phasma Schwengel, 1940	A	
Crassispira rhythmica Melvill, 1927	A	
Crassispira sanibelensis Bartsch and Rehder, 1939	A	
Crassispira semiinflata (Grant and Gale, 1931)	P	
Cryoturris cerinella (Dall, 1889)	A	waxy mangelia
Cryoturris citronella (Dall, 1889)	A	yellow mangelia
Cryoturris fargoi McGinty, 1955	A	
Cryoturris filifera (Dall, 1881)	A	
Cryoturris quadrilineata (C. B. Adams, 1850)	A	
Cymakra aspera (Carpenter, 1864)	P	
Cymakra gracilior (Tryon, 1884)	P	
Cymatosyrinx hemphilli (Stearns, 1871)	P	

SCIENTIFIC NAME	OCCURRENCE	COMMON NAME
Cymatosyrinx johnsoni Arnold, 1903	P	
Cymatosyrinx pagodula (Dall, 1889)	A	
Daphnella clathrata (Gabb, 1865)	P	
Daphnella corbicula Dall, 1889	A	
Daphnella lymneiformis (Kiener, 1840)	A	volute daphnelle
Daphnella margaretae Lyons, 1972	A	
Daphnella morra (Dall, 1881)	A	Morro daphnelle
Daphnella reticulosa Dall, 1889	A	
Daphnella retifera Dall, 1889	A	
Daphnella stegeri McGinty, 1955	A	
Drillia acurugata (Dall, 1890)	A	rough drillia
Drillia albicoma (Dall, 1889)	A	
Drillia canna (Dall, 1889)	A	
Drillia cydia (Bartsch, 1943)	A	
Drilliola loprestiana (Calcara, 1841)	A	
Elaeocyma empyrosia (Dall, 1899)	P	
Eubela macgintyi Schwengel, 1943	A	
Fenimorea fucata (Reeve, 1845)	A	
Fenimorea halidorema (Schwengel, 1940)	A	
Fenimorea janetae Bartsch, 1934	A	
Gemmula periscelida (Dall, 1889)	A	Atlantic gem-turris
Glyphostoma adria Dall, 1919	P	
Glyphostoma cymodoce Dall, 1919	P	
Glyphostoma dentiferum Gabb, 1872	A	
Glyphostoma elsae Bartsch, 1934	A	
Glyphostoma gabbii Dall, 1889	A	
Glyphostoma golfoyaquense Maury, 1917	A	
Glyphostoma hesione (Dall, 1919)	P	
Glyphostoma pilsbryi Schwengel, 1940	A	
Glyphostomops hendersoni Bartsch, 1934	A	
Glyphoturris quadrata (Reeve, 1845)	A	
Glyphoturris diminuta (C. B. Adams, 1850)	A	
Glyphoturris eritima (Bush, 1885)	A	
Glyphoturris rugirima (Dall, 1889)	A	
Granoturris presleyi Lyons, 1972	A	
Gymnobela agassizii (A. E. Verrill and S. Smith, 1880)	A	
Hindsiclava alesidota (Dall, 1889)	A	
Inodrillia acova Bartsch, 1943	A	
Inodrillia aepynota (Dall, 1889)	A	
Inodrillia avira Bartsch, 1943	A	
Inodrillia carpenteri (A. E. Verrill and S. Smith 1880, non de Folin, 1867)	A	
Inodrillia dalli (A. E. Verrill and S. Smith, 1882)	A	
Inodrillia dido Bartsch, 1943	A	
Inodrillia gibba Bartsch, 1943	A	
Inodrillia hatterasensis Bartsch, 1943	A	
Inodrillia ino Bartsch, 1943	A	
Inodrillia martha Bartsch, 1943	A	
Inodrillia miamia Bartsch, 1943	A	
Inodrillia vetula Bartsch, 1943	A	
Ithycythara cymella (Dall, 1889)	A	
Ithycythara lanceolata (C. B. Adams, 1850)	A	spear mangelia

SCIENTIFIC NAME	OCCURRENCE	COMMON NAME
Ithycythara parkeri Abbott, 1958	A	
Ithycythara pentagonalis (Reeve, 1845)	A	
Ithycythara psila (Bush, 1885)	A	
Kurtzia arteaga (Dall and Bartsch, 1910)	P	beaded mangelia
Kurtziella accincta (Montagu, 1808)	A	
Kurtziella atrostyla (Tryon, 1884)	A	brown-tip mangelia
Kurtziella cerina (Kurtz and Stimpson, 1851)	A	
Kurtziella diomedea Bartsch and Rehder, 1939	A	
Kurtziella limonitella (Dall, 1883)	A	punctate mangelia
Kurtziella newcombei (Dall, 1919)	P	
Kurtziella perryae Bartsch and Rehder, 1939	A	
Kurtziella plumbea (Hinds, 1843)	P	violet-band mangelia
Kurtziella rubella (Kurtz and Stimpson, 1851)	A	reddish mangelia
Kurtziella serga (Dall, 1881)	A	
Kurtziella variegata (Carpenter, 1864)	P	tan mangelia
Kylix halocydne (Dall, 1919)	P	
Leptadrillia cookei (E. A. Smith, 1882)	A	
Leptadrillia splendida (Bartsch, 1934)	A	
Leucosyrinx kincaidi Dall, 1919	P	
Mangelia aculea (Dall, 1919)	P	
Mangelia amatula (Dall, 1919)	P	
Mangelia astricta Reeve, 1846	A	
Mangelia bandella (Dall, 1881)	A	
Mangelia ceroplasta Bush, 1885	A	
Mangelia cesta Dall, 1919	P	
Mangelia constricta Gabb, 1865	P	
Mangelia densilineata (Dall, 1921)	P	
Mangelia eriphyle Dall, 1919	P	
Mangelia evadne Dall, 1919	P	
Mangelia hexagona Gabb, 1865	P	
Mangelia hooveri Arnold, 1903	P	
Mangelia louisa (Dall, 1919)	P	
Mangelia merita Hinds, 1843	P	
Mangelia painei Arnold, 1903	P	
Mangelia perattenuata Dall, 1905	P	
Mangelia philodice Dall, 1919	P	
Mangelia pomara (Dall, 1919)	P	
Mangelia victoriana (Dall, 1897)	P	
Megasurcula carpenteriana (Gabb, 1865)	P	
Megasurcula remondii (Gabb, 1866)	P	
Megasurcula stearnsiana (Raymond, 1904)	P	
Miraclathurella herminea (Bartsch, 1934)	A	
Mitrolumna biplicata (Dall, 1889)	A	
Mitromorpha carpenteri Glibert, 1954	P	
Nannodiella melanitica (Bush, 1885)	A	
Nannodiella oxia (Bush, 1885)	A	
Nannodiella vespuciana (d'Orbigny, 1842)	A	
Nodotoma impressa (Mörch, 1869)	P, Ac	
Obesotoma arctica (A. Adams, 1855)	P	
Obesotoma lawrenciana (Dall, 1919)	P	
Oenopota alaskensis (Dall, 1871)	P	
Oenopota albrechti (Krause, 1885)	P	
Oenopota aleuticus (Dall, 1871)	P, Ac	
Oenopota alitakensis (Dall, 1919)	P	

SCIENTIFIC NAME	OCCURRENCE	COMMON NAME
Oenopota althorpensis (Dall, 1919)	P	
Oenopota althorpi (Dall, 1919)	P	
Oenopota amiantus (Dall, 1919)	P	
Oenopota angulosus (G. O. Sars, 1878)	A	
Oenopota babylonius (Dall, 1919)	P	
Oenopota beckii (Möller, 1842)	Ac	
Oenopota bicarinatus (Couthuoy, 1833)	A	
Oenopota blaneyi (Bush, 1909)	A	
Oenopota cancellatus (Mighels and C. B. Adams, 1842)	A	
Oenopota chiachianus (Dall, 1919)	P	
Oenopota concinnulus (A. E. Verrill, 1882)	A	
Oenopota decussatus (Couthouy, 1839)	A, Ac	
Oenopota elegans (Möller, 1842)	A, P	
Oenopota eriopis (Dall, 1919)	P	
Oenopota exaratus (Möller, 1842)	A	
Oenopota excurvatus (Carpenter, 1865)	P	
Oenopota fidicula (Gould, 1849)	P	
Oenopota fiora (Dall, 1919)	P	
Oenopota galgana (Dall, 1919)	P	
Oenopota gouldii (A. E. Verrill, 1882)	A	
Oenopota granticus (Dall, 1919)	P	
Oenopota harpa (Dall, 1885)	P, Ac	
Oenopota harpularius (Couthouy, 1838)	A, P, Ac	
Oenopota healyi (Dall, 1919)	P	
Oenopota hebes (A. E. Verrill, 1880)	A	
Oenopota incisulus (A. E. Verrill, 1882)	A, Ac	
Oenopota inequita (Dall, 1919)	P	
Oenopota krausei (Dall, 1886)	P	
Oenopota kyskanus (Dall, 1919)	P	
Oenopota laevigatus (Dall, 1871)	P	
Oenopota levidensis (Carpenter, 1864)	P	
Oenopota lotta (Dall, 1919)	P	
Oenopota luetkeni (Dall, 1919)	P	
Oenopota lutkeanus (Krause, 1885)	P	
Oenopota maurellei (Dall and Bartsch, 1910)	P	
Oenopota metschigmensis (Krause, 1885)	P	
Oenopota mitratus (Dall, 1919)	P	
Oenopota morchi (Leche, 1878)	P, Ac	
Oenopota murdochianus (Dall, 1885)	P, Ac	
Oenopota nazanensis (Dall, 1919)	P	
Oenopota nodulosus (Krause, 1885)	P	
Oenopota novajasemljensis (Leche, 1878)	P, Ac	
Oenopota nunivakensis (Dall, 1919)	P	
Oenopota pavlova (Dall, 1919)	P	
Oenopota pingelii (Möller, 1842)	A	
Oenopota pleurotomaris (Couthuoy, 1838)	A	
Oenopota popovius (Dall, 1919)	P	
Oenopota pribilovus (Dall, 1919)	P, Ac	
Oenopota pyramidalis (Strøm, 1788)	A, P, Ac	
Oenopota rassinus (Dall, 1919)	P	
Oenopota regulus (Dall, 1919)	P	
Oenopota reticulatus (T. Brown, 1827)	A, P, Ac	
Oenopota roseus (Lovén, 1846)	P	

SCIENTIFIC NAME	OCCURRENCE	COMMON NAME
Oenopota roseus (G. O. Sars, 1878)	A	
Oenopota sarsii (A. E. Verrill, 1880)	A	
Oenopota scalaris (Möller, 1842)	A, Ac	
Oenopota sculpturatus (Dall, 1886)	P	
Oenopota simplex (Middendorff, 1849)	P, Ac	
Oenopota solidus (Dall, 1887)	P	
Oenopota subvitreus (A. E. Verrill, 1884)	A	
Oenopota tabulatus (Carpenter, 1865)	P	
Oenopota tenuiliratus cymatus (Dall, 1919)	P	
Oenopota tenuissimus (Dall, 1919)	P	
Oenopota turricula (Montagu, 1803)	A, P, Ac	
Oenopota violaceus (Mighels and C. B. Adams, 1842)	A, P, Ac	
Oenopota woodianus (Möller, 1842)	A, P, Ac	
Ophiodermella cancellata (Carpenter, 1864)	P	cancellate snakeskin-snail
Ophiodermella inermis (Hinds, 1843)	P	gray snakeskin-snail
Ophiodermella montereyensis Bartsch, 1944	P	
Pilsbryspira albomaculata (d'Orbigny, 1842)	A	white-band drillia
Pilsbryspira leucocyma (Dall, 1883)	A	white-knob drillia
Pilsbryspira monilis (Bartsch and Rehder, 1939)	A	
Pleurotomella blakeana (Dall, 1889)	A	
Pleurotomella packardii (A. E. Verrill, 1872)	A	
Polystira albida (G. Perry, 1811)	A	white giant-turris
Polystira tellea (Dall, 1889)	A	delicate giant-turris
Polystira vibex (Dall, 1889)	A	twilled giant-turris
Pseudomelatoma grippi (Dall, 1919)	P	
Pseudomelatoma penicillata (Carpenter, 1864)	P	
Pseudomelatoma sticta S. S. Berry, 1956	P	
Pseudomelatoma torosa (Carpenter, 1865)	P	
Pyrgocythara balteata (Reeve, 1846)	A	
Pyrgocythara candidissima (C. B. Adams, 1845)	A	
Pyrgocythara filosa Rehder, 1943	A	filose mangelia
Pyrgocythara hemphilli Bartsch and Rehder, 1939	A	
Pyrgocythara plicosa (C. B. Adams, 1850)	A	plicate mangelia
Pyrgospira ostrearum (Stearns, 1872)	A	
Pyrgospira tampaensis (Bartsch and Rehder, 1939)	A	
Rhodopetoma rhodope (Dall, 1919)	P	
Saccharoturris monocingulata (Dall, 1889)	A	
Splendrillia lissotropis (Dall, 1881)	A	
Splendrillia moseri brunnescens Rehder, 1943	A	
Splendrillia moseri moseri (Dall, 1889)	A	
Splendrillia woodringi (Bartsch, 1934)	A	
Stellatoma stellata (Stearns, 1872)	A	
Strictispira solida (C. B. Adams, 1850)	A	solid drillia
Suavodrillia kennicottii (Dall, 1871)	P	
Suavodrillia willetti Dall, 1919	P	
Tenaturris bartletti (Dall, 1889)	A	
Tenaturris janira (Dall, 1919)	P	
Typhlomangelia nivalis (Lovén, 1846)	A	
Viridrillia cervina Bartsch, 1943	A	
Viridrillia hendersoni Bartsch, 1943	A	
Viridrillia williami Bartsch, 1943	A	
Vitricythara elata (Dall, 1889)	A	

SCIENTIFIC NAME	OCCURRENCE	COMMON NAME
Vitricythara metria (Dall, 1903)	A	

ORDER PYRAMIDELLOIDA

Pyramidellidae

SCIENTIFIC NAME	OCCURRENCE	COMMON NAME
Boonea bisuturalis (Say, 1822)	A	two-groove odostome
Boonea impressa (Say, 1822)	A	impressed odostome
Boonea seminuda (C. B. Adams, 1839)	A	half-smooth odostome
Cyclostremella californica Bartsch, 1907	P	
Cyclostremella conradia Bartsch, 1920	P	
Cyclostremella humilis Bush, 1897	A	
Fargoa bartschi (Winkley, 1909)	A	
Fargoa bushiana (Bartsch, 1909)	A	
Fargoa dianthophila (H. W. Wells and M. J. Wells, 1961)	A	serpulid odostome
Fargoa gibbosa (Bush, 1909)	A	
Iselica anomala (C. B. Adams, 1850)	A	
Iselica obtusa (Carpenter, 1864	P	
Iselica ovoidea (Gould, 1853)	P	
Kleinella cedrosa (Dall, 1884)	A	
Miralda havanensis (Pilsbry and Aguayo, 1933)	A	
Odostomia aepynota Dall and Bartsch, 1909	P	
Odostomia aequisculpta Carpenter, 1864	P	
Odostomia aleutica Dall and Bartsch, 1909	P	
Odostomia altina Dall and Bartsch, 1909	P	
Odostomia americana Dall and Bartsch, 1904	P	
Odostomia amiantus Dall and Bartsch, 1907	P	
Odostomia amilda Dall and Bartsch, 1909	P	
Odostomia angularis Dall and Bartsch, 1907	P	
Odostomia arctica Dall and Bartsch, 1909	P	
Odostomia astricta Dall and Bartsch, 1907	P	
Odostomia atossa Dall, 1908	P	
Odostomia bachia Bartsch, 1927	P	
Odostomia baldridgae Bartsch, 1912	P	
Odostomia barkleyensis Dall and Bartsch, 1910	P	
Odostomia beringi Dall, 1871	P	
Odostomia calcarella Bartsch, 1912	P	
Odostomia callimene Bartsch, 1912	P	
Odostomia callimorpha Dall and Bartsch, 1909	P	
Odostomia calliope Bartsch, 1912	P	
Odostomia cancellata d'Orbigny, 1842	A	
Odostomia canfieldi Dall, 1908	P	
Odostomia capitana Dall and Bartsch, 1909	P	
Odostomia cassandra Bartsch, 1912	P	
Odostomia catalinensis Bartsch, 1927	P	
Odostomia chinooki Bartsch, 1927	P	
Odostomia churchi A. G. Smith and Gordon, 1948	P	
Odostomia cincta (Carpenter, 1864)	P	
Odostomia clementensis Bartsch, 1927	P	
Odostomia clementina Dall and Bartsch, 1909	P	
Odostomia columbiana Dall and Bartsch, 1907	P	

SCIENTIFIC NAME	OCCURRENCE	COMMON NAME
Odostomia cookeana Bartsch, 1910	P	
Odostomia cumshewaensis Bartsch, 1921	P	
Odostomia cypria Dall and Bartsch, 1912	P	
Odostomia dicella Bartsch, 1912	P	
Odostomia dinella Dall and Bartsch, 1909	P	
Odostomia edmondi Jordan, 1920	P	
Odostomia eldorana Bartsch, 1912	P	
Odostomia elsa Dall and Bartsch, 1909	P	
Odostomia enbergi Bartsch, 1920	P	
Odostomia engonia Bush, 1885	A	
Odostomia enora Dall and Bartsch, 1909	P	
Odostomia esilda Dall and Bartsch, 1909	P	
Odostomia eucosmia Dall and Bartsch, 1909	P	
Odostomia eugena Dall and Bartsch, 1909	P	
Odostomia euglypta Jordan, 1920	P	
Odostomia exara Dall and Bartsch, 1907	P	
Odostomia excisa Bartsch, 1912	P	
Odostomia eyerdami Bartsch, 1927	P	
Odostomia farallonensis Dall and Bartsch, 1909	P	
Odostomia farma Dall and Bartsch, 1909	P	
Odostomia fetella Dall and Bartsch, 1909	P	
Odostomia franciscana Bartsch, 1917	P	
Odostomia gloriosa Bartsch, 1912	P	
Odostomia gracilentis (Keep, 1887)	P	
Odostomia gravida Gould, 1852	P	
Odostomia grippiana Bartsch, 1912	P	
Odostomia hartfordensis Dall and Bartsch, 1907	P	
Odostomia heathi A. G. Smith and Gordon, 1948	P	
Odostomia helena Bartsch, 1912	P	
Odostomia helga Dall and Bartsch, 1909	P	
Odostomia hemphilli Dall and Bartsch, 1909	P	
Odostomia hendersoni Bartsch, 1909	A	
Odostomia herilda Dall and Bartsch, 1909	P	
Odostomia heterocincta Bartsch, 1912	P	
Odostomia hypatia Dall and Bartsch, 1912	P	
Odostomia hypocurta Dall and Bartsch, 1909	P	
Odostomia iliuliukensis Dall and Bartsch, 1909	P	
Odostomia inflata Carpenter, 1864	P	
Odostomia kelseyi Bartsch, 1912	P	
Odostomia kennerleyi Dall and Bartsch, 1907	P	
Odostomia killisnooensis Dall and Bartsch, 1909	P	
Odostomia krausei Clessin, 1900	P	
Odostomia laevigata (d'Orbigny, 1842)	A	
Odostomia lastra Dall and Bartsch, 1909	P	
Odostomia laxa Dall and Bartsch, 1909	P	
Odostomia martensi Dall and Bartsch, 1906	P	
Odostomia moratora Dall and Bartsch, 1909	P	
Odostomia movilla Dall and Bartsch, 1909	P	
Odostomia navisa Dall and Bartsch, 1907	P	
Odostomia nemo Dall and Bartsch, 1909	P	
Odostomia nota Dall and Bartsch, 1909	P	
Odostomia nuciformis Carpenter, 1865	P	
Odostomia nunivakensis Dall and Bartsch, 1909	P	
Odostomia oregonensis Dall and Bartsch, 1909	P	

SCIENTIFIC NAME	OCCURRENCE	COMMON NAME
Odostomia ornatissima (Haas, 1943)	P	
Odostomia phanella Dall and Bartsch, 1909	P	
Odostomia pharcida Dall and Bartsch, 1907	P	
Odostomia pocahontasae J. B. Henderson and Bartsch, 1914	A	
Odostomia pratoma Dall and Bartsch, 1909	P	
Odostomia producta (C. B. Adams, 1840)	A	
Odostomia profundicola Dall and Bartsch, 1909	P	
Odostomia pulcherrima Dall and Bartsch, 1909	P	
Odostomia pulcia Dall and Bartsch, 1909	P	
Odostomia quadrae Dall and Bartsch, 1910	P	
Odostomia resina Dall and Bartsch, 1909	P	
Odostomia richi Dall and Bartsch, 1909	P	
Odostomia ritteri Dall and Bartsch, 1909	P	
Odostomia sanjuanensis Bartsch, 1920	P	
Odostomia sapia Dall and Bartsch, 1909	P	
Odostomia satura Carpenter, 1865	P	
Odostomia septentrionalis Dall and Bartsch, 1909	P	
Odostomia sillana Dall and Bartsch, 1909	P	
Odostomia skidegatensis Bartsch, 1912	P	
Odostomia smithii A. E. Verrill, 1880	A	
Odostomia spreadboroughi Dall and Bartsch, 1910	P	
Odostomia strongi Bartsch, 1927	P	
Odostomia subglobosa Bartsch, 1912	P	
Odostomia subturrita Dall and Bartsch, 1909	P	
Odostomia sulcosa (Mighels, 1843)	A	
Odostomia tacomaensis Dall and Bartsch, 1907	P	
Odostomia tenuisculpta Carpenter, 1864	P	
Odostomia teres Bush, 1885	A	
Odostomia thalia Bartsch, 1912	P	
Odostomia thea Bartsch, 1912	P	
Odostomia tornata A. E. Verrill, 1884	A	
Odostomia trachis Dall and Bartsch, 1909	P	
Odostomia tremperi Bartsch, 1927	P	
Odostomia turricula Dall and Bartsch, 1903	P	
Odostomia unalaskensis Dall and Bartsch, 1909	P	
Odostomia unidentata Fleming, 1813	A	
Odostomia vancouverensis Dall and Bartsch, 1910	P	
Odostomia vicola Dall and Bartsch, 1909	P	
Odostomia vincta Dall and Bartsch, 1909	P	
Odostomia virginica J. B. Henderson and Bartsch, 1914	A	
Odostomia washingtonia Bartsch, 1920	P	
Odostomia whitei Bartsch, 1927	P	
Odostomia willetti Bartsch, 1917	P	
Odostomia winkleyi Bartsch, 1909	A	
Odostomia youngi Dall and Bartsch, 1910	P	
Peristichia agria Dall, 1889	A	
Peristichia pedroana (Dall and Bartsch, 1909)	P	
Peristichia toreta Dall, 1889	A	
Pyramidella adamsi Carpenter, 1864	P	
Pyramidella candida Mörch, 1875	A	
Pyramidella crassula Forbes, 1843	A	

SCIENTIFIC NAME	OCCURRENCE	COMMON NAME
Pyramidella crenulata (Holmes, 1859)	A	
Pyramidella dolabrata (Linnaeus, 1758)	A	
Pyramidella mazatlanticus Dall and Bartsch, 1909	P	
Pyramidella mexicana Dall and Bartsch, 1909	P	
Pyramidella resticula (Dall, 1889)	A	
Pyramidella unifasciata Forbes, 1843	A	
Pyramidella ventricosa Forbes, 1843	A	
Sayella chesapeakea Morrison, 1939	A	
Sayella crosseana (Dall, 1885)	A	
Sayella fusca (C. B. Adams, 1839)	A	
Sayella hemphilli (Dall, 1889)	A	
Sayella livida Rehder, 1935	A	
Triptychus niveus Mörch, 1875	A	
Turbonilla acra Dall and Bartsch, 1909	P	
Turbonilla adusta Dall and Bartsch, 1909	P	
Turbonilla aepynota Dall and Bartsch, 1909	P	
Turbonilla aequalis (Say, 1827)	A	
Turbonilla alaskana Dall and Bartsch, 1909	P	
Turbonilla almo Dall and Bartsch, 1909	P	
Turbonilla antestriata Dall and Bartsch, 1907	P	
Turbonilla aragoni Dall and Bartsch, 1909	P	
Turbonilla aresta Dall and Bartsch, 1909	P	
Turbonilla attrita Dall and Bartsch, 1909	P	
Turbonilla aurantia (Carpenter, 1865)	P	
Turbonilla auricoma Dall and Bartsch, 1903	P	
Turbonilla bakeri Bartsch, 1912	P	
Turbonilla barcleyensis Bartsch, 1917	P	
Turbonilla belotheca Dall, 1889	A	
Turbonilla burchi Gordon, 1938	P	
Turbonilla callia Dall and Bartsch, 1909	P	
Turbonilla callimene Bartsch, 1912	P	
Turbonilla canadensis Bartsch, 1917	P	
Turbonilla canfieldi Dall and Bartsch, 1907	P	
Turbonilla carpenteri Dall and Bartsch, 1909	P	
Turbonilla castanea Keep, 1887	P	
Turbonilla castanella Dall, 1908	P	
Turbonilla chocolata (Carpenter, 1864)	P	
Turbonilla clarinda Bartsch, 1912	P	
Turbonilla clementina Bartsch, 1927	P	
Turbonilla conradi Bush, 1899	A	
Turbonilla curta Dall, 1889	A	
Turbonilla dalli Bush, 1899	A	
Turbonilla delmontana Bartsch, 1937	P	
Turbonilla diegensis Dall and Bartsch, 1909	P	
Turbonilla dinora Bartsch, 1912	P	
Turbonilla dora Bartsch, 1917	P	
Turbonilla dracona Bartsch, 1912	P	
Turbonilla edwardensis Bartsch, 1909	P	
Turbonilla elegantula branfordensis Bartsch, 1909	A	
Turbonilla elegantula elegantula A. E. Verrill, 1882	A	
Turbonilla encella Bartsch, 1912	P	

SCIENTIFIC NAME	OCCURRENCE	COMMON NAME
Turbonilla engbergi Bartsch, 1920	P	
Turbonilla enna Bartsch, 1927	P	
Turbonilla eschscholtzi Dall and Bartsch, 1907	P	
Turbonilla eucosmobasis Dall and Bartsch, 1907	P	
Turbonilla eva Bartsch, 1917	P	
Turbonilla exilis (C. B. Adams, 1850)	A	
Turbonilla eyerdami Bartsch, 1927	P	
Turbonilla fackenthallae A. G. Smith and Gordon, 1948	P	
Turbonilla franciscana Bartsch, 1917	P	
Turbonilla gabbiana (J. G. Cooper, 1867)	P	
Turbonilla gilli delmontensis Dall and Bartsch, 1907	P	
Turbonilla gilli gilli Dall and Bartsch, 1907	P	
Turbonilla gloriosa Bartsch, 1912	P	
Turbonilla grippi Bartsch, 1912	P	
Turbonilla halibrecta Dall and Bartsch, 1909	P	
Turbonilla halistrepta Dall and Bartsch, 1909	P	
Turbonilla hecuba Dall and Bartsch, 1913	A	
Turbonilla hemphilli Bush, 1899	A	
Turbonilla hypolispa Dall and Bartsch, 1909	P	
Turbonilla ilfa Bartsch, 1927	P	
Turbonilla ina Bartsch, 1917	P	
Turbonilla incisa constricta Bush, 1899	A	
Turbonilla incisa incisa Bush, 1899	A	
Turbonilla interrupta (Totten, 1835)	A	
Turbonilla ista Bartsch, 1917	P	
Turbonilla jewetti Dall and Bartsch, 1909	P	
Turbonilla kelseyi Dall and Bartsch, 1909	P	
Turbonilla kincaidi Bartsch, 1921	P	
Turbonilla kurtzii Mazyck, 1913	A	
Turbonilla laevis (C. B. Adams, 1850)	A	
Turbonilla laminata (Carpenter, 1865)	P	
Turbonilla lituyana Dall and Bartsch, 1909	P	
Turbonilla lordi (E. A. Smith, 1880)	P	
Turbonilla louiseae Clarke, 1954	A	
Turbonilla lyalli Dall and Bartsch, 1907	P	
Turbonilla macouni Dall and Bartsch, 1910	P	
Turbonilla middendorffi Bartsch, 1927	P	
Turbonilla mighelsi Bartsch, 1909	A	
Turbonilla morchi Dall and Bartsch, 1907	P	
Turbonilla multicostata (C. B. Adams, 1850)	A	
Turbonilla muricatoides Dall and Bartsch, 1907	P	
Turbonilla nereia Dall and Bartsch, 1909	P	
Turbonilla newcombei Dall and Bartsch, 1907	P	
Turbonilla nivea (Stimpson, 1851)	A	
Turbonilla nuttingi Dall and Bartsch, 1909	P	
Turbonilla obeliscus (C. B. Adams, 1850)	A	
Turbonilla obesa Dall and Bartsch, 1909	P	
Turbonilla oregonensis Dall and Bartsch, 1907	P	
Turbonilla pauli A. G. Smith and Gordon, 1958	P	
Turbonilla pentalopha Dall and Bartsch, 1903	P	
Turbonilla perlepida A. E. Verrill, 1885	A	

SCIENTIFIC NAME	OCCURRENCE	COMMON NAME
Turbonilla pesa Dall and Bartsch, 1910	P	
Turbonilla pluto Dall and Bartsch, 1909	P	
Turbonilla pocahontasae J. B. Henderson and Bartsch, 1914	A	
Turbonilla polita (A. E. Verrill, 1872)	A	
Turbonilla powhatani J. B. Henderson and Bartsch, 1914	A	
Turbonilla protracta Dall, 1892	A	
Turbonilla pugetensis Bartsch, 1917	P	
Turbonilla puncta (C. B. Adams, 1850)	A	
Turbonilla punicea Dall, 1884	A	
Turbonilla pusilla (C. B. Adams, 1850)	A	
Turbonilla rathbuni A. E. Verrill and S. Smith, 1880	A	
Turbonilla raymondi Dall and Bartsch, 1909	P	
Turbonilla recta Dall and Bartsch, 1909	P	
Turbonilla reticulata (C. B. Adams, 1850)	A	
Turbonilla ridgwayi Dall and Bartsch, 1909	P	
Turbonilla rinella Dall and Bartsch, 1910	P	
Turbonilla santarosana Dall and Bartsch, 1909	P	
Turbonilla serrae Dall and Bartsch, 1907	P	
Turbonilla shuyakensis Bartsch, 1927	P	
Turbonilla signae Dall and Bartsch, 1909	P	
Turbonilla simpsoni Dall and Bartsch, 1909	P	
Turbonilla stelleri Bartsch, 1927	P	
Turbonilla stillmani A. G. Smith and Gordon, 1948	P	
Turbonilla stimpsoni Bush, 1899	A	
Turbonilla strongi Willett, 1931	P	
Turbonilla styliformis Mörch, 1875	A	
Turbonilla stylina (Carpenter, 1865)	P	
Turbonilla subulata (C. B. Adams, 1850)	A	
Turbonilla sumneri Bartsch, 1909	A	
Turbonilla swani Dall and Bartsch, 1909	P	
Turbonilla talma Dall and Bartsch, 1910	P	
Turbonilla taylori Dall and Bartsch, 1907	P	
Turbonilla tenuicula (Gould, 1853)	P	
Turbonilla textilis (Kurtz, 1860)	A	
Turbonilla torquata (Gould, 1852)	P	
Turbonilla toyatani J. B. Henderson and Bartsch, 1914	A	
Turbonilla tremperi Bartsch, 1917	P	
Turbonilla tridentata (Carpenter, 1864)	P	
Turbonilla unilirata Bush, 1899	A	
Turbonilla verrillii Bartsch, 1909	A	
Turbonilla vexativa Dall and Bartsch, 1909	P	
Turbonilla virga Dall, 1884	A	
Turbonilla virgata Dall, 1892	A	
Turbonilla virginica J. B. Henderson and Bartsch, 1914	A	
Turbonilla virgo (Carpenter, 1864)	P	
Turbonilla viridaria Dall, 1884	A	
Turbonilla weldi Dall and Bartsch, 1909	P	
Turbonilla whiteavesi Bartsch, 1909	A	

SCIENTIFIC NAME	OCCURRENCE	COMMON NAME
Turbonilla wickhami Dall and Bartsch, 1909	P	
Turbonilla willetti A. G. Smith and Gordon, 1948	P	

Order Cephalaspidea

Acteonidae

Acteon candens Rehder, 1939	A	
Acteon finlayi McGinty, 1955	A	
Acteon traskii Stearns, 1898	P	
Japonacteon pusillus (Forbes, 1843)	A	miniature baby-bubble
Microglyphis brevicula (Dall, 1902)	P	short baby-bubble
Microglyphis estuarina (Dall, 1908)	P	estuarine baby-bubble
Mysouffa cumingii (A. Adams, 1854)	A	
Rictaxis painei Dall, 1903	P	
Rictaxis punctocaelatus (Carpenter, 1864)	P	
Rictaxis punctostriatus (C. B. Adams, 1840)	A	pitted baby-bubble

Bullinidae

Bullina exquisita McGinty, 1955	A	exquisite bubble

Hydatinidae

Hydatina physis (Linnaeus, 1758)	A	brown-line paperbubble
Micromelo undatus (Bruguière, 1792)	A	miniature melo

Ringiculidae

Ringicula nitida A. E. Verrill, 1873	A	
Ringicula semistriata d'Orbigny, 1842	A	striate helmet-bubble

Scaphandridae

Acteocina atrata P. S. Mikkelsen and P. M. Mikkelsen, 1984	A	blackback barrel-bubble
Acteocina bidentata (d'Orbigny, 1841)	A	two-tooth barrel-bubble
Acteocina bullata (Kiener, 1834)	A	striate barrel-bubble
Acteocina canaliculata (Say, 1826)	A	channeled barrel-bubble
Acteocina candei (d'Orbigny, 1841)	A	
Acteocina cerealis (Gould, 1853)	P	grain barrel-bubble
Acteocina culcitella (Gould, 1853)	P	pillow barrel-bubble
Acteocina eburnea A. E. Verrill, 1885	A	ivory barrel-bubble
Acteocina harpa (Dall, 1871)	P	
Acteocina inculta (Gould, 1855)	P	rude barrel-bubble
Acteocina intermedia Willett, 1928	P	intermediate barrel-bubble
Acteocina oldroydi Dall, 1925	P	
Acteocina oryza (Totten, 1835)	A	rice barrel-bubble
Acteocina recta (d'Orbigny, 1841)	A	straight barrel-bubble
Acteocina smirna Dall, 1919	P	
Scaphander pilsbryi McGinty, 1955	A	
Scaphander punctostriatus (Mighels, 1841)	A, Ac	giant canoe bubble

SCIENTIFIC NAME	OCCURRENCE	COMMON NAME
Scaphander watsoni Dall, 1881	A	
Scaphander willetti Dall, 1919	P	

Cylichnidae

SCIENTIFIC NAME	OCCURRENCE	COMMON NAME
Cylichna alba (T. Brown, 1827)	A, P, Ac	white chalice-bubble
Cylichna attonsa (Carpenter, 1865)	P	
Cylichna diegensis (Dall, 1919)	P	San Diego chalice-bubble
Cylichna eburnea A. E. Verrill, 1885	A	ivory chalice-bubble
Cylichna gouldii (Couthouy, 1839)	A, Ac	
Cylichna linearis Jeffreys, 1867	A	lined chalice-bubble
Cylichna nucleolus (Reeve, 1855)	P, Ac	kernel chalice-bubble
Cylichna occulta (Mighels and C. B. Adams, 1842)	A, P, Ac	concealed chalice-bubble
Cylichna verrillii Dall, 1889	A	

Aglajidae

SCIENTIFIC NAME	OCCURRENCE	COMMON NAME
Aglaja ocelligera (Bergh, 1894)	P	eyespot aglaja
Chelidonura hirundinina (Quoy and Gaimard, 1833)	A	leech aglaja
Chelidonura sabina Ev. Marcus and Er. Marcus, 1970	A	Sabine Island aglaja
Melanochlamys diomedea (Bergh, 1894)	P	albatross aglaja
Navanax aenigmaticus (Bergh, 1893)	A, P	mysterious aglaja
Navanax inermis (J. G. Cooper, 1863)	P	California aglaja
Philinopsis pusa (Ev. Marcus and Er. Marcus, 1967)	A	pusa aglaja

Philinidae

SCIENTIFIC NAME	OCCURRENCE	COMMON NAME
Philine alba Mattox, 1958	A, P	white paperbubble
Philine angulata Jeffreys, 1867	A	angled paperbubble
Philine bakeri Dall, 1919	P	
Philine californica Willett, 1944	P	
Philine cingulata G. O. Sars, 1878	A	girdled paperbubble
Philine finmarchia M. Sars in G. O. Sars, 1878	A	Finmark paperbubble
Philine fragilis G. O. Sars, 1878	A	fragile paperbubble
Philine lima (T. Brown, 1825)	A, Ac	file paperbubble
Philine polaris Aurivillius, 1885	P, Ac	axial paperbubble
Philine quadrata (S. V. Wood, 1839)	A, Ac	quadrate paperbubble
Philine sagra (d'Orbigny, 1841)	A	crenulate paperbubble
Philine sinuata Stimpson, 1850	A, P	sinuate paperbubble
Philine tincta A. E. Verrill, 1882	A	tinted paperbubble

Gastropteridae

SCIENTIFIC NAME	OCCURRENCE	COMMON NAME
Gastropteron cinereum Dall, 1925	P	gray batwing seaslug
Gastropteron pacificum Bergh, 1894	P	Pacific batwing seaslug
Gastropteron rubrum (Rafinesque, 1814)	A	batwing seaslug
Gastropteron vespertilium Gosliner and Armes, 1984	A	flapping dingbat

SCIENTIFIC NAME	OCCURRENCE	COMMON NAME
Diaphanidae		
Diaphana brunnea Dall, 1919	P	brown. paperbubble
Diaphana californica Dall, 1919	P	
Diaphana debilis (Gould, 1839)	A	weak paperbubble
Diaphana minuta (T. Brown, 1827)	A, P, Ac	Arctic paperbubble
Woodbridgea polystrigma (Dall, 1908)	P	furrowed paperbubble
Runcinidae		
Runcina divae (Ev. Marcus and Er. Marcus, 1963)	A	
Bullidae		
Bulla clausa Dall, 1889	A	imperforate bubble
Bulla gemma A. E. Verrill, 1880	A	jewel bubble
Bulla gouldiana Pilsbry, 1895	P	California bubble
Bulla solida Gmelin, 1791	A	solid bubble
Bulla striata Bruguière, 1792	A	striate bubble
Atyidae		
Atys caribaeus (d'Orbigny, 1841)	A	Caribbean glassy-bubble
Atys castus Carpenter, 1864	P	clean glassy-bubble
Atys nonscriptus (A. Adams, 1850)	P	clean-slate glassy-bubble
Atys obscuratus Dall, 1896	A	obscure glassy-bubble
Atys riiseanus Mörch, 1875	A	
Atys sandersoni Dall, 1881	A	
Haminoea antillarum (d'Orbigny, 1841)	A	Antilles glassy-bubble
Haminoea elegans (J. E. Gray, 1825)	A	elegant glassy-bubble
Haminoea petitii (d'Orbigny, 1841)	A	straight glassy-bubble
Haminoea solitaria (Say, 1822)	A	solitary glassy-bubble
Haminoea succinea (Conrad, 1846)	A	amber glassy-bubble
Haminoea vesicula Gould, 1855	P	blister glassy-bubble
Haminoea virescens (Sowerby, 1833)	P	green glassy-bubble
Retusidae		
Coleophysis carinata (Carpenter, 1857)	P	keeled barrel-bubble
Coleophysis harpa (Dall, 1871)	P	harp barrel-bubble
Pyrunculus caelatus (Bush, 1885)	A	engraved barrel-bubble
Pyrunculus ovatus (Jeffreys, 1870)	A	ovate barrel-bubble
Retusa mayoi (Dall, 1889)	A	
Retusa montereyensis A. G. Smith and Gordon, 1948	P	Monterey barrel-bubble
Retusa obtusa (Montagu, 1803)	A, P, Ac	Arctic barrel-bubble
Retusa pertenuis Mighels, 1843	Ac	
Retusa semen (Reeve, 1856)	Ac	
Retusa sulcata (d'Orbigny, 1841)	A	sulcate barrel-bubble
Retusa umbilicata (Montagu, 1803)	Ac	
Sulcoretusa xystrum (Dall, 1919)	P	polished barrel-bubble
Volvulella californica Dall, 1919	P	California spindle-bubble
Volvulella cylindrica (Carpenter, 1864)	P	cylindrical spindle-bubble
Volvulella panamica Dall, 1919	P	Panamic spindle-bubble

SCIENTIFIC NAME	OCCURRENCE	COMMON NAME
Volvulella paupercula (Watson, 1883)	A	spineless spindle-bubble
Volvulella persimilis (Mörch, 1875)	A	southern spindle-bubble
Volvulella recta (Mörch, 1875)	A	spined spindle-bubble
Volvulella texasiana Harry, 1967	A	Texas spindle-bubble

ORDER ACOCHLIDIOIDEA

Microhedylidae

Unela remanei Er. Marcus, 1953	A	sand nudibranch

ORDER THECOSOMATA

Limacinidae

Limacina bulimoides (d'Orbigny, 1836)	A	bulimoid pteropod
Limacina helicina (Phipps, 1774)	A, P, Ac	helicid pteropod
Limacina inflata (d'Orbigny, 1836)	A	planorbid pteropod
Limacina lesueurii (d'Orbigny, 1836)	A	
Limacina retroversa (Fleming, 1823)	A, Ac	retrovert pteropod
Limacina trochiformis (d'Orbigny, 1836)	A	trochiform pteropod

Cavoliniidae

Cavolinia gibbosa (d'Orbigny, 1836)	A, P	gibbose cavoline
Cavolinia inflexa (Lesueur, 1813)	A, P	inflexed cavoline
Cavolinia longirostris (de Blainville, 1821)	A, P	longsnout cavoline
Cavolinia tridentata (Niebuhr, 1775)	A, P	threetooth cavoline
Cavolinia uncinata (Rang, 1829)	A	uncinate cavoline
Clio chaptalii J. E. Gray, 1850	A	
Clio cuspidata (Bosc, 1801)	A	cuspidate clio
Clio polita (Pelseneer, 1888)	A	two-keel clio
Clio pyramidata Linnaeus, 1767	A, P	pyramid clio
Clio recurva (Children, 1823)	A, P	wavy clio
Creseis acicula (Rang, 1828)	A, P	straight needle-pteropod
Creseis virgula (Rang, 1828)	A, P	curved needle-pteropod
Cuvierina columnella (Rang, 1827)	A, P	cigar pteropod
Diacria quadridentata (de Blainville, 1821)	A, P	fourtooth cavoline
Diacria trispinosa (de Blainville, 1821)	A, P	threespine cavoline
Hyalocylis striata (Rang, 1828)	A	striate clio
Styliola subula (Quoy and Gaimard, 1827)	A, P	keeled clio

Peraclididae

Peracle apicifulva Meisenheimer, 1906	A, P	
Peracle bispinosa Pelseneer, 1888	A	two-spine pteropod
Peracle reticulata (d'Orbigny, 1836)	A, P	reticulate pteropod
Peracle triacantha P. Fischer, 1882	A	

Cymbuliidae

Cleba cordata Niebuhr, 1776	A	
Corolla calceola A. E. Verrill, 1880	A	Atlantic corolla

SCIENTIFIC NAME	OCCURRENCE	COMMON NAME
Corolla intermedia (Tesch, 1903)	A	
Corolla ovata (Quoy and Gaimard, 1832)	A	
Corolla spectabilis Dall, 1871	P	spectacular corolla
Cymbulia peronii de Blainville, 1818	A	

Desmopteridae

Desmopterus pacificus Essenberg, 1919	P	
Desmopterus papilio Chun, 1889	A	

ORDER GYMNOSOMATA

Clionidae

Clione limacina (Phipps, 1774)	A, P, Ac	
Paedoclione doliiformis Danforth, 1907	A	
Paraclione longicaudata (Souleyet, 1852)	A	

Cliopsidae

Cliopsis krohni Troschel, 1854	A, P	

Hydromylidae

Hydromyles globulosus (Rang, 1825)	A, P	

Notobranchaeidae

Notobranchaea macdonaldi Pelseneer, 1886	A	
Prionoglossa tetrabranchiata (Bonnevie, 1913)	A	

Pneumodermatidae

Pneumoderma atlanticum atlanticum (Oken, 1815)	A	
Pneumoderma atlanticum pacificum Dall, 1872	P	
Pneumoderma mediterraneum (van Beneden, 1838)	A	
Pneumodermopsis macrochira (Meisenheimer, 1905)	A, P	
Pneumodermopsis paucidens Boas, 1886	A	

Thliptodontidae

Thliptodon diaphanus (Meisenheimer, 1903)	A, P	

ORDER ANASPIDEA

Akeridae

Akera thompsoni Olsson and McGinty, 1951	A	

SCIENTIFIC NAME	OCCURRENCE	COMMON NAME
Aplysiidae		
Aplysia brasiliana Rang, 1828	A	sooty seahare
Aplysia californica J. G. Cooper, 1863	P	California seahare
Aplysia cervina (Dall and Simpson, 1901)	A	
Aplysia dactylomela Rang, 1828	A	spotted seahare
Aplysia donca Ev. Marcus and Er. Marcus, 1959	A	
Aplysia geographica (A. Adams and Reeve, 1850)	A	
Aplysia juliana Quoy and Gaimard, 1832	A	walking seahare
Aplysia morio A. E. Verrill, 1901	A	Atlantic black seahare
Aplysia parvula Guilding in Mörch, 1863	A	
Aplysia reticulopoda Beeman, 1960	P	net-foot seahare
Aplysia vaccaria Winkler, 1955	P	California black seahare
Aplysia willcoxi Heilprin, 1886	A	
Bursatella leachii pleii Rang, 1828	A	ragged seahare
Dolabrifera dolabrifera (Rang, 1828)	A	warty seacat
Notarchus punctatus Philippi, 1836	A	
Petalifera petalifera (Rang, 1828)	A	
Petalifera ramosa Baba, 1959	A	
Phyllaplysia cymodacea K. B. Clark, 1976	A	
Phyllaplysia engeli Er. Marcus, 1955	A	
Phyllaplysia smaragda K. B. Clark, 1977	A	emerald leafslug
Phyllaplysia taylori Dall, 1900	P	zebra leafslug
Stylocheilus citrinus (Rang, 1828)	A	yellow seahare
Stylocheilus longicauda (Quoy and Gaimard, 1825)	A	blue-ring seahare
ORDER SACOGLOSSA		
Boselliidae		
Bosellia corinneae Ev. Marcus, 1973	A	
Bosellia marcusi Ev. Marcus, 1972	A	
Bosellia mimetica Trinchese, 1891	A	
Caliphyllidae		
Caliphylla mediterranea A. Costa, 1867	A	
Cyerce antillensis Engel, 1927	A	Antilles glass-slug
Cyerce cristallina (Trinchese, 1881)	A	harlequin glass-slug
Mourgona germaineae Ev. Marcus and Er. Marcus, 1970	A	
Phyllobranchillus viridis (Deshayes, 1857)	A, P	
Costasiellidae		
Costasiella ocellifera (Simroth, 1895)	A	eyespot costasiella
Cylindrobullidae		
Ascobulla ulla (Ev. Marcus and Er. Marcus, 1970)	A	
Cylindrobulla beauii P. Fischer, 1856	A	

SCIENTIFIC NAME	OCCURRENCE	COMMON NAME
Elysiidae		
Elysia canguzua Er. Marcus, 1955	A	
Elysia catulus Gould, 1870	A	
Elysia chlorotica (Gould, 1870)	A	emerald elysia
Elysia evelinae Er. Marcus, 1957	A	
Elysia hedgpethi Er. Marcus, 1961	P	
Elysia ornata (Swainson, 1840)	A	ornate elysia
Elysia papillosa A. E. Verrill, 1901	A	papillose elysia
Elysia patina Ev. Marcus, 1980	A	
Elysia picta A. E. Verrill, 1901	A	painted elysia
Elysia serca Er. Marcus, 1955	A	seagrass elysia
Elysia subornata A. E. Verrill, 1901	A	
Elysia tuca Ev. Marcus and Er. Marcus, 1967	A	
Tridachia crispata (Mörch, 1863)	A	lettuce slug
Hermaeidae		
Aplysiopsis enteromorphae (Cockerell and Eliot, 1905)	P	
Aplysiopsis smithi (Er. Marcus, 1961)	P	
Aplysiopsis zebra K. B. Clark, 1982	A	
Hermaea cruciata Gould, 1870	A	
Hermaea olivae MacFarland, 1966	P	
Hermaea vancouverensis (O'Donoghue, 1924)	A, P	
Juliidae		
Berthelinia caribbea Edmunds, 1963	A	Caribbean bivalved snail
Lobigeridae		
Lobiger souverbii P. Fischer, 1856	A	
Oxynoidae		
Oxynoe antillarum Mörch, 1863	A	Antilles oxynoe
Oxynoe azuropunctata K. R. Jensen, 1980	A	blue-spot oxynoe
Stiligeridae		
Alderia modesta (Lovén, 1844)	A, P	modest alderia
Ercolania coerulea Trinchese, 1893	A	blue stiliger
Ercolania costai Pruvot-Fol, 1951	A	
Ercolania funerea (A. Costa, 1867)	A	
Ercolania fuscata (Gould, 1870)	A	dusky stiliger
Limapontia zonata Girard, 1852	A	
Olea hansineensis Agersborg, 1923	P	Hansine seaslug
Placida dendritica (Alder and Hancock, 1843)	A, P	
Placida kingstoni T. E. Thompson, 1977	A	
Stiliger fuscovittatus Lance, 1962	A, P	brown-streak stiliger
Stiliger vossi Ev. Marcus and Er. Marcus, 1960	A	

SCIENTIFIC NAME	OCCURRENCE	COMMON NAME

ORDER NOTASPIDEA

Tylodinidae

Tylodina americana Dall, 1890	A	
Tylodina fungina Gabb, 1865	P	yellow umbrella slug
Tylodinella spongotheras Bertsch, 1980	P	

Umbraculidae

Umbraculum umbraculum (Lightfoot, 1786)	A	Atlantic umbrella slug

Pleurobranchidae

Berthella agassizii (MacFarland, 1909)	A	
Berthella californica (Dall, 1900)	P	California sidegill slug
Berthella sideralis (Lovén, 1847)	P	
Berthella tupala Er. Marcus, 1957	A	
Berthellina citrina (Rüppell and Leuckart, 1828)	P	
Berthellina engeli Gardiner, 1936	A, P	
Pleurobranchus areolatus Mörch, 1863	A, P	Atlantic sidegill slug
Pleurobranchus reesi White, 1952	A	
Pleurobranchus strongi MacFarland, 1966	P	

Pleurobranchaeidae

Pleurobranchaea bonnieae Ev. Marcus and Gosliner, 1984	A	
Pleurobranchaea californica MacFarland, 1966	P	
Pleurobranchaea inconspicua Bergh, 1897	A	
Pleurobranchaea tarda A. E. Verrill, 1880	A	

ORDER NUDIBRANCHIA

Corambidae

Corambe pacifica MacFarland and O'Donoghue, 1929	P	frost-spot corambe
Doridella burchi Ev. Marcus and Er. Marcus, 1967	A	
Doridella obscura A. E. Verrill, 1870	A	obscure corambe
Doridella steinbergae (Lance, 1962)	P	

Goniodorididae

Ancula evelinae Er. Marcus, 1961	A	
Ancula gibbosa (Risso, 1818)	A, Ac	Atlantic ancula
Ancula lentiginosa Farmer in Farmer and Sloan, 1964	P	freckled ancula
Ancula pacifica MacFarland, 1905	P	Pacific ancula
Hopkinsia rosacea MacFarland, 1905	P	Hopkins rose

SCIENTIFIC NAME	OCCURRENCE	COMMON NAME
Okenia angelensis Lance, 1966	P	Angeles okenia
Okenia ascidicola M. P. Morse, 1972	A	
Okenia cupella (Vogel and Schultz, 1970)	A	
Okenia impexa (Er. Marcus, 1957)	A	
Okenia modesta A. E. Verrill, 1875	A	
Okenia plana Baba, 1960	P	flat okenia
Okenia pulchella Alder and Hancock, 1854	A, Ac	
Okenia sapelona (Ev. Marcus and Er. Marcus, 1967)	A	Sapelo okenia
Okenia vancouverensis (O'Donoghue, 1921)	P	Vancouver okenia
Okenia zoobotryon (Smallwood, 1910)	A	
Trapania dalva Ev. Marcus, 1972	A	
Trapania velox (Cockerell, 1901)	P	

Onchidorididae

SCIENTIFIC NAME	OCCURRENCE	COMMON NAME
Acanthodoris armata O'Donoghue, 1927	P	
Acanthodoris atrogriseata O'Donoghue, 1927	P	
Acanthodoris brunnea MacFarland, 1905	P	brown spiny doris
Acanthodoris caerulescens Bergh, 1880	P	
Acanthodoris hudsoni MacFarland, 1905	P	
Acanthodoris lutea MacFarland, 1925	P	orange-peel doris
Acanthodoris nanaimoensis O'Donoghue, 1921	P	wine-plumed spiny doris
Acanthodoris pilosa (Abildgaard in Müller, 1789)	A, P, Ac	hairy spiny doris
Acanthodoris rhodoceras Cockerell in Cockerell and Eliot, 1905	P	black-tipped spiny doris
Adalaria albopapillosa (Dall, 1871)	P	
Adalaria pacifica Bergh, 1880	P, Ac	
Adalaria proxima (Alder and Hancock, 1854)	A, Ac	yellow false doris
Adalaria tschuktschica Krause, 1885	Ac	
Adalaria virescens Bergh, 1880	Ac	
Akiodoris lutescens Bergh, 1880	Ac	
Diaphorodoris lirulatocauda Millen, 1985	P	
Onchidoris bilamellata (Linnaeus, 1767)	A, P, Ac	barnacle-eating onchidoris
Onchidoris diademata Gould, 1870	A	
Onchidoris grisea Gould, 1870	A	
Onchidoris muricata (Müller, 1776)	A, P, Ac	fuzzy onchidoris
Onchidoris tenella Gould, 1870	A	

Triophidae

SCIENTIFIC NAME	OCCURRENCE	COMMON NAME
Crimora coneja Er. Marcus, 1961	P	rabbit doris
Triopha catalinae (J. G. Cooper, 1863)	P	sea-clown triopha
Triopha maculata MacFarland, 1905	P	maculated triopha
Triopha occidentalis (Fewkes, 1889)	P	grand triopha

Heterodorididae

SCIENTIFIC NAME	OCCURRENCE	COMMON NAME
Heterodoris robusta A. E. Verrill and Emerton, 1882	A	

SCIENTIFIC NAME	OCCURRENCE	COMMON NAME
Aegiretidae		
Aegires albopunctatus MacFarland, 1905.........	P	salt-and-pepper doris
Gymnodorididae		
Nembrotha gratiosa Bergh, 1890	A	
Polyceridae		
Issena pacifica (Bergh, 1894)	A, P	
Issena ramosa (A. E. Verrill and Emerton, 1881)..	A	
Laila cockerelli MacFarland, 1905	P	
Palio pallida Bergh, 1880	Ac	
Polycera atra MacFarland, 1905.................	P	orange-spike polycera
Polycera aurisula Er. Marcus, 1957	A	
Polycera chilluna Er. Marcus, 1961	A	
Polycera dubia (M. Sars, 1829).................	A	
Polycera hedgpethi Er. Marcus, 1964...........	P	
Polycera hummi Abbott, 1952...................	A	
Polycera odhneri Er. Marcus, 1955.............	A	
Polycera rycia Ev. Marcus, 1970	A	
Polycera tricolor Robilliard, 1971	P	three-color polycera
Polycera zosterae O'Donoghue, 1924............	P	eelgrass polycera
Polycerella conyna Er. Marcus, 1957............	A	
Polycerella davenportii Balch, 1899	A	
Polycerella emertoni A. E. Verrill, 1881..........	A	
Cadlinidae		
Cadlina flavomaculata MacFarland, 1905.........	P	yellow-spot cadlina
Cadlina laevis (Linnaeus, 1767)	A, Ac	white Atlantic cadlina
Cadlina limbaughi Lance, 1962.................	P	
Cadlina luteomarginata MacFarland, 1966........	P	
Cadlina marginata MacFarland, 1905............	P	yellow-rim cadlina
Cadlina modesta MacFarland, 1966	P	
Cadlina pacifica Bergh, 1879....................	P	
Cadlina rumia Er. Marcus, 1955.................	A	
Cadlina scabriuscula (Bergh, 1890)..............	A	
Cadlina sparsa Odhner, 1921	P	dark-spot cadlina
Chromodorididae		
Chromodoris aila Er. Marcus, 1961	A	
Chromodoris clenchi (Russell, 1935).............	A	harlequin blue doris
Chromodoris dalli (Bergh, 1879)................	P	
Chromodoris macfarlandi Cockerell, 1901	P	three-stripe doris
Chromodoris nyalya Ev. Marcus and Er. Marcus, 1967............	A	red-line blue doris
Chromodoris roseopicta (A.E. Verrill, 1900)	A	
Felimare bayeri Ev. Marcus and Er. Marcus, 1967.	A	
Hypselodoris californiensis (Bergh, 1879).........	P	California blue doris
Hypselodoris edenticulata (White, 1952)	A	Florida regal doris
Mexichromis porterae (Cockerell, 1901)..........	P	

SCIENTIFIC NAME	OCCURRENCE	COMMON NAME
Asteronotidae		
Aphelodoris antillensis (Bergh, 1879)	A	
Sclerodoris tanya (Ev. Marcus, 1971)	P	feline doris
Actinocyclidae		
Hallaxa chani Gosliner and Williams, 1975	P	
Conualeviidae		
Conualevia alba Collier and Farmer, 1964	P	white smooth-horn doris
Calycidorididae		
Calycidoris guentheri Abraham, 1876	Ac	
Aldisidae		
Aldisa cooperi Robilliard and Baba, 1972	P	
Aldisa sanguinea (J. G. Cooper, 1863)	P	blood-spot doris
Aldisa zetlandica (Alder and Hancock, 1854)	A, Ac	
Rostangidae		
Rostanga pulchra MacFarland, 1905	P	red sponge doris
Dorididae		
Doris odonoghuei Steinberg, 1963	P	
Doris verrucosa Linnaeus, 1758	A	
Siraius kyolis Ev. Marcus and Er. Marcus, 1967	A	
Dendrodorididae		
Dendrodoris krebsii (Mörch, 1863)	A	
Dendrodoris warta Ev. Marcus and Gallagher, 1976	A	
Doriopsilla albopunctata (J. G. Cooper, 1863)	P	salted yellow doris
Doriopsilla areolata (Bergh, 1880)	A	
Doriopsilla leia Er. Marcus, 1961	A	
Doriopsilla nigromaculata (Cockerell in Cockerell and Eliot, 1905)	P	tiny black-spot doris
Doriopsilla pharpa Er. Marcus, 1961	A	
Phyllidiidae		
Phyllidiopsis papilligera Bergh, 1890	A	
Archidorididae		
Archidoris montereyensis (J. G. Cooper, 1863)	P	Monterey sea-lemon
Archidoris odhneri (MacFarland, 1966)	P	white night doris
Atagema alba (O'Donoghue, 1927)	P	hunchback doris

SCIENTIFIC NAME	OCCURRENCE	COMMON NAME
Atagema prea (Ev. Marcus and Er. Marcus, 1967).	A	

Discodorididae

Anisodoris lentiginosa Millen, 1982	P	mottled pale sea-lemon
Anisodoris nobilis (MacFarland, 1905)	P	Pacific sea-lemon
Anisodoris prea Ev. Marcus and Er. Marcus, 1967.	A	
Anisodoris worki Ev. Marcus and Er. Marcus, 1967.	A	
Diaulula sandiegensis (J. G. Cooper, 1863)	P	ringed doris
Discodoris alba White, 1952	A	
Discodoris heathi MacFarland, 1905	P	gritty doris
Discodoris evelinae Er. Marcus, 1955	A	
Discodoris phoca Ev. Marcus and Er. Marcus, 1959	A	
Discodoris purcina Ev. Marcus and Er. Marcus, 1967	A	
Discodoris pusae Er. Marcus, 1955	A	
Geitodoris complanata (A. E. Verrill, 1880)	A	
Peltodoris greeleyi MacFarland, 1909	A	
Taringa aivica timia Ev. Marcus and Er. Marcus, 1967	P	dusky brown taringa
Taringa telopia disa Ev. Marcus and Er. Marcus, 1967	A	
Thordisa bimaculata Lance, 1966	P	two-spot thordis
Thordisa rubescens Behrens and R. Henderson, 1981	P	red thordis

Kentrodorididae

Jorunna pardus Behrens and R. Henderson, 1981	P	leopard jorunna

Platydorididae

Platydoris angustipes (Mörch, 1863)	A	
Platydoris macfarlandi Hanna, 1951	P	California flat doris

Tritoniidae

Tochuina tetraquetra (Pallas, 1788)	P	giant orange tochui
Tritonia bayeri bayeri Ev. Marcus, 1978	A	
Tritonia bayeri misa Er. Marcus, 1967	A	
Tritonia diomedea Bergh, 1894	A, P	rosy tritonia
Tritonia festiva (Stearns, 1873)	P	diamondback tritonia
Tritonia palmeri (J. G. Cooper, 1862)	P	
Tritonia wellsi Er. Marcus, 1961	A	

Hancockiidae

Hancockia californica MacFarland, 1923	P	

Dendronotidae

Dendronotus albopunctatus Robilliard, 1972	P	

SCIENTIFIC NAME	OCCURRENCE	COMMON NAME
Dendronotus albus MacFarland, 1966	P	white frond-aeolis
Dendronotus dalli Bergh, 1879	P	
Dendronotus diversicolor Robilliard, 1970	P	multicolor frond-aeolis
Dendronotus frondosus (Ascanius, 1774)	A, P, Ac	frond-aeolis
Dendronotus iris J. G. Cooper, 1863	P	giant frond-aeolis
Dendronotus robustus A. E. Verrill, 1870	A	robust frond-aeolis
Dendronotus rufus O'Donoghue, 1921	P	red frond-aeolis
Dendronotus subramosus MacFarland, 1966	P	stubby frond-aeolis

Tethyidae

Melibe leonina (Gould, 1852)	P	lion nudibranch

Lomanotidae

Lomanotus stauberi K. B. Clark and Goetzfried, 1976	A	

Scyllaeidae

Scyllaea pelagica Linnaeus, 1758	A	sargassum nudibranch

Phylliroidae

Phylliroe atlantica Bergh, 1871	A	
Phylliroe bucephalum Péron and Lesueur, 1810	A	

Dotoidae

Doto amyra Er. Marcus, 1961	P	hammerhead doto
Doto chica Er. Marcus, 1960	A	
Doto columbiana O'Donoghue, 1921	P	British Columbia doto
Doto coronata (Gmelin, 1791)	A	crown doto
Doto divae Er. Marcus, 1960	A	
Doto doerga Ev. Marcus and Er. Marcus, 1963	A	
Doto formosa A. E. Verrill, 1875	A	
Doto kya Er. Marcus, 1961	P	dark doto
Doto pita Er. Marcus, 1955	A	
Doto uva Er. Marcus, 1955	A	
Miesea evelinae (Er. Marcus, 1957)	A	

Arminidae

Armina californica (J. G. Cooper, 1863)	P	California armina
Armina tigrina Rafinesque, 1814	A	tiger armina

Dironidae

Dirona albolineata MacFarland in Cockerell and Eliot, 1905	P	white-line dirona
Dirona aurantia Hurst, 1966	P	golden dirona
Dirona picta MacFarland in Cockerell and Eliot, 1905	P	painted dirona

SCIENTIFIC NAME	OCCURRENCE	COMMON NAME
Janolidae		
Antiopella barbarensis (J. G. Cooper, 1863)	P	cockscomb nudibranch
Coryphellidae		
Coryphella cooperi Cockerell, 1901	P	blue-patch aeolis
Coryphella diversa (Couthouy, 1839)	A	
Coryphella fusca O'Donoghue, 1921	P	predaceous aeolis
Coryphella iodinea (J. G. Cooper, 1863)	P	purple aeolis
Coryphella longicaudata O'Donoghue, 1922	P	long-tail aeolis
Coryphella nobilis A. E. Verrill, 1880	A	
Coryphella pellucida (Alder and Hancock, 1843)	A	pellucid aeolis
Coryphella pricei MacFarland, 1966	P	smooth-tooth aeolis
Coryphella salmonacea (Couthouy, 1839)	A, P, Ac	salmon aeolis
Coryphella subrosacea (Eschscholtz, 1831)	P	
Coryphella trilineata O'Donoghue, 1921	P	threeline aeolis
Coryphella trophina (Bergh, 1894)	P	
Coryphella verrucosa rufibranchialis (Johnston, 1832)	A, Ac	red-finger aeolis
Eubranchidae		
Eubranchus conicla (Er. Marcus, 1958)	A	
Eubranchus columbianus (O'Donoghue, 1922)	P	
Eubranchus exiguus (Alder and Hancock, 1848)	A, Ac	dwarf balloon aeolis
Eubranchus misakiensis Baba, 1960	P	Misaki balloon aeolis
Eubranchus olivaceus (O'Donoghue, 1922)	P	green balloon aeolis
Eubranchus pallidus (Alder and Hancock, 1842)	A	
Eubranchus rustyus (Er. Marcus, 1961)	P	
Eubranchus sanjuanensis Roller, 1972	P	
Eubranchus tricolor (Forbes, 1838)	A, Ac	painted balloon aeolis
Cumanotidae		
Cumanotus beaumonti (Eliot, 1906)	P	polyp aeolis
Tergipedidae		
Catriona columbiana (O'Donoghue, 1922)	P	red-tentacle cuthona
Catriona gymnota (Couthouy, 1838)	A	
Catriona maua Ev. Marcus and Er. Marcus, 1960	A	
Catriona rickettsi Behrens, 1984	P	
Cuthona abronia (MacFarland, 1966)	P	colorful cuthona
Cuthona albocrusta (MacFarland, 1966)	P	white-crust cuthona
Cuthona aurantia (Alder and Hancock, 1842)	A, Ac	orange-tip cuthona
Cuthona cocoachroma Williams and Gosliner, 1979	P	brown cuthona
Cuthona concina (Alder and Hancock, 1843)	A, P	
Cuthona flavovulta (MacFarland, 1966)	P	yellowish cuthona
Cuthona fulgens (MacFarland, 1966)	P	black-and-yellow cuthona
Cuthona lagunae (O'Donoghue, 1926)	P	orange-face cuthona
Cuthona nana (Alder and Hancock, 1842)	A	
Cuthona perca (Er. Marcus, 1958)	A, P	Lake Merritt cuthona
Cuthona phoenix Gosliner, 1981	P	bornagain cuthona

SCIENTIFIC NAME	OCCURRENCE	COMMON NAME
Cuthona pustulata (Alder and Hancock, 1845)	A	
Cuthona stimpsoni A. E. Verrill, 1880	A	
Cuthona tina (Er. Marcus, 1957)	A	
Cuthona veronica (A. E. Verrill, 1880)	A	
Cuthona virens (MacFarland, 1966)	P	green cuthona
Precuthona divae Er. Marcus, 1961	P	rose-pink cuthona
Tenellia adspersa (Nordmann, 1845)	P	miniature aeolis
Tenellia fuscata Gould, 1870	A	
Tenellia ventilabrum (Dalyell, 1853)	A	
Tergipes tergipes (Forskål, 1775)	A, Ac	

Fionidae

Fiona pinnata (Eschscholtz in Rathke, 1831)	A, P	fiona

Babakinidae

Babakina festiva (Roller, 1972)	P	single-stalk aeolis

Facelinidae

Austraeolis catina Ev. Marcus and Er. Marcus, 1967	A	
Cratena kaoruae Er. Marcus, 1955	A	
Cratena pilata (Gould in W. G. Binney, 1870)	A	
Dondice occidentalis (Engel, 1925)	A	
Emarcusia morroensis Roller, 1972	P	western dondice
Facelina bostoniensis (Couthouy, 1838)	A	Boston facelina
Facelina stearnsi Cockerell, 1901	P	scarlet-tip aeolis
Favorinus auritulus Er. Marcus, 1955	A	
Godiva rubrolineata Edmunds, 1964	A	
Hermissenda crassicornis (Eschscholtz in Rathke, 1831)	P	hermissenda
Learchis poica Ev. Marcus and Er. Marcus, 1960	A	
Phidiana hiltoni (O'Donoghue, 1927)	P	pugnaceous aeolis
Phidiana lynceus Bergh, 1867	A	
Sakuraeolis enosimensis (Baba, 1930)	P	white-tentacle Japanese aeolis

Aeolidiidae

Aeolidia papillosa (Linnaeus, 1761)	A, P, Ac	shag-rug aeolis
Aeolidiella takanosimensis Baba, 1930	P	vermillion Japanese aeolis
Cerberilla mosslandica McDonald and Nybakken, 1975	P	brown burrowing aeolis
Cerberilla tanna Er. Marcus, 1959	A	

Spurillidae

Baeolidia benteva Er. Marcus, 1958	A	
Berghia verrucicornis (O. G. Costa, 1864)	A	
Spurilla chromosoma Cockerell in Cockerell and Eliot, 1905	P	frosted spurilla
Spurilla neapolitana (delle Chiaje, 1823)	A	neapolitan spurilla
Spurilla oliviae (MacFarland, 1966)	P	red-tentacle spurilla

SCIENTIFIC NAME	OCCURRENCE	COMMON NAME
Glaucidae		
Glaucus atlanticus Forster, 1777	A	blue glaucus

ORDER ARCHAEOPULMONATA

Melampodidae

SCIENTIFIC NAME	OCCURRENCE	COMMON NAME
Apodopsis novimundi Pilsbry and McGinty, 1949	A	
Blauneria heteroclita (Montagu, 1808)	A	left-hand melampus
Detracia bulloides (Montagu, 1808)	A	bubble melampus
Detracia clarki Morrison, 1951	A	
Detracia floridana (Pfeiffer, 1856)	A	Florida melampus
Ellobium pellucens (Menke, 1830)	A	white melampus
Laemodonta cubensis (Pfeiffer, 1854)	A	Cuba dwarf pedipes
Marinula succinea (Pfeiffer, 1854)	A	
Melampus bidentatus Say, 1822	A	eastern melampus
Melampus coffeus (Linnaeus, 1758)	A	coffee melampus
Melampus monilis (Bruguière, 1789)	A	Caribbean melampus
Melampus olivaceus Carpenter, 1857	P	California melampus
Microtralia occidentalis (Pfeiffer, 1854)	A	
Ovatella myosotis (Draparnaud, 1801)	A, P	
Pedipes angulatus C. B. Adams, 1852	P	angular pedipes
Pedipes mirabilis (Mühlfeld, 1816)	A	miraculous pedipes
Pedipes unisulcatus J. G. Cooper, 1866	P	one-groove pedipes
Tralia ovula (Bruguière, 1789)	A	egg melampus

ORDER BASOMMATOPHORA

Acroloxidae

SCIENTIFIC NAME	OCCURRENCE	COMMON NAME
Acroloxus coloradensis (J. Henderson, 1930)	F	Rocky Mountain capshell

Lymnaeidae

SCIENTIFIC NAME	OCCURRENCE	COMMON NAME
Acella haldemani (W. G. Binney, 1867)	F	spindle lymnaea
Bulimnaea megasoma (Say, 1824)	F	mammoth lymnaea
Fisherola nuttalli (Haldeman, 1841)	F	shortface lanx
Fossaria alberta F. C. Baker, 1919	F	*classification uncertain*
Fossaria bulimoides (I. Lea, 1841)	F	prairie fossaria
Fossaria cockerelli Pilsbry and Ferriss, 1906	F	*classification uncertain*
Fossaria cubensis (Pfeiffer, 1839)	F	Carib fossaria
Fossaria cyclostoma (Walker, 1808)	F	bugle fossaria
Fossaria dalli (F. C. Baker, 1907)	F	dusky fossaria
Fossaria exigua (I. Lea, 1841)	F	*classification uncertain*
Fossaria galbana (Say, 1825)	F	boreal fossaria
Fossaria hendersoni F. C. Baker, 1909	F	*classification uncertain*
Fossaria humilis (Say, 1822)	F	marsh fossaria
Fossaria modicella (Say, 1825)	F	rock fossaria
Fossaria obrussa (Say, 1825)	F	golden fossaria
Fossaria parva (I. Lea, 1841)	F	pygmy fossaria
Fossaria peninsulae (Walker, 1908)	F	*classification uncertain*

SCIENTIFIC NAME	OCCURRENCE	COMMON NAME
Fossaria perplexa F. C. Baker and J. Henderson, 1929	F	*classification uncertain*
Fossaria perpolita (Dall, 1905)	F	glossy fossaria
Fossaria rustica (I. Lea, 1841)	F	*classification uncertain*
Fossaria sonomaensis Hemphill, 1906	F	Sonoma fossaria
Fossaria tazewelliana (Wolf, 1869)	F	Tazwell fossaria
Fossaria techella Haldeman, 1867	F	*classification uncertain*
Fossaria truncatula (Müller, 1774)	F	attenuate fossaria
Fossaria vancouverensis F. C. Baker, 1939	F	*classification uncertain*
Lanx alta (Tryon, 1865)	F	highcap lanx
Lanx klamathensis Hannibal, 1912	F	scale lanx
Lanx patelloides (I. Lea, 1856)	F	kneecap lanx
Lanx subrotunda (Tryon, 1865)	F	rotund lanx
Lymnaea atkaensis Dall, 1884	F	frigid lymnaea
Lymnaea stagnalis Linnaeus, 1758	F	swamp lymnaea
Pseudosuccinea columella (Say, 1817)	F	mimic lymnaea
Radix auricularia (Linnaeus, 1758)	F [I]	big-ear radix
Stagnicola apicina (I. Lea, 1838)	F	abbreviate pondsnail
Stagnicola arcticus (I. Lea, 1864)	F	arctic pondsnail
Stagnicola bonnevillensis (Call, 1884)	F	fat-whorled pondsnail
Stagnicola caperatus (Say, 1829)	F	wrinkled marshsnail
Stagnicola catascopium (Say, 1867)	F	woodland pondsnail
Stagnicola contractus (Currier, 1881)	F	deepwater pondsnail
Stagnicola elodes (Say, 1821)	F	marsh pondsnail
Stagnicola elrodi (F. C. Baker and J. Henderson, 1933)	F	flathead pondsnail
Stagnicola elrodianus F. C. Baker, 1935	F	longmouth pondsnail
Stagnicola emarginatus (Say, 1821)	F	St. Lawrence pondsnail
Stagnicola exilis (I. Lea, 1834)	F	flat-whorled pondsnail
Stagnicola gabbi (Tryon, 1865)	F	striate pondsnail
Stagnicola hinkleyi (F. C. Baker, 1906)	F	rustic pondsnail
Stagnicola idahoensis (J. Henderson, 1931)	F	shortspire pondsnail
Stagnicola mighelsi (W. G. Binney, 1865)	F	bigmouth pondsnail
Stagnicola montanensis (F. C. Baker, 1913)	F	mountain marshsnail
Stagnicola neopalustris (F. C. Baker, 1911)	F	Piedmont pondsnail
Stagnicola oronoensis (F. C. Baker, 1904)	F	obese pondsnail
Stagnicola petoskeyensis (Walker, 1908)	F	Petoskey pondsnail
Stagnicola pilsbryi (Hemphill, 1890)	F	Fish Springs marshsnail
Stagnicola traski (Tryon, 1863)	F	widelip pondsnail
Stagnicola utahensis (Call, 1884)	F	thickshell pondsnail
Stagnicola walkerianus F. C. Baker, 1926	F	calabash pondsnail
Stagnicola woodruffi (F. C. Baker, 1901)	F	coldwater pondsnail

Physidae

SCIENTIFIC NAME	OCCURRENCE	COMMON NAME
Aplexa elongata (Say, 1821)	F	lance aplexa
Physa jennessi Dall, 1919	F	obtuse physa
Physa skinneri Taylor, 1954	F [I]	glass physa
Physella acuta (Draparnaud, 1805)	F	European physa
Physella ancillaria (Say, 1825)	F	pumpkin physa
Physella bermudezi (Aguayo, 1935)	F	lowdome physa
Physella bottimeri (Clench, 1924)	F	Comanche physa
Physella boucardi (Crosse and P. Fischer, 1881)	F	desert physa

SCIENTIFIC NAME	OCCURRENCE	COMMON NAME
Physella columbiana (Hemphill, 1890)	F	rotund physa
Physella conoidea (P. Fischer and Crosse, 1886)	F	Texas physa
Physella cooperi (Tryon, 1865)	F	olive physa
Physella costata (Newcomb, 1861)	F	ornate physa
Physella cubensis (Pfeiffer, 1839)	F	Carib physa
Physella globosa (Haldeman, 1841)	F	globose physa
Physella gyrina (Say, 1821)	F	tadpole physa
Physella hendersoni (Clench, 1925)	F	bayou physa
Physella heterostropha (Say, 1817)	F	pewter physa
Physella hordacea (I. Lea, 1864)	F	grain physa
Physella humerosa (Gould, 1855)	F	corkscrew physa
Physella integra (Haldeman, 1841)	F	ashy physa
Physella johnsoni (Clench, 1926)	F	striate physa
Physella lordi (Baird, 1863)	F	twisted physa
Physella magnalacustris (Walker, 1901)	F	Great Lakes physa
Physella microstriata (Chamberlain and E. G. Berry, 1930)	F	Fish Lake physa
Physella parkeri (Currier, 1881)	F	broadshoulder physa
Physella propinqua (Tryon, 1865)	F	Rocky Mountain physa
Physella osculans (Haldeman, 1841)	F	cayuse physa
Physella spelunca (Turner and Clench, 1974)	F	cave physa
Physella squalida (Morelet, 1851)	F	squalid physa
Physella traski (I. Lea, 1864)	F	sculpted physa
Physella utahensis (Clench, 1925)	F	Utah physa
Physella vinosa (Gould, 1847)	F	banded physa
Physella virgata (Gould, 1855)	F	protean physa
Physella virginea (Gould, 1847)	F	sunset physa
Physella wrighti Te and Clarke, 1985	F	hotwater physa
Physella zionis (Pilsbry, 1926)	F	wet-rock physa
Stenophysa maugeriae (J. E. Gray, 1837)	F	tawny aplexa
Stenophysa marmorata (Guilding, 1828)	F	marbled aplexa
Planorbidae		
Amphigyra alabamensis Pilsbry, 1906	F	shoal sprite
Biomphalaria glabrata (Say, 1818)	F [I]	bloodfluke planorb
Biomphalaria havanensis (Pfeiffer, 1839)	F	ghost rams-horn
Drepanotrema aeruginosum (Morelet, 1851)	F [I]	rusty rams-horn
Drepanotrema cimex (Moricand, 1839)	F [I]	ridged rams-horn
Drepanotrema kermatoides (d'Orbigny, 1835)	F [I]	crested rams-horn
Gyraulus circumstriatus (Tryon, 1866)	F	disc gyro
Gyraulus crista (Linnaeus, 1858)	F	star gyro
Gyraulus deflectus (Say, 1824)	F	flexed gyro
Gyraulus hornensis F. C. Baker, 1934	F	tuba gyro
Gyraulus parvus (Say, 1817)	F	ash gyro
Helisoma anceps (Menke, 1830)	F	two-ridge rams-horn
Helisoma newberryi (I. Lea, 1858)	F	Great Basin rams-horn
Menetus opercularis (Gould, 1847)	F	button sprite
Micromenetus alabamensis (Pilsbry, 1895)	F	marsh sprite
Micromenetus brogniartianus (I. Lea, 1842)	F	disc sprite
Micromenetus dilatatus (Gould, 1841)	F	bugle sprite
Micromenetus floridensis (F. C. Baker, 1945)	F	penny sprite
Micromenetus sampsoni (Ancey, 1885)	F	*classification uncertain*

SCIENTIFIC NAME	OCCURRENCE	COMMON NAME
Neoplanorbis carinatus Walker, 1908	F	*classification uncertain*
Neoplanorbis smithi Walker, 1908	F	*classification uncertain*
Neoplanorbis tantillus Pilsbry, 1906	F	*classification uncertain*
Neoplanorbis umbilicatus Walker, 1908	F	*classification uncertain*
Planorbella ammon (Gould, 1855)	F	Jupiter rams-horn
Planorbella armigera (Say, 1821)	F	thicklip rams-horn
Planorbella binneyi (Tryon, 1867)	F	coarse rams-horn
Planorbella campanulata (Say, 1821)	F	bellmouth rams-horn
Planorbella campestris (Dawson, 1875)	F	meadow rams-horn
Planorbella columbiensis (F. C. Baker, 1945)	F	caribou rams-horn
Planorbella corpulenta (Say, 1824)	F	corpulent rams-horn
Planorbella duryi (Wetherby, 1879)	F	Seminole rams-horn
Planorbella magnifica (Pilsbry, 1903)	F	magnificent rams-horn
Planorbella multivolvis (Case, 1847)	F	acorn rams-horn
Planorbella occidentalis (J. G. Cooper, 1870)	F	fine-lined rams-horn
Planorbella oregonensis (Tryon, 1865)	F	lamb rams-horn
Planorbella pilsbryi (F. C. Baker, 1926)	F	file rams-horn
Planorbella scalaris (Jay, 1839)	F	mesa rams-horn
Planorbella tenuis (Dunker, 1850)	F	Mexican rams-horn
Planorbella traski (I. Lea, 1856)	F	keeled rams-horn
Planorbella trivolvis (Say, 1817)	F	marsh rams-horn
Planorbella truncata (M. Miles, 1861)	F	druid rams-horn
Promenetus exacuous (Say, 1821)	F	sharp sprite
Promenetus umbilicatellus (Cockerell, 1887)	F	umbilicate sprite
Vorticifex effusus (I. Lea, 1856)	F	Artemesian rams-horn
Vorticifex solidus (Dall, 1870)	F	*classification uncertain*

Ancylidae

SCIENTIFIC NAME	OCCURRENCE	COMMON NAME
Ferrissia fragilis (Tryon, 1863)	F	fragile ancylid
Ferrissia hendersoni Walker, 1925	F	blackwater ancylid
Ferrissia mcneili Walker, 1925	F	hood ancylid
Ferrissia parallela (Haldeman, 1841)	F	oblong ancylid
Ferrissia rivularis (Say, 1817)	F	creeping ancylid
Ferrissia walkeri (Pilsbry and Ferriss, 1907)	F	cloche ancylid
Hebetoncylus excentricus (Morelet, 1851)	F	excentric ancylid
Laevapex diaphanus (Haldeman, 1841)	F	cymbal ancylid
Laevapex fuscus (C. B. Adams, 1841)	F	dusky ancylid
Laevapex peninsulae (Pilsbry, 1903)	F	peninsula ancylid
Rhodacme elatior (Anthony, 1855)	F	domed ancylid
Rhodacme filosa (Conrad, 1834)	F	wicker ancylid
Rhodacme hinkleyi (Walker, 1908)	F	knobby ancylid

Carychiidae

SCIENTIFIC NAME	OCCURRENCE	COMMON NAME
Carychium clappi Hubricht, 1959	T	Appalachian thorn
Carychium exiguum (Say, 1822)	T	obese thorn
Carychium exile I. Lea, 1842	T	ice thorn
Carychium mexicanum Pilsbry, 1891	T	southern thorn
Carychium minimum Müller, 1774	T [I]	herald snail
Carychium nannodes G. H. Clapp, 1905	T	file thorn
Carychium occidentale Pilsbry, 1891	T	western thorn
Carychium riparium Hubricht, 1978	T	floodplain thorn
Carychium stygium Call, 1897	T	cave thorn

SCIENTIFIC NAME	OCCURRENCE	COMMON NAME
Siphonariidae		
Siphonaria alternata Say, 1826	A	
Siphonaria brannani Stearns, 1872	P	
Siphonaria pectinata (Linnaeus, 1758)	A	striped falselimpet
Siphonaria thersites Carpenter, 1864	P	
Williamia krebsii (Mörch, 1877)	A	
Williamia peltoides (Carpenter, 1864)	P	shield falselimpet
Trimusculidae		
Trimusculus carinatus (Dall, 1870)	A	carinate gadinia
Trimusculus reticulatus (Sowerby, 1835)	P	reticulate gadinia
ORDER STYLOMMATOPHORA		
Cochlicopidae		
Cochlicopa lubrica (Müller, 1774)	T	glossy pillar
Cochlicopa lubricella (Porro, 1838)	T	thin pillar
Cochlicopa morseana (Doherty, 1878)	T	Appalachian pillar
Cochlicopa nitens (Gallenstein, 1848)	T	robust pillar
Pupillidae		
Bothriopupa variolosa (Gould, 1848)	T	pitted birddrop
Chaenaxis tuba (Pilsbry and Ferriss, 1906)	T	hollow tuba
Columella simplex (Gould, 1841)	T	high-spire column
Columella edentula (Draparnaud, 1805)	T	toothless column
Gastrocopta abbreviata (Sterki, 1909)	T	plains snaggletooth
Gastrocopta armifera (Say, 1821)	T	armed snaggletooth
Gastrocopta ashmuni (Sterki, 1898)	T	sluice snaggletooth
Gastrocopta carnegiei (Sterki, 1916)	T	Erie snaggletooth
Gastrocopta clappi (Sterki, 1909)	T	bluegrass snaggletooth
Gastrocopta cochisensis (Pilsbry and Ferriss, 1910)	T	Apache snaggletooth
Gastrocopta contracta (Say, 1822)	T	bottleneck snaggletooth
Gastrocopta corticaria (Say, 1816)	T	bark snaggletooth
Gastrocopta cristata (Pilsbry and Vanatta, 1900)	T	crested snaggletooth
Gastrocopta dalliana (Sterki, 1898)	T	shortneck snaggletooth
Gastrocopta holzingeri (Sterki, 1889)	T	lambda snaggletooth
Gastrocopta pellucida (Pfeiffer, 1841)	T	slim snaggletooth
Gastrocopta pentodon (Say, 1821)	T	comb snaggletooth
Gastrocopta pilsbryana (Sterki, 1890)	T	montane snaggletooth
Gastrocopta procera (Gould, 1840)	T	wing snaggletooth
Gastrocopta prototypus (Pilsbry, 1899)	T	Sonoran snaggletooth
Gastrocopta quadridens Pilsbry, 1916	T	cross snaggletooth
Gastrocopta riograndensis (Pilsbry and Vanatta, 1900)	T	Rio Grande snaggletooth
Gastrocopta riparia Hubricht, 1978	T	Gulf Coast snaggletooth
Gastrocopta ruidosensis (Cockerell, 1899)	T	Ruidoso snaggletooth
Gastrocopta rupicola (Say, 1821)	T	tapered snaggletooth
Gastrocopta similis (Sterki, 1909)	T	Great Lakes snaggletooth
Gastrocopta tappaniana (C. B. Adams, 1842)	T	white snaggletooth

SCIENTIFIC NAME	OCCURRENCE	COMMON NAME
Pupilla blandi E. S. Morse, 1865	T	Rocky Mountain column
Pupilla hebes (Ancey, 1881)	T	crestless column
Pupilla muscorum (Linnaeus, 1758)	T	widespread column
Pupilla sonorana (Sterki, 1899)	T	threetooth column
Pupilla syngenes (Pilsbry, 1890)	T	top-heavy column
Pupisoma dioscoricola (C. B. Adams, 1845)	T	yam babybody
Pupisoma macneilli (G. H. Clapp, 1918)	T	Gulf babybody
Pupisoma minus Pilsbry, 1920	T	*classification uncertain*
Pupoides albilabris (C. B. Adams, 1841)	T	white-lip dagger
Pupoides hordaceus (Gabb, 1866)	T	ribbed dagger
Pupoides inornatus Vanatta, 1915	T	Rocky Mountain dagger
Pupoides modicus (Gould, 1848)	T	island dagger
Sterkia calamitosa (Pilsbry, 1889)	T	Mexican birddrop
Sterkia clementina (Sterki, 1890)	T	insular birddrop
Sterkia eryiesi rhoadsi (Pilsbry, 1899)	T	Caribbean birddrop
Sterkia hemphilli (Sterki, 1890)	T	California birddrop
Vertigo alabamensis G. H. Clapp, 1915	T	Alabama vertigo
Vertigo allyniana S. S. Berry, 1919	T	smalltooth vertigo
Vertigo alpestris Alder, 1838	T	tundra vertigo
Vertigo andrusiana (Pilsbry, 1899)	T	Pacific vertigo
Vertigo arthuri von Martens, 1882	T	*classification uncertain*
Vertigo berryi Pilsbry, 1919	T	rotund vertigo
Vertigo binneyana Sterki, 1890	T	cylindrical vertigo
Vertigo bollesiana (E. S. Morse, 1865)	T	delicate vertigo
Vertigo californica (Rowell, 1861)	T	ribbed vertigo
Vertigo clappi Brooks and Hunt, 1936	T	cupped vertigo
Vertigo columbiana Sterki, 1892	T	Columbia vertigo
Vertigo concinnula Cockerell, 1897	T	mitered vertigo
Vertigo conecuhensis G. H. Clapp, 1915	T	Conecuh vertigo
Vertigo dalliana Sterki, 1890	T	horseshoe vertigo
Vertigo elatior Sterki, 1894	T	tapered vertigo
Vertigo gouldi (A. Binney, 1843)	T	variable vertigo
Vertigo hebardi Vanatta, 1912	T	Keys vertigo
Vertigo hinkleyi Pilsbry, 1921	T	heart vertigo
Vertigo idahoensis Pilsbry, 1934	T	Idaho vertigo
Vertigo meramecensis Van Devender, 1979	T	bluff vertigo
Vertigo milium (Gould, 1840)	T	blade vertigo
Vertigo modesta (Say, 1824)	T	cross vertigo
Vertigo morsei Sterki, 1894	T	six-whorl vertigo
Vertigo nylanderi Sterki, 1909	T	deep-throat vertigo
Vertigo occidentalis Sterki, 1907	T	cone vertigo
Vertigo oralis Sterki, 1898	T	palmetto vertigo
Vertigo oscariana Sterki, 1890	T	capital vertigo
Vertigo ovata Say, 1822	T	ovate vertigo
Vertigo paradoxa Sterki, 1900	T	*classification uncertain*
Vertigo parvula Sterki, 1890	T	smallmouth vertigo
Vertigo perryi Sterki, 1905	T	olive vertigo
Vertigo pygmaea (Draparnaud, 1801)	T	crested vertigo
Vertigo rowelli (Newcomb, 1860)	T	threaded vertigo
Vertigo rugosula Sterki, 1890	T	striate vertigo
Vertigo sterkii (Pilsbry, 1919)	T	chestnut vertigo
Vertigo teskeyae Hubricht, 1961	T	swamp vertigo

SCIENTIFIC NAME	OCCURRENCE	COMMON NAME
Vertigo tridentata Wolf, 1870	T	honey vertigo
Vertigo ventricosa (E. S. Morse, 1865)	T	five-tooth vertigo
Vertigo wheeleri Pilsbry, 1928	T	Monte Sano vertigo

Valloniidae

Planogyra asteriscus (E. S. Morse, 1857)	T	eastern flat-whorl
Planogyra clappi (Pilsbry, 1898)	T	western flat-whorl
Vallonia albula Sterki, 1893	T	indecisive vallonia
Vallonia costata (Müller, 1774)	T	costate vallonia
Vallonia cyclophorella (Sterki, 1892)	T	silky vallonia
Vallonia excentrica Sterki, 1893	T	Iroquois vallonia
Vallonia gracilicosta Reinhardt, 1883	T	multirib vallonia
Vallonia parvula Sterki, 1893	T	trumpet vallonia
Vallonia perspectiva Sterki, 1892	T	thin-lip vallonia
Vallonia pulchella (Müller, 1774)	T	lovely vallonia
Zoogenetes harpa (Say, 1824)	T	boreal top

Strobilopsidae

Strobilops aeneus Pilsbry, 1926	T	bronze pinecone
Strobilops affinis Pilsbry, 1893	T	eightfold pinecone
Strobilops hubbardi A. D. Brown, 1861	T	flattened pinecone
Strobilops labyrinthicus (Say, 1817)	T	maze pinecone
Strobilops texasianus Pilsbry and Ferriss, 1906	T	southern pinecone

Ceriidae

Cerion casablancae Bartsch, 1920	T [I]	*classification uncertain*
Cerion chrysalis fastigatum Maynard, 1896	T [I]	*classification uncertain*
Cerion incanum (A. Binney, 1851)	T	gray peanut-snail
Cerion sculptum marielinum Torre, 1927	T [I]	*classification uncertain*
Cerion tridentatum Pilsbry and Vanatta, 1895	T [I]	*classification uncertain*
Cerion viaregis Bartsch, 1920	T [I]	*classification uncertain*

Ferussaciidae

Cecilioides acicula (Müller, 1774)	T [I]	blind awlsnail
Cecilioides apertus (Swainson, 1840)	T [I]	obtuse awlsnail

Subulinidae

Lamellaxis clavulinus (Potiez and Michaud, 1838)	T [I]	spike awlsnail
Lamellaxis gracilis (Hutton, 1834)	T [I]	graceful awlsnail
Lamellaxis mauritianus (Pfeiffer, 1852)	T [I]	Mauritian awlsnail
Lamellaxis micrus (d'Orbigny, 1835)	T [I]	tiny awlsnail
Opeas pumilum (Pfeiffer, 1840)	T [I]	dwarf awlsnail
Opeas pyrgula Schmacker and Boettger, 1891	T [I]	sharp awlsnail
Rumina decollata (Linnaeus, 1758)	T [I]	decollate snail
Subulina octona (Bruguière, 1792)	T [I]	miniature awlsnail

SCIENTIFIC NAME	OCCURRENCE	COMMON NAME
Spiraxidae		
Euglandina rosea (Férussac, 1818)	T	rosey wolfsnail
Euglandina singleyana W. G. Binney, 1892	T	striate wolfsnail
Euglandina texasiana (Pfeiffer 1857)	T	glossy wolfsnail
Melaniella gracillima (Pfeiffer, 1839)	T	brown foxsnail
Pseudosubulina cheatumi Pilsbry, 1950	T	Chisos foxsnail
Achatinidae		
Achatina fulica (Férussac, 1821)	T [I]	giant African snail
Streptaxidae		
Huttonella bicolor (Hutton, 1834)	T [I]	two-tone gulella
Haplotrematidae		
Haplotrema alameda Pilsbry, 1930	T	Alameda lancetooth
Haplotrema caelatum Mazyck, 1886	T	slotted lancetooth
Haplotrema catalinense (Hemphill, 1890)	T	Catalina lancetooth
Haplotrema concavum (Say, 1821)	T	gray-foot lancetooth
Haplotrema continentis H. B. Baker, 1930	T	grizzly lancetooth
Haplotrema costatum A. G. Smith, 1957	T	costate lancetooth
Haplotrema duranti (Newcomb, 1864)	T	ribbed lancetooth
Haplotrema keepi (Hemphill, 1890)	T	glassy lancetooth
Haplotrema kendeighi Webb, 1951	T	blue-foot lancetooth
Haplotrema minimum (Ancey, 1888)	T	California lancetooth
Haplotrema sportella (Gould, 1846)	T	beaded lancetooth
Haplotrema transfuga (Hemphill, 1892)	T	striate lancetooth
Haplotrema vancouverense (I. Lea, 1839)	T	robust lancetooth
Haplotrema voyanum (Newcomb, 1865)	T	hooded lancetooth
Urocoptidae		
Bostrychocentrum metcalfi F. G. Thompson, 1974	T	Metcalf holospira
Cochlodinella poeyana (d'Orbigny, 1841)	T	truncate urocoptid
Holospira arizonensis Stearns, 1890	T	Arizona holospira
Holospira bilamellata Dall, 1895	T	bilamellate holospira
Holospira chiricahuana Pilsbry, 1905	T	Cave Creek holospira
Holospira cockerelli Dall, 1897	T	Cockerell holospira
Holospira crossei Dall, 1895	T	Cross holospira
Holospira danielsi Pilsbry and Ferriss, 1915	T	strongrib holospira
Holospira ferrissi Pilsbry, 1905	T	stocky holospira
Holospira montivaga Pilsbry, 1946	T	vagabond holospira
Holospira oritis Pilsbry and Cheatum, 1951	T	mountain holospira
Holospira pityis Pilsbry and Cheatum, 1951	T	pinecone holospira
Holospira regis Pilsbry and Cockerell, 1905	T	royal holospira
Holospira tantalus Bartsch, 1906	T	teasing holospira
Holospira whetstonensis Pilsbry and Ferriss, 1923	T	Whetstone holospira
Holospira yucatanensis Bartsch, 1906	T	Bartsch holospira
Metastoma goldfussi (Menke, 1847)	T	New Braunfels metastoma
Metastoma hamiltoni Dall, 1897	T	Hamilton metastoma

SCIENTIFIC NAME	OCCURRENCE	COMMON NAME
Metastoma mesolia Pilsbry, 1912	T	widemouth metastoma
Metastoma pasonis Dall, 1895	T	robust metastoma
Metastoma riograndense Pilsbry, 1946	T	Rio Grande metastoma
Metastoma roemeri (Pfeiffer, 1848)	T	distorted metastoma
Microceramus floridanus (Pilsbry, 1898)	T	Florida urocoptid
Microceramus pontificus (Gould, 1848)	T	pontiff urocoptid
Microceramus texanus (Pilsbry, 1898)	T	Texas urocoptid

Bulimulidae

SCIENTIFIC NAME	OCCURRENCE	COMMON NAME
Bulimulus guadalupensis (Bruguière, 1789)	T [I]	West Indian bulimulus
Drymaeus dominicus (Reeve, 1850)	T	master treesnail
Drymaeus dormani (W. G. Binney, 1857)	T	manatee treesnail
Drymaeus multilineatus (Say, 1825)	T	lined treesnail
Liguus fasciatus (Müller, 1774)	T	Florida treesnail
Orthalicus floridensis Pilsbry, 1899	T	banded treesnail
Orthalicus reses nesodryas Pilsbry, 1946	T	Florida Keys treesnail
Orthalicus reses reses (Say, 1830)	T	Stock Island treesnail
Rabdotus alternatus (Say, 1830)	T	striped rabdotus
Rabdotus dealbatus (Say, 1821)	T	whitewashed rabdotus
Rabdotus mooreanus (Pfeiffer, 1868)	T	prairie rabdotus
Rabdotus nigromontanus (Dall, 1897)	T	Black Mountain rabdotus
Rabdotus pilsbryi (Ferriss, 1925)	T	elongate rabdotus

Punctidae

SCIENTIFIC NAME	OCCURRENCE	COMMON NAME
Punctum blandianum Pilsbry, 1900	T	brown spot
Punctum californicum Pilsbry, 1898	T	ribbed spot
Punctum conspectum (Bland, 1865)	T	striate spot
Punctum minutissimum (I. Lea, 1841)	T	small spot
Punctum randolphi (Dall, 1895)	T	conical spot
Punctum smithi Morrison, 1935	T	lamellate spot
Punctum vitreum (H. B. Baker, 1930)	T	glass spot
Zonites diegoensis Hemphill, 1892	T	classification uncertain

Charopidae

SCIENTIFIC NAME	OCCURRENCE	COMMON NAME
Radiodiscus abietum H. B. Baker, 1930	T	fir pinwheel
Radiodiscus millicostatus Pilsbry and Ferriss, 1906	T	ribbed pinwheel

Helicodiscidae

SCIENTIFIC NAME	OCCURRENCE	COMMON NAME
Helicodiscus aldrichianus (G. H. Clapp, 1907)	T	burrowing coil
Helicodiscus barri Hubricht, 1962	T	raccoon coil
Helicodiscus bonamicus Hubricht, 1978	T	spiral coil
Helicodiscus diadema Grimm, 1967	T	shaggy coil
Helicodiscus eigenmanni Pilsbry, 1900	T	Mexican coil
Helicodiscus enneodon Hubricht, 1967	T	bluff coil
Helicodiscus fimbriatus Wetherby, 1881	T	fringed coil
Helicodiscus hadenoecus Hubricht, 1962	T	cricket coil
Helicodiscus hexodon Hubricht, 1966	T	toothy coil
Helicodiscus inermis H. B. Baker, 1929	T	oldfield coil
Helicodiscus lirellus Hubricht, 1975	T	rubble coil

SCIENTIFIC NAME	OCCURRENCE	COMMON NAME
Helicodiscus multidens Hubricht, 1962	T	twilight coil
Helicodiscus notius Hubricht, 1962	T	tight coil
Helicodiscus nummus (Vanatta, 1899)	T	wax coil
Helicodiscus parallelus (Say, 1817)	T	compound coil
Helicodiscus punctatellus Morrison, 1942	T	punctate coil
Helicodiscus salmonaceus W. G. Binney, 1886	T	salmon coil
Helicodiscus saludensis (Morrison, 1937)	T	corncob snail
Helicodiscus shimeki Hubricht, 1962	T	temperate coil
Helicodiscus singleyanus (Pilsbry, 1890)	T	smooth coil
Helicodiscus tridens (Morrison, 1935)	T	crosstimbers coil
Helicodiscus triodus Hubricht, 1958	T	tallus coil
Polygyriscus virginicus P. R. Burch, 1947	T	Virginia coil
Speleodiscoides spirellum A. G. Smith, 1957	T	glass coil

Discidae

Anguispira alabama (Clapp, 1920)	T	Alabama disc
Anguispira alternata (Say, 1816)	T	flamed disc
Anguispira cumberlandiana (I. Lea, 1840)	T	Cumberland disc
Anguispira fergusoni (Bland, 1861)	T	coastal-plain disc
Anguispira jessica Kutchka, 1938	T	mountain disc
Anguispira knoxensis (Pilsbry, 1899)	T	rustic disc
Anguispira kochi (Pfeiffer, 1821)	T	banded globe
Anguispira mordax (Shuttleworth, 1852)	T	Appalachian disc
Anguispira nimapuna H. B. Baker, 1932	T	Nimapuna disc
Anguispira picta (G. H. Clapp, 1920)	T	painted disc
Anguispira rugoderma Hubricht, 1938	T	Pine Mountain disc
Anguispira strongyloides (Pfeiffer, 1854)	T	southeastern disc
Discus bryanti (Harper, 1881)	T	saw-tooth disc
Discus catskillensis (Pilsbry, 1896)	T	angular disc
Discus clappi (Pilsbry, 1924)	T	channelled disc
Discus cronkhitei (Newcomb, 1865)	T	forest disc
Discus macclintocki (F. C. Baker, 1928)	T	Pleistocene disc
Discus marmorensis H. B. Baker, 1932	T	marbled disc
Discus nigrimontanus (Pilsbry, 1924)	T	Black Mountain disc
Discus patulus (Deshayes, 1830)	T	domed disc
Discus rotundatus (Müller, 1776)	T [I]	rotund disc
Discus selinitoides (Pilsbry, 1890)	T	file disc
Discus shimeki (Pilsbry, 1890)	T	striate disc

Arionidae

Anadenulus cockerelli (Hemphill, 1890)	T	American keeledslug
Arion ater (Linnaeus, 1758)	T [I]	black arion
Arion circumscriptus Johnston, 1828	T [I]	brown-banded arion
Arion distinctus Mabille, 1868	T [I]	darkface arion
Arion fasciatus (Nilsson, 1822)	T [I]	orange-banded arion
Arion hortensis Férussac, 1819	T [I]	garden arion
Arion intermedius (Normand, 1852)	T [I]	hedgehog arion
Arion owenii Davies, 1979	T [I]	warty arion
Arion rufus (Linnaeus, 1758)	T [I]	chocolate arion
Arion subfuscus (Draparnaud, 1805)	T [I]	dusky arion
Arion sylvaticus Lohmander, 1937	T [I]	forest arion
Ariolimax californicus J. G. Cooper, 1872	T	California bananaslug

SCIENTIFIC NAME	OCCURRENCE	COMMON NAME
Ariolimax columbianus (Gould, 1851)	T	Pacific bananaslug
Ariolimax dolichophallus Mead, 1943	T	slender bananaslug
Binneya guadalupensis Pilsbry, 1927	T	Guadalupe shelledslug
Binneya notabilis J. G. Cooper, 1863	T	Santa Barbara shelledslug
Hemphillia burringtoni Pilsbry, 1948	T	keeled jumping-slug
Hemphillia camelus Pilsbry and Vanatta, 1897	T	pale jumping-slug
Hemphillia danielsi Vanatta, 1914	T	marbled jumping-slug
Hemphillia dromedarius Branson, 1972	T	dromedary jumping-slug
Hemphillia glandulosa Bland and W. G. Binney, 1872	T	warty jumping-slug
Hemphillia malonei Pilsbry, 1917	T	Malone jumping-slug
Hemphillia pantherina Branson, 1975	T	panther jumping-slug
Hesperarion hemphilli (W. G. Binney, 1875)	T	Hemphill slug
Hesperarion niger (J. G. Cooper, 1872)	T	black slug
Magnipelta mycophaga Pilsbry, 1953	T	spotted slug
Prophysaon andersoni (J. G. Cooper, 1872)	T	reticulate taildropper
Prophysaon boreale Pilsbry, 1946	T	northern taildropper
Prophysaon coeruleum Cockerell, 1890	T	blue-gray taildropper
Prophysaon dubium Cockerell, 1890	T	papillose taildropper
Prophysaon fasciatum Cockerell, 1890	T	banded taildropper
Prophysaon foliolatum (Gould, 1851)	T	yellow-bordered taildropper
Prophysaon humile Cockerell, 1890	T	smoky taildropper
Prophysaon obscurum Cockerell, 1890	T	mottled taildropper
Prophysaon vanattae Pilsbry, 1946	T	scarletback taildropper

Philomycidae

SCIENTIFIC NAME	OCCURRENCE	COMMON NAME
Megapallifera mutabilis (Hubricht, 1951)	T	changeable mantleslug
Megapallifera ragsdalei (Webb, 1950)	T	Ozark mantleslug
Megapallifera wetherbyi (W. G. Binney, 1874)	T	blotchy mantleslug
Pallifera dorsalis (A. Binney, 1842)	T	pale mantleslug
Pallifera fosteri F. C. Baker, 1939	T	Foster mantleslug
Pallifera hemphilli (W. G. Binney, 1885)	T	black mantleslug
Pallifera marmorea Pilsbry, 1948	T	marbled mantleslug
Pallifera ohioensis (Sterki, 1908)	T	redfoot mantleslug
Pallifera pilsbryi C. D. Miles and Mead, 1960	T	Arizona mantleslug
Pallifera secreta Cockerell, 1900	T	severed mantleslug
Pallifera varia Hubricht, 1953	T	variable mantleslug
Philomycus carolinianus (Bosc, 1802)	T	Carolina mantleslug
Philomycus flexuolaris Rafinesque, 1820	T	winding mantleslug
Philomycus sellatus Hubricht, 1972	T	Alabama mantleslug
Philomycus togatus (Gould, 1841)	T	toga mantleslug
Philomycus venustus Hubricht, 1953	T	brown-spotted mantleslug
Philomycus virginicus Hubricht, 1953	T	Virginia mantleslug
Zacoleus idahoensis Pilsbry, 1903	T	sheathed slug

Succineidae

SCIENTIFIC NAME	OCCURRENCE	COMMON NAME
Catinella aprica Hubricht, 1968	T	diurnal ambersnail
Catinella avara (Say, 1824)	T	suboval ambersnail
Catinella hubrichti Grimm, 1960	T	Snowhill ambersnail
Catinella oklahomarum (Webb, 1953)	T	detritis ambersnail
Catinella pugilator Hubricht, 1961	T	weedpatch ambersnail

SCIENTIFIC NAME	OCCURRENCE	COMMON NAME
Catinella rehderi (Pilsbry, 1948)	T	chrome ambersnail
Catinella stretchiana Bland, 1865	T	Sierra ambersnail
Catinella vagans (Pilsbry, 1900)	T	mudbank ambersnail
Catinella wandae (Webb, 1953)	T	slope ambersnail
Oxyloma decampi (Tryon, 1866)	T	Marshall ambersnail
Oxyloma effusum (Pfeiffer, 1853)	T	coastal plain ambersnail
Oxyloma groenlandicum (Möller, 1842)	T	ruddy ambersnail
Oxyloma hawkinsi (Baird, 1863)	T	boundary ambersnail
Oxyloma haydeni (W. G. Binney, 1858)	T	Niobrara ambersnail
Oxyloma kanabense Pilsbry, 1948	T	Kanab ambersnail
Oxyloma missoula Hubricht, 1982	T	Ninepipes ambersnail
Oxyloma nuttallianum (I. Lea, 1841)	T	oblique ambersnail
Oxyloma peoriense (Wolf, 1874)	T	depressed ambersnail
Oxyloma retusum (I. Lea, 1834)	T	blunt ambersnail
Oxyloma salleanum (Pfeiffer, 1849)	T	Louisiana ambersnail
Oxyloma sillimani (Bland, 1865)	T	Humboldt ambersnail
Oxyloma subeffusum Pilsbry, 1948	T	Chesapeake ambersnail
Succinea barberi (W. B. Marshall, 1926)	T	Sanibel ambersnail
Succinea californica P. Fischer and Crosse, 1878	T	San Tomas ambersnail
Succinea campestris Say, 1817	T	crinkled ambersnail
Succinea chittenangoensis Pilsbry, 1908	T	Appalachian ambersnail
Succinea floridana Pilsbry, 1905	T	Florida chalksnail
Succinea forsheyi I. Lea, 1864	T	spotted ambersnail
Succinea gabbi Tryon, 1866	T	riblet ambersnail
Succinea greeri Tryon, 1866	T	dryland ambersnail
Succinea grosvenori I. Lea, 1864	T	Santa Rita ambersnail
Succinea indiana Pilsbry, 1905	T	xeric ambersnail
Succinea luteola Gould, 1848	T	Mexico ambersnail
Succinea oregonensis I. Lea, 1841	T	Oregon ambersnail
Succinea ovalis Say, 1817	T	oval ambersnail
Succinea paralia Hubricht, 1983	T	saltmarsh ambersnail
Succinea putris (Linnaeus, 1758)	T	European ambersnail
Succinea rusticana Gould, 1846	T	rustic ambersnail
Succinea solastra Hubricht, 1961	T	Lone Star ambersnail
Succinea strigata Pfeiffer, 1855	T	striate ambersnail
Succinea unicolor Tryon, 1866	T	squatty ambersnail
Succinea urbana Hubricht, 1961	T	urban ambersnail
Succinea wilsoni I. Lea, 1864	T	golden ambersnail

Helicarionidae

SCIENTIFIC NAME	OCCURRENCE	COMMON NAME
Dryachloa dauca F. G. Thompson and Lee, 1980	T	carrot glass
Euconulus chersinus (Say, 1821)	T	wild hive
Euconulus dentatus (Sterki, 1893)	T	toothed hive
Euconulus fulvus (Müller, 1774)	T	brown hive
Euconulus polygyratus (Pilsbry, 1899)	T	fat hive
Euconulus trochulus (Reinhardt, 1883)	T	silk hive
Guppya gundlachi (Pfeiffer, 1839)	T	glossy granule
Guppya miamiensis Pilsbry, 1903	T	*classification uncertain*
Guppya sterkii (Dall, 1888)	T	*classification uncertain*

SCIENTIFIC NAME	OCCURRENCE	COMMON NAME
Zonitidae		
Gastrodonta fonticula Wurtz, 1948	T	Appalachia bellytooth
Gastrodonta interna (Say, 1822)	T	brown bellytooth
Glyphyalinia carolinensis (Cockerell, 1890)	T	spiral mountain glyph
Glyphyalinia clingmani (Dall, 1890)	T	fragile glyph
Glyphyalinia cryptomphala (G. H. Clapp, 1915)	T	thin glyph
Glyphyalinia cumberlandiana (G. H. Clapp, 1919)	T	hill glyph
Glyphyalinia floridana (Morrison, 1937)	T	Ocala glyph
Glyphyalinia indentata (Say, 1823)	T	carved glyph
Glyphyalinia junaluskana (Clench and Banks, 1932)	T	dark glyph
Glyphyalinia latebricola Hubricht, 1968	T	stone glyph
Glyphyalinia lewisiana (G. H. Clapp, 1908)	T	pale glyph
Glyphyalinia luticola Hubricht, 1966	T	furrowed glyph
Glyphyalinia ocoae Hubricht, 1978	T	blue-gray glyph
Glyphyalinia pecki Hubricht, 1966	T	blind glyph
Glyphyalinia pentadelphia (Pilsbry, 1900)	T	pink glyph
Glyphyalinia picea Hubricht, 1976	T	rust glyph
Glyphyalinia praecox (H. B. Baker, 1930)	T	brilliant glyph
Glyphyalinia raderi (Dall, 1898)	T	Maryland glyph
Glyphyalinia rhoadsi (Pilsbry, 1899)	T	sculpted glyph
Glyphyalinia rimula Hubricht, 1968	T	tongued glyph
Glyphyalinia roemeri (Pilsbry and Ferriss, 1906)	T	pretty glyph
Glyphyalinia sculptilis (Bland, 1858)	T	suborb glyph
Glyphyalinia specus Hubricht, 1965	T	hollow glyph
Glyphyalinia umbilicata (Cockerell, 1893)	T	Texas glyph
Glyphyalinia vanattai (Pilsbry and Walker, 1902)	T	honey glyph
Glyphyalinia virginica (Morrison, 1937)	T	depressed glyph
Glyphyalinia wheatleyi (Bland, 1883)	T	bright glyph
Hawaiia alachuana (Dall, 1885)	T	southeastern gem
Hawaiia minuscula (A. Binney, 1840)	T	minute gem
Hawaiia neomexicana (Cockerell and Pilsbry, 1900)	T	striate gem
Mesomphix andrewsae (Pilsbry, 1895)	T	mountain button
Mesomphix anurus Hubricht, 1962	T	frog button
Mesomphix capnodes (W. G. Binney, 1857)	T	dusky button
Mesomphix cupreus (Rafinesque, 1831)	T	copper button
Mesomphix friabilis (W. G. Binney, 1857)	T	brittle button
Mesomphix globosus (MacMillan, 1940)	T	globose button
Mesomphix inornatus (Say, 1821)	T	plain button
Mesomphix latior (Pilsbry, 1900)	T	broad button
Mesomphix perfragilis (Wetherby, 1894)	T	fragile button
Mesomphix perlaevis (Pilsbry, 1900)	T	smooth button
Mesomphix pilsbryi (G. H. Clapp, 1904)	T	striate button
Mesomphix rugeli (W. G. Binney, 1879)	T	wrinkled button
Mesomphix subplanus (A. Binney, 1842)	T	flat button
Mesomphix vulgatus H. B. Baker, 1933	T	common button
Nesovitrea binneyana (E. S. Morse, 1864)	T	blue glass
Nesovitrea dalliana (Pilsbry and Simpson, 1889)	T	depressed glass
Nesovitrea electrina (Gould, 1841)	T	amber glass

SCIENTIFIC NAME	OCCURRENCE	COMMON NAME
Nesovitrea susannae Pratt, 1978	T	live oak glass
Oxychilus alliarius (J. S. Miller, 1822)	T [I]	garlic glass-snail
Oxychilus cellarius (Müller, 1774)	T [I]	cellar glass-snail
Oxychilus draparnaudi (Beck, 1837)	T [I]	dark-bodied glass-snail
Oxychilus helveticus (Blum, 1881)	T [I]	Swiss glass-snail
Paravitrea alethia Hubricht, 1978	T	goddess supercoil
Paravitrea amicalola Hubricht, 1976	T	tight supercoil
Paravitrea andrewsae (W. G. Binney, 1879)	T	high mountain supercoil
Paravitrea aulacogyra (Pilsbry and Ferris, 1906)	T	classification uncertain
Paravitrea bellona Hubricht, 1978	T	club supercoil
Paravitrea bidens Hubricht, 1963	T	gray supercoil
Paravitrea blarina Hubricht, 1963	T	shrew supercoil
Paravitrea calcicola H. B. Baker, 1931	T	pearl supercoil
Paravitrea capsella (Gould, 1851)	T	dimple supercoil
Paravitrea ceres Hubricht, 1978	T	sidelong supercoil
Paravitrea clappi (Pilsbry, 1898)	T	Mirey Ridge supercoil
Paravitrea conecuhensis (G. H. Clapp, 1917)	T	triangular supercoil
Paravitrea dentilla Hubricht, 1978	T	comb supercoil
Paravitrea diana Hubricht, 1983	T	hunted supercoil
Paravitrea grimmi Hubricht, 1968	T	buff supercoil
Paravitrea hera Hubricht, 1983	T	spirit supercoil
Paravitrea lacteodens (Pilsbry, 1903)	T	Ramp Cove supercoil
Paravitrea lamellidens (Pilsbry, 1898)	T	lamellate supercoil
Paravitrea lapilla Hubricht, 1965	T	gem supercoil
Paravitrea metallacta Hubricht, 1963	T	Caneyfork supercoil
Paravitrea mira Hubricht, 1975	T	funnel supercoil
Paravitrea multidentata (A. Binney, 1840)	T	dentate supercoil
Paravitrea petrophila (Bland, 1883)	T	Cherokee supercoil
Paravitrea pilsbryana (G. H. Clapp, 1919)	T	translucent supercoil
Paravitrea placentula (Shuttleworth, 1852)	T	glossy supercoil
Paravitrea pontis H. B. Baker, 1931	T	Natural Bridge supercoil
Paravitrea reesi Morrison, 1937	T	round supercoil
Paravitrea septadens Hubricht, 1978	T	brown supercoil
Paravitrea seradens Hubricht, 1972	T	barred supercoil
Paravitrea significans (Bland, 1866)	T	domed supercoil
Paravitrea simpsoni (Pilsbry, 1889)	T	amber supercoil
Paravitrea subtilis Hubricht, 1978	T	slender supercoil
Paravitrea tantilla Hubricht, 1963	T	teasing supercoil
Paravitrea ternaria Hubricht, 1978	T	sculpted supercoil
Paravitrea tiara Hubricht, 1978	T	crowned supercoil
Paravitrea toma Hubricht, 1975	T	sharp supercoil
Paravitrea tridens Pilsbry, 1946	T	white-foot supercoil
Paravitrea umbilicaris (Ancey, 1887)	T	open supercoil
Paravitrea variabilis H. B. Baker, 1929	T	variable supercoil
Paravitrea varidens Hubricht, 1978	T	roan supercoil
Pilsbryna aurea H. B. Baker, 1929	T	ornate bud
Pilsbryna castanea H. B. Baker, 1931	T	prominent bud
Pristiloma arcticum (Lehnert, 1884)	T	northern tightcoil
Pristiloma chersinellum (Dall, 1886)	T	black-foot tightcoil
Pristiloma gabrielinum (S. S. Berry, 1924)	T	waxy tightcoil
Pristiloma idahoense Pilsbry, 1902	T	thinlip tightcoil
Pristiloma johnsoni (Dall, 1895)	T	broadwhorl tightcoil
Pristiloma juniperum A. G. Smith, 1957	T	cedar tightcoil

SCIENTIFIC NAME	OCCURRENCE	COMMON NAME
Pristiloma lansingi (Bland, 1875)	T	denticulate tightcoil
Pristiloma nicholsoni H. B. Baker, 1930	T	riparian tightcoil
Pristiloma orotis (S. S. Berry, 1930)	F	minute tightcoil
Pristiloma pilsbryi Vanatta, 1899	T	crowned tightcoil
Pristiloma shepardae (Hemphill, 1892)	T	island tightcoil
Pristiloma stearnsi (Bland, 1875)	T	striate tightcoil
Pristiloma subrupicola (Dall, 1877)	T	southern tightcoil
Pristiloma wascoense (Hemphill, 1911)	T	shiny tightcoil
Striatura exigua (Stimpson, 1850)	T	ribbed striate
Striatura ferrea E. S. Morse, 1864	T	black striate
Striatura meridionalis (Pilsbry and Ferriss, 1906)	T	median striate
Striatura milium (E. S. Morse, 1859)	T	fine-ribbed striate
Striatura pugentensis (Dall, 1895)	T	northwest striate
Ventridens acerra (J. Lewis, 1870)	T	glossy dome
Ventridens arcellus Hubricht, 1976	T	golden dome
Ventridens brittsi (Pilsbry, 1892)	T	western dome
Ventridens cerinoideus (Anthony, 1865)	T	wax dome
Ventridens coelaxis (Pilsbry, 1899)	T	bidentate dome
Ventridens collisella (Pilsbry, 1896)	T	sculptured dome
Ventridens decussatus (Walker and Pilsbry, 1902)	T	crossed dome
Ventridens demissus (A. Binney, 1843)	T	perforate dome
Ventridens eutropis Pilsbry, 1946	T	carinate dome
Ventridens gularis (Say, 1822)	T	throaty dome
Ventridens intertextus (A. Binney, 1841)	T	pyramid dome
Ventridens lasmodon (Phillips, 1841)	T	hollow dome
Ventridens lawae (W. G. Binney, 1892)	T	rounded dome
Ventridens ligera (Say, 1821)	T	globose dome
Ventridens monodon Hubricht, 1964	T	blade dome
Ventridens percallosus (Pilsbry, 1898)	T	Tennessee dome
Ventridens pilsbryi Hubricht, 1964	T	yellow dome
Ventridens suppressus (Say, 1829)	T	flat dome
Ventridens theloides (Walker and Pilsbry, 1902)	T	copper dome
Ventridens virginicus (Vanatta, 1936)	T	split-tooth dome
Ventridens volusiae (Pilsbry, 1900)	T	Seminole dome
Vitrea contracta (Westerlund, 1871)	T [I]	contracted glass-snail
Vitrizonites latissimus (J. Lewis, 1875)	T	glassy grapeskin
Zonitoides arboreus (Say, 1816)	T	quick gloss
Zonitoides elliotti (Redfield, 1856)	T	green dome
Zonitoides kirbyi R. W. Fullington, 1974	T	shadow gloss
Zonitoides lateumbilicatus (Pilsbry, 1895)	T	striate gloss
Zonitoides limatulus (A. Binney, 1840)	T	dull gloss
Zonitoides nitidus (Müller, 1774)	T	black gloss
Zonitoides patuloides (Pilsbry, 1895)	T	Appalachian gloss

Vitrinidae

Vitrina angelicae Beck, 1837	T	eastern glass-snail
Vitrina pellucida (Müller, 1774)	T	western glass-snail

Limacidae

Deroceras caruanae (Pollonera, 1891)	T [I]	longneck fieldslug

SCIENTIFIC NAME	OCCURRENCE	COMMON NAME
Deroceras hesperium Pilsbry, 1944.............	T	.... evening fieldslug
Deroceras heterura (Pilsbry, 1944)	T	.... marsh slug
Deroceras laeve (Müller, 1774)	T	.... meadow slug
Deroceras monentolophus Pilsbry, 1944..........	T	.... one-ridge fieldslug
Deroceras reticulatum (Müller, 1774)	T [I]	.... gray fieldslug
Limax flavus Linnaeus, 1758...................	T [I]	.... yellow gardenslug
Limax marginatus Müller, 1774	T [I]	.... tree slug
Limax maximus (Linnaeus, 1758)...............	T [I]	.... giant gardenslug
Limax nyctelius Bourguignat, 1861	T [I]	.... vine slug
Limax valentianus Férussac, 1823	T [I]	.... threeband gardenslug

Milacidae

Milax gagates (Draparnaud, 1801)	T [I]	.... greenhouse slug

Testacellidae

Testacella haliotidea Draparnaud, 1801	T [I]	.... earshell slug

Polygyridae

Allogona lombardii A. G. Smith, 1943	T	.... Selway forestsnail
Allogona profunda (Say, 1821)	T	.... broad-banded forestsnail
Allogona ptychophora (A. D. Brown, 1870).......	T	.... Idaho forestsnail
Allogona townsendiana (I. Lea, 1838)	T	.... Oregon forestsnail
Ashmunella amblya Pilsbry, 1940	T	..... Pine Springs woodlandsnail
Ashmunella angulata Pilsbry, 1905..............	T	.... angulate woodlandsnail
Ashmunella animasensis Vagvolgyi, 1974.........	T	..... Animas Peak woodlandsnail
Ashmunella ashmuni (Dall, 1896)	T	.... Jemez woodlandsnail
Ashmunella auriculata Vagvolgyi, 1974	T	.. Boulder Canyon woodlandsnail
Ashmunella bequaerti Clench and W. B. Miller, 1966	T	.... Goat Cave woodlandsnail
Ashmunella binneyi Pilsbry and Ferriss, 1917	T	..... Silver Creek woodlandsnail
Ashmunella carlsbadensis Pilsbry, 1932	T	.... Guadalupe woodlandsnail
Ashmunella chiricahuana (Dall, 1895)	T	...... Cave Creek woodlandsnail
Ashmunella cockerelli Pilsbry and Ferriss, 1917 ...	T	..... Black Range woodlandsnail
Ashmunella danielsi Pilsbry and Ferriss, 1915	T	. Whitewater Creek woodlandsnail
Ashmunella escuritor Pilsbry, 1915..............	T	.... Barefoot woodlandsnail
Ashmunella ferrissi Pilsbry, 1905	T	.. Reed's Mountain woodlandsnail
Ashmunella harrisi Metcalf and Smartt, 1977......	T	... Goat Mountain woodlandsnail
Ashmunella hebardi Pilsbry and Vanatta, 1923	T	.. Hacheta Grande woodlandsnail
Ashmunella kochi G. H. Clapp, 1908	T	..... San Andreas woodlandsnail
Ashmunella lepiderma Pilsbry and Ferriss, 1910 ...	T	.... Whitetail woodlandsnail
Ashmunella levettei (Bland, 1882)...............	T	.... Huachuca woodlandsnail
Ashmunella macromphala Vagvolgyi, 1974	T	..... Cook's Peak woodlandsnail
Ashmunella mearnsi (Dall, 1895)................	T	...... Big Hatchet woodlandsnail
Ashmunella mendax Pilsbry and Ferriss, 1917.....	T	.... Iron Creek woodlandsnail
Ashmunella mogollonensis Pilsbry, 1905	T	.... Mogollon woodlandsnail
Ashmunella mudgei Cheatham, 1971	T	. Sawtooth Mountain woodlandsnail
Ashmunella organensis Pilsbry, 1936	T	.. Organ Mountain woodlandsnail
Ashmunella pasonis (Drake, 1951)	T	. Franklin Mountain woodlandsnail
Ashmunella pilsbryana Ferriss, 1914	T	... Blue Mountain woodlandsnail
Ashmunella proxima Pilsbry, 1915	T	.... Chiricahua woodlandsnail
Ashmunella pseudodonta (Dall, 1897)............	T	.... Capitan woodlandsnail

SCIENTIFIC NAME	OCCURRENCE	COMMON NAME
Ashmunella rhyssa (Dall, 1897)	T	Sierra Blanca woodlandsnail
Ashmunella rileyensis Metcalf and Hurley, 1971	T	Mount Riley woodlandsnail
Ashmunella salinasensis Vagvolgyi, 1974	T	Salinas Peak woodlandsnail
Ashmunella sprouli R. W. Fullington and K. E. Fullington, 1978	T	Hell's Canyon woodlandsnail
Ashmunella tetrodon Pilsbry and Ferriss, 1915	T	Dry Creek woodlandsnail
Ashmunella thompsoniana (Ancey, 1887)	T	Sangre de Cristo woodlandsnail
Ashmunella todseni Metcalf and Smartt, 1977	T	Maple Canyon woodlandsnail
Ashmunella varicifera (Ancey, 1897)	T	Miller Canyon woodlandsnail
Ashmunella walkeri Ferriss, 1904	T	Florida Mountain woodlandsnail
Cryptomastix devia (Gould, 1846)	T	Puget oregonian
Cryptomastix germana (Gould, 1851)	T	pygmy oregonian
Cryptomastix harfordiana (W. G. Binney, 1878)	T	classification uncertain
Cryptomastix hendersoni (Pilsbry, 1928)	T	Columbia oregonian
Cryptomastix magnidentata (Pilsbry, 1940)	T	Mission Creek oregonian
Cryptomastix mullani (Bland and J. G. Cooper, 1861)	T	Coeur d'Alene oregonian
Cryptomastix sanburni (W. G. Binney, 1886)	T	Kingston oregonian
Euchemotrema cheatumi (R. W. Fullington, 1974)	T	palmetto pillsnail
Euchemotrema fasciatum (Pilsbry, 1940)	T	mountain pillsnail
Euchemotrema fraternum (Say, 1824)	T	upland pillsnail
Euchemotrema hubrichti (Pilsbry, 1940)	T	carinate pillsnail
Euchemotrema leai (A. Binney, 1841)	T	lowland pillsnail
Euchemotrema wichitorum (Branson, 1972)	T	Wichita Mountain pillsnail
Mesodon andrewsae W. G. Binney, 1879	T	balsam globe
Mesodon appressus (Say, 1821)	T	flat bladetooth
Mesodon approximans (G. H. Clapp, 1905)	T	tight-gapped shagreen
Mesodon archeri Pilsbry, 1940	T	Ocoee covert
Mesodon binneyanus (Pilsbry, 1899)	T	half-lidded oval
Mesodon chilhoweensis (J. Lewis, 1870)	T	queen crater
Mesodon christyi (Bland, 1860)	T	glossy covert
Mesodon clarki clarki (I. Lea, 1858)	T	dwarf proud globe
Mesodon clarki nantahala Clench, 1933	T	noonday globe
Mesodon clausus (Say, 1821)	T	yellow globelet
Mesodon clenchi (Rehder, 1932)	T	Calico Rock oval
Mesodon downieanus (Bland, 1861)	T	dwarf globelet
Mesodon edentatus (Sampson, 1889)	T	smooth-lip shagreen
Mesodon elevatus (Say, 1821)	T	proud globe
Mesodon ferrissi (Pilsbry, 1897)	T	Smoky Mountain covert
Mesodon indianorum (Pilsbry, 1899)	T	lidded oval
Mesodon inflectus (Say, 1821)	T	shagreen
Mesodon jonesianus (Archer, 1938)	T	big-tooth covert
Mesodon kalmianus Hubricht, 1965	T	brown globelet
Mesodon kiowaensis (Simpson, 1888)	T	drywoods oval
Mesodon leatherwoodi Pratt, 1971	T	Pedernales oval
Mesodon magazinensis (Pilsbry and Ferriss, 1906)	T	Magazine Mountain shagreen
Mesodon mitchellianus (I. Lea, 1839)	T	sealed globelet
Mesodon normalis (Pilsbry, 1900)	T	grand globe
Mesodon orestes Hubricht, 1975	T	engraved covert
Mesodon panselenus Hubricht, 1976	T	Virginia bladetooth
Mesodon pennsylvanicus (Green, 1827)	T	proud globelet
Mesodon perigraptus Pilsbry, 1894	T	engraved bladetooth
Mesodon roemeri (Pfeiffer, 1848)	T	Texas oval

SCIENTIFIC NAME	OCCURRENCE	COMMON NAME
Mesodon rugeli (Shuttleworth, 1852)	T	deep-tooth shagreen
Mesodon sanus (Clench and Archer, 1933)	T	squat globelet
Mesodon sargentianus (C. W. Johnson and Pilsbry, 1892)	T	grand bladetooth
Mesodon sayanus (Pilsbry, 1906)	T	spike-lip crater
Mesodon smithi (G. H. Clapp, 1905)	T	Alabama shagreen
Mesodon subpalliatus (Pilsbry, 1893)	T	velvet covert
Mesodon thyroidus (Say, 1816)	T	white-lip globe
Mesodon wetherbyi (Bland, 1873)	T	clifty covert
Mesodon wheatleyi (Bland, 1860)	T	cinnamon covert
Mesodon zaletus (A. Binney, 1837)	T	toothed globe
Polygyra auriculata Say, 1818	T	Ocala liptooth
Polygyra auriformis (Bland, 1859)	T	rockpile liptooth
Polygyra avara Say, 1818	T	Florida liptooth
Polygyra cereolus (Mühlfeld, 1818)	T	southern flatcoil
Polygyra chisosensis (Pilsbry, 1936)	T	Chisos liptooth
Polygyra delecta Hubricht, 1976	T	Gulf Hammock liptooth
Polygyra deltoidea (Simpson, 1889)	T	Oklahoma liptooth
Polygyra dorfueilliana (I. Lea, 1838)	T	oakwood liptooth
Polygyra fastigiata (Say, 1829)	T	bluegrass liptooth
Polygyra gracilis Hubricht, 1961	T	Edwards Plateau liptooth
Polygyra hausmani Jackson, 1948	T	Dixie liptooth
Polygyra hippocrepis (Pfeiffer, 1848)	T	horseshoe liptooth
Polygyra jacksoni (Bland, 1866)	T	Ozark liptooth
Polygyra leporina (Gould, 1848)	T	Gulf coast liptooth
Polygyra mooreana (W. G. Binney, 1858)	T	grassland liptooth
Polygyra peninsulae Pilsbry, 1940	T	St. Johns liptooth
Polygyra peregrina (Rehder, 1932)	T	White liptooth
Polygyra plana bahamensis Vanatta, 1919	T	Bahama flatcoil
Polygyra plicata (Say, 1821)	T	Cumberland liptooth
Polygyra polita Pilsbry and Hinkley, 1907	T	Tamaulipas liptooth
Polygyra postelliana (Bland, 1859)	T	coastal liptooth
Polygyra pustula (Say, 1821)	T	grooved liptooth
Polygyra pustuloides (Bland, 1858)	T	tiny liptooth
Polygyra septemvolva (Say, 1818)	T	Florida flatcoil
Polygyra simpsoni (Pilsbry and Ferriss, 1907)	T	Wyandotte liptooth
Polygyra subclausa Pilsbry, 1940	T	Suwannee liptooth
Polygyra texasiana (Moricand, 1833)	T	Texas liptooth
Polygyra troostiana (I. Lea, 1839)	T	Nashville liptooth
Polygyra uvulifera (Shuttleworth, 1852)	T	peninsula liptooth
Praticolella bakeri Vanatta, 1915	T	ridge scrubsnail
Praticolella berlandieriana (Moricand, 1833)	T	banded scrubsnail
Praticolella candida Hubricht, 1983	T	white scrubsnail
Praticolella griseola (Pfeiffer, 1841)	T	vagrant scrubsnail
Praticolella jujuna (Say, 1821)	T	Florida scrubsnail
Praticolella lawae (J. Lewis, 1874)	T	Appalachian scrubsnail
Praticolella mobiliana (I. Lea, 1841)	T	Choctaw scrubsnail
Praticolella pachyloma (Menke, 1847)	T	sandyland scrubsnail
Praticolella taeniata Pilsbry, 1940	T	striped scrubsnail
Praticolella trimatris Hubricht, 1983	T	Hidalgo scrubsnail
Stenotrema angellum Hubricht, 1958	T	Kentucky slitmouth
Stenotrema altispira (Pilsbry, 1894)	T	highland slitmouth
Stenotrema barbatum (G. H. Clapp, 1904)	T	bristled slitmouth
Stenotrema barbigerum (Redfield, 1856)	T	fringed slitmouth

SCIENTIFIC NAME	OCCURRENCE	COMMON NAME
Stenotrema blandianum (Pilsbry, 1903)	T	Missouri slitmouth
Stenotrema brevipila (G. H Clapp, 1907)	T	Talladega slitmouth
Stenotrema calvescens Hubricht, 1961	T	Chattanooga slitmouth
Stenotrema cohuttense (G. H. Clapp, 1914)	T	Cohutta slitmouth
Stenotrema deceptum (G. H. Clapp, 1905)	T	Monte Sano slitmouth
Stenotrema depilatum (Pilsbry, 1895)	T	Great Smoky slitmouth
Stenotrema edgarianum (I. Lea, 1841)	T	Sequatchie slitmouth
Stenotrema edvardsi (Bland, 1856)	T	ridge-and-valley slitmouth
Stenotrema exodon (Pilsbry, 1900)	T	Alabama slitmouth
Stenotrema florida Pilsbry, 1940	T	Apalachicola slitmouth
Stenotrema hirsutum (Say, 1817)	T	hairy slitmouth
Stenotrema labrosum (Bland, 1862)	T	Ozark slitmouth
Stenotrema magnafumosum (Pilsbry, 1900)	T	Appalachian slitmouth
Stenotrema maxillatum (Gould, 1848)	T	ridge-lip slitmouth
Stenotrema pilsbryi (Ferriss, 1900)	T	Rich Mountain slitmouth
Stenotrema pilula (Pilsbry, 1900)	T	pygmy slitmouth
Stenotrema simile Grimm, 1971	T	Bear Creek slitmouth
Stenotrema spinosum (I. Lea, 1830)	T	carinate slitmouth
Stenotrema stenotrema (Pfeiffer, 1842)	T	inland slitmouth
Stenotrema unciferum (Pilsbry, 1900)	T	Oachita slitmouth
Stenotrema waldense Archer, 1938	T	Doaks Creek slitmouth
Triodopsis alabamensis (Pilsbry, 1902)	T	Alabama threetooth
Triodopsis albolabris (Say, 1816)	T	whitelip
Triodopsis alleni (Sampson, 1883)	T	western whitelip
Triodopsis burchi Hubricht, 1950	T	Pittsylvania threetooth
Triodopsis caroliniensis (I. Lea, 1834)	T	blunt wedge
Triodopsis chadwicki (Ferriss, 1907)	T	Kaw whitelip
Triodopsis claibornensis Lutz, 1950	T	Claiborne threetooth
Triodopsis complanata (Pilsbry, 1898)	T	glossy threetooth
Triodopsis cragini Call, 1886	T	Post Oak threetooth
Triodopsis denotata (Férussac, 1821)	T	velvet wedge
Triodopsis dentifera (A. Binney, 1837)	T	big-tooth whitelip
Triodopsis discoidea Pilsbry, 1894	T	rivercliff threetooth
Triodopsis divesta (Gould, 1848)	T	Ozark whitelip
Triodopsis fallax (Say, 1825)	T	mimic threetooth
Triodopsis fosteri (F. C. Baker, 1921)	T	bladetooth wedge
Triodopsis fradulenta (Pilsbry, 1894)	T	baffled threetooth
Triodopsis fulcidens Hubricht, 1952	T	dwarf threetooth
Triodopsis henriettae (Mazyck, 1877)	T	pineywood threetooth
Triodopsis hopetonensis (Shuttleworth, 1852)	T	magnolia threetooth
Triodopsis juxtidens (Pilsbry, 1894)	T	Atlantic threetooth
Triodopsis lioderma (Pilsbry, 1902)	T	Tulsa whitelip
Triodopsis major (A. Binney, 1837)	T	southeastern whitelip
Triodopsis maratima (Pilsbry, 1890)	T	coastal whitelip
Triodopsis messana Hubricht, 1952	T	pinhole threetooth
Triodopsis multilineata (Say, 1821)	T	striped whitelip
Triodopsis neglecta (Pilsbry, 1899)	T	Ozark threetooth
Triodopsis obsoleta (Pilsbry, 1894)	T	nubbin threetooth
Triodopsis obstricta (Say, 1821)	T	sharp wedge
Triodopsis occidentalis (Pilsbry and Ferriss, 1894)	T	Arkansas wedge
Triodopsis palustris Hubricht, 1958	T	Santee threetooth
Triodopsis pendula Hubricht, 1952	T	Hanging Rock threetooth
Triodopsis picea Hubricht, 1958	T	Spruce Knob threetooth

SCIENTIFIC NAME	OCCURRENCE	COMMON NAME
Triodopsis platysayoides (Brooks, 1933)	T	Cheat threetooth
Triodopsis rugosa Brooks and McMillan, 1940	T	buttress threetooth
Triodopsis soelneri (J. B. Henderson, 1907)	T	Cape Fear threetooth
Triodopsis tridentata (Say, 1816)	T	northern threetooth
Triodopsis tennesseensis (Walker, 1902)	T	budded threetooth
Triodopsis vannostrandi (Bland, 1875)	T	coiled threetooth
Triodopsis vulgata Pilsbry, 1940	T	dished threetooth
Triodopsis vultuosa (Gould, 1848)	T	Texas threetooth
Trilobopsis loricata (Gould, 1846)	T	scaley chaparral
Trilobopsis penitens (Hanna and Rixford, 1923)	T	Mormon Island chaparral
Trilobopsis roperi (Pilsbry, 1889)	T	Shasta chaparral
Trilobopsis tehamana (Pilsbry, 1928)	T	Tehama chaparral
Trilobopsis trachypepla (S. S. Berry, 1933)	T	Bridge Creek chaparral
Vespericola armiger (Ancey, 1887)	T	Santa Cruz hesperian
Vespericola columbianus (I. Lea, 1838)	T	northwest hesperian
Vespericola haplus (S. S. Berry, 1933)	T	Butte Creek hesperian
Vespericola karakorum Talmadge, 1962	T	Karok hesperian
Vespericola megasoma (Pilsbry, 1928)	T	redwood hesperian
Vespericola pinicola (S. S. Berry, 1916)	T	Monterey hesperian
Vespericola shasta (S. S. Berry, 1921)	T	Shasta hesperian
Vespericola sierranus (S. S. Berry, 1921)	T	Siskiyou hesperian

Sagdidae

SCIENTIFIC NAME	OCCURRENCE	COMMON NAME
Hojeda inaguensis (Weinland, 1880)	T	Keys mudcloak
Lacteoluna selenina (Gould, 1848)	T	moonlight mudcloak

Thysanophoridae

SCIENTIFIC NAME	OCCURRENCE	COMMON NAME
Microphysula cookei (Pilsbry, 1922)	T	Vancouver snail
Microphysula ingersolli (Bland, 1874)	T	spruce snail
Thysanophora horni (Gabb, 1866)	T	southwestern fringed-snail
Thysanophora plagioptycha (Shuttleworth, 1854)	T	lyrate fringed-snail

Camaenidae

SCIENTIFIC NAME	OCCURRENCE	COMMON NAME
Zachrysia auricoma (Férussac, 1821)	T [I]	golden zachrysia
Zachrysia provisoria (Pfeiffer, 1858)	T [I]	garden zachrysia

Ammonitellidae

SCIENTIFIC NAME	OCCURRENCE	COMMON NAME
Ammonitella yatesi (J. G. Cooper, 1868)	T	tight coin
Glyptostoma gabrielense Pilsbry, 1938	T	San Gabriel chestnut
Glyptostoma newberryanum (W. G. Binney, 1858)	T	San Diego chestnut
Megomphix californicus A. G. Smith, 1960	T	Natural Bridge megomphix
Megomphix hemphilli (W. G. Binney, 1879)	T	Oregon megomphix
Megomphix leutarius H. B. Baker, 1932	T	Umatilla megomphix
Polygyrella polygyrella (Bland and W. Cooper, 1861)	T	humped coin
Polygyroidea harfordiana (J. G. Cooper, 1870)	T	toothed coin

Oreohelicidae

SCIENTIFIC NAME	OCCURRENCE	COMMON NAME
Oreohelix alpina (Elrod, 1901)	T	alpine mountainsnail

SCIENTIFIC NAME	OCCURRENCE	COMMON NAME
Oreohelix amariradix Pilsbry, 1902	T	Bitter Root mountainsnail
Oreohelix anchana Gregg, 1953	T	Ancha mountainsnail
Oreohelix barbata Pilsbry, 1905	T	bearded mountainsnail
Oreohelix californica S. S. Berry, 1931	T	Clark mountainsnail
Oreohelix carinifera Pilsbry, 1912	T	keeled mountainsnail
Oreohelix concentrata (Dall, 1890)	T	Huachuca mountainsnail
Oreohelix confragosa Metcalf, 1974	T	Pinos Altos mountainsnail
Oreohelix elrodi (Pilsbry, 1900)	T	carinate mountainsnail
Oreohelix eurekensis J. Henderson and Daniels, 1916	T	Eureka mountainsnail
Oreohelix florida Pilsbry, 1939	T	Florida mountainsnail
Oreohelix grahamensis Gregg and W. B. Miller, 1974	T	Pinaleno mountainsnail
Oreohelix handi Pilsbry and Ferriss, 1918	T	Spring mountainsnail
Oreohelix haydeni (Gabb, 1869)	T	lyrate mountainsnail
Oreohelix hemphilli (Newcomb, 1869)	T	Whitepine mountainsnail
Oreohelix hendersoni Pilsbry, 1912	T	pallid mountainsnail
Oreohelix houghi W. B. Marshall, 1929	T	Diablo mountainsnail
Oreohelix howardi Jones, 1944	T	Mill Creek mountainsnail
Oreohelix idahoensis (Newcomb, 1866)	T	costate mountainsnail
Oreohelix intersum (Hemphill, 1890)	T	Deep Slide mountainsnail
Oreohelix jaegeri S. S. Berry, 1931	T	Kyle Canyon mountainsnail
Oreohelix jugalis (Hemphill, 1890)	T	Boulder Pile mountainsnail
Oreohelix junii Pilsbry, 1934	T	Grand Coulee mountainsnail
Oreohelix littoralis Crews and Metcalf, 1982	T	San Augustin mountainsnail
Oreohelix magdalenae Pilsbry, 1939	T	Magdalena mountainsnail
Oreohelix metcalfei Cockerell, 1905	T	Black Range mountainsnail
Oreohelix nevadensis S. S. Berry, 1932	T	Schell Creek mountainsnail
Oreohelix peripherica (Ancey, 1881)	T	Deseret mountainsnail
Oreohelix pilsbryi Ferriss, 1917	T	Mineral Creek mountainsnail
Oreohelix pygmaea Pilsbry, 1913	T	pygmy mountainsnail
Oreohelix socorroensis Pilsbry, 1905	T	Socorro mountainsnail
Oreohelix strigosa (Gould, 1846)	T	Rocky mountainsnail
Oreohelix subrudis (Reeve, 1854)	T	subalpine mountainsnail
Oreohelix swopei Pilsbry and Ferriss, 1917	T	Morgan Creek mountainsnail
Oreohelix tenuistriata J. Henderson and Daniels, 1916	T	thin-ribbed mountainsnail
Oreohelix uinta Brooks, 1939	T	Uinta mountainsnail
Oreohelix vortex S. S. Berry, 1932	T	whorled mountainsnail
Oreohelix waltoni Solem, 1975	T	lava rock mountainsnail
Oreohelix yavapai Pilsbry, 1905	T	Yavapai mountainsnail
Radiocentrum avalonense (Hemphill, 1905)	T	Catalina mountainsnail
Radiocentrum chiricahuanum (Pilsbry, 1905)	T	Chiricahuana mountainsnail
Radiocentrum clappi (Ferriss, 1904)	T	Cave Creek mountainsnail
Radiocentrum ferrissi (Pilsbry, 1915)	T	fringed mountainsnail
Radiocentrum hachetanum (Pilsbry, 1915)	T	Hacheta mountainsnail
Udosarx lyrata Webb, 1959	T	lyre mantleslug

Bradybaenidae

SCIENTIFIC NAME	OCCURRENCE	COMMON NAME
Bradybaena similaris (Férussac, 1821)	T [I]	Asian trampsnail

SCIENTIFIC NAME	OCCURRENCE	COMMON NAME
Helminthoglyptidae		
Eremarionta aquaealbae S. S. Berry, 1922	T	Whitewater desertsnail
Eremarionta brunnea Willett, 1935	T	Chuckwalla Spring desertsnail
Eremarionta greggi W. B. Miller, 1981	T	Panamint desertsnail
Eremarionta immaculata Willett, 1937	T	white desertsnail
Eremarionta indioensis (Yates, 1890)	T	Coachella desertsnail
Eremarionta millepalmarum S. S. Berry, 1930	T	Thousand Palms desertsnail
Eremarionta morongoana S. S. Berry, 1929	T	Morongo desertsnail
Eremarionta newcombi Pilsbry and Ferriss, 1923	T	*classification uncertain*
Eremarionta orocopia Willett, 1939	T	Orocopia desertsnail
Eremarionta rowelli (Newcomb, 1865)	T	eastern desertsnail
Eremariontoides argus (Edson, 1912)	T	Argus desertsnail
Gliabates oregonius Webb, 1959	T	salamander slug
Helminthoglypta allyniana (S. S. Berry, 1920)	T	Sierra shoulderband
Helminthoglypta allynsmithi Pilsbry, 1939	T	Merced Canyon shoulderband
Helminthoglypta arrosa (W. G. Binney, 1858)	T	bronze shoulderband
Helminthoglypta ayresiana Newcomb, 1861	T	San Miguel shoulderband
Helminthoglypta benitoensis Lowe, 1930	T	Pinnacles shoulderband
Helminthoglypta berryi Hanna, 1929	T	Tehachapi shoulderband
Helminthoglypta californiensis (I. Lea, 1838)	T	Point Pinos shoulderband
Helminthoglypta callistoderma (Pilsbry and Ferriss, 1918)	T	Kern shoulderband
Helminthoglypta carpenteri Newcomb, 1861	T	San Juaquin shoulderband
Helminthoglypta caruthersi Willett, 1934	T	Morris Canyon shoulderband
Helminthoglypta contracosta (Pilsbry, 1895)	T	Contracosta shoulderband
Helminthoglypta crotalina S. S. Berry, 1928	T	sidewinder shoulderband
Helminthoglypta cuyama Hanna and A. G. Smith, 1937	T	Cuyama shoulderband
Helminthoglypta cuyamacensis (Bartsch, 1895)	T	Cuyamaca shoulderband
Helminthoglypta cypreophila (W. G. Binney and Bland, 1869)	T	foothill shoulderband
Helminthoglpta diabloensis (J. G. Cooper, 1868)	T	silky shoulderband
Helminthoglypta dupetithoursi (Deshayes, 1840)	T	cypress shoulderband
Helminthoglypta edwardsi Gregg and W. B. Miller, 1976	T	Pine Valley shoulderband
Helminthoglypta euomphalodes S. S. Berry, 1938	T	Greenhorn shoulderband
Helminthoglypta exarata (Pfeiffer, 1857)	T	Santa Cruz shoulderband
Helminthoglypta expansilabris (Pilsbry, 1898)	T	Mendocino shoulderband
Helminthoglypta ferrissi Pilsbry, 1924	T	Kings shoulderband
Helminthoglypta fieldi Pilsbry, 1930	T	surf shoulderband
Helminthoglypta fisheri Bartsch, 1904	T	Panamint shoulderband
Helminthoglypta fontiphila Gregg, 1931	T	Soledad shoulderband
Helminthoglypta graniticola S. S. Berry, 1926	T	granite shoulderband
Helminthoglypta greggi Willett, 1931	T	Mohave shoulderband
Helminthoglypta hertleini Hanna and A. G. Smith, 1937	T	Oregon shoulderband
Helminthoglypta inglesi S. S. Berry, 1938	T	Horse Meadows shoulderband
Helminthoglypta isabella S. S. Berry, 1938	T	yucca shoulderband
Helminthoglypta jaegeri S. S. Berry, 1928	T	Sweetwater shoulderband
Helminthoglypta liodoma S. S. Berry, 1938	T	Cottonwood shoulderband
Helminthoglypta micrometalleoides W. B. Miller, 1970	T	mimic shoulderband

SCIENTIFIC NAME	OCCURRENCE	COMMON NAME
Helminthoglypta mohaveana S. S. Berry, 1927	T	Victorville shoulderband
Helminthoglypta napaea S. S. Berry, 1938	T	bigtree shoulderband
Helminthoglypta nickliniana (I. Lea, 1838)	T	Coast Range shoulderband
Helminthoglypta orina S. S. Berry, 1938	T	Breckenridge shoulderband
Helminthoglypta petricola (S. S. Berry, 1916)	T	Transverse Range shoulderband
Helminthoglypta proles (Hemphill, 1892)	T	Yosemite shoulderband
Helminthoglypta reediana Willett, 1932	T	Lowe Canyon shoulderband
Helminthoglypta sequoicola (J. G. Cooper, 1866)	T	redwood shoulderband
Helminthoglypta similans Hanna and A. G. Smith, 1937	T	Coalinga shoulderband
Helminthoglypta sonoma Pilsbry, 1937	T	Sonoma shoulderband
Helminthoglypta stageri Willett, 1938	T	Paiute shoulderband
Helminthoglypta stiversiana (J. G. Cooper, 1875)	T	Point Reyes shoulderband
Helminthoglypta thermimontis S. S. Berry, 1953	T	coyote shoulderband
Helminthoglypta traski (Newcomb, 1861)	T	Peninsular Range shoulderband
Helminthoglypta tudiculata (A. Binney, 1843)	T	southern shoulderband
Helminthoglypta tularensis (Hemphill, 1892)	T	Tulare shoulderband
Helminthoglypta umbilicata (Pilsbry, 1897)	T	Big Sur shoulderband
Helminthoglypta walkeriana (Hemphill, 1911)	T	Morro shoulderband
Helminthoglypta waltoni Gregg and W. B. Miller, 1976	T	Laguna shoulderband
Hemitrochus varians (Menke, 1829)	T	seagrape snail
Humboldtiana agavophila Pratt, 1971	T	agave threeband
Humboldtiana cheatumi Pilsbry, 1935	T	Davis Mountain threeband
Humboldtiana chisosensis Pilsbry, 1927	T	Chisos threeband
Humboldtiana edithae Parodíz, 1954	T	boulder slide threeband
Humboldtiana ferrissiana Pilsbry, 1928	T	Mitre Peak threeband
Humboldtiana fullingtoni Cheatham, 1972	T	Capote threeband
Humboldtiana hoegiana praesidii Pilsbry, 1939	T	San Carlos threeband
Humboldtiana palmeri Clench, 1930	T	Mt. Livermore threeband
Humboldtiana texana Pilsbry, 1891	T	Stockton Plateau threeband
Humboldtiana ultima Pilsbry, 1927	T	northern threeband
Micrarionta facta (Newcomb, 1864)	T	Santa Barbara islandsnail
Micrarionta feralis (Hemphill, 1901)	T	San Nicolas islandsnail
Micrarionta gabbi (Newcomb, 1864)	T	San Clemente islandsnail
Micrarionta opuntia Roth, 1975	T	pricklypear islandsnail
Micrarionta rufocincta (Newcomb, 1864)	T	Santa Catalina islandsnail
Mohavelix micrometaleus (S. S. Berry, 1930)	T	El Paso shoulderband
Monadenia callipeplus S. S. Berry, 1940	T	classification uncertain
Monadenia churchi Hanna and A. G. Smith, 1933	T	Klamath sideband
Monadenia circumcarinata (Stearns, 1879)	T	keeled sideband
Monadenia cristulata S. S. Berry, 1940	T	classification uncertain
Monadenia fidelis (J. E. Gray, 1834)	T	Pacific sideband
Monadenia hillebrandi (Newcomb, 1864)	T	Mariposa sideband
Monadenia infumata (Gould, 1855)	T	redwood sideband
Monadenia marmoratis S. S. Berry, 1940	T	classification uncertain
Monadenia mormonum (Pfeiffer, 1857)	T	Sierra sideband
Monadenia rotifera S. S. Berry, 1940	T	classification uncertain
Monadenia setosa Talmadge, 1952	T	Trinity bristlesnail
Monadenia troglodytes Hanna and A. G. Smith, 1933	T	Shasta sideband
Monadenia tuolumnea S. S. Berry, 1955	T	Tuolumne sideband
Plesarionta stearnsiana (Gabb, 1867)	T	speckled cactussnail

SCIENTIFIC NAME	OCCURRENCE	COMMON NAME
Sonorelix angelus (Gregg, 1949)	T	Soledad desertsnail
Sonorelix avawatzica (S. S. Berry, 1930)	T	Avawatz desertsnail
Sonorelix baileyi (Bartsch, 1904)	T	Resting Spring desertsnail
Sonorelix borregoensis (S. S. Berry, 1929)	T	Borrego desertsnail
Sonorelix harperi (Bryant, 1900)	T	Mountain Spring desertsnail
Sonorelix melanophylon (S. S. Berry, 1930)	T	Black Rock desertsnail
Sonorelix rixfordi (Pilsbry, 1919)	T	Joshua Tree desertsnail
Sonorella allynsmithi Gregg and W. B. Miller, 1969	T	Squaw Park talussnail
Sonorella ambigua Pilsbry and Ferriss, 1915	T	Papago talussnail
Sonorella anchana S. S. Berry, 1948	T	Sierra Ancha talussnail
Sonorella animasensis Pilsbry, 1939	T	Animas talussnail
Sonorella apache Pilsbry and Ferriss, 1915	T	Apache talussnail
Sonorella ashmuni Bartsch, 1904	T	Richinbar talussnail
Sonorella baboquivariensis Pilsbry and Ferriss, 1915	T	Baboquivari talussnail
Sonorella bagnarai W. B. Miller, 1967	T	Rincon talussnail
Sonorella bartschi Pilsbry and Ferriss, 1915	T	Escabrosa talussnail
Sonorella bequaerti W. B. Miller, 1976	T	Happy Valley talussnail
Sonorella bicipitis Pilsbry and Ferriss, 1910	T	Dos Cabezas talussnail
Sonorella binneyi Pilsbry and Ferriss, 1910	T	Horseshoe Canyon talussnail
Sonorella bowiensis Pilsbry, 1905	T	Quartzite Hill talussnail
Sonorella caerulifluminis Pilsbry and Ferriss, 1919	T	Blue talussnail
Sonorella christenseni Fairbanks and Reeder, 1980	T	Clark Peak talussnail
Sonorella clappi Pilsbry and Ferriss, 1915	T	Madera talussnail
Sonorella coloradoensis (Stearns, 1890)	T	Grand Canyon talussnail
Sonorella coltoniana Pilsbry, 1939	T	Walnut Canyon talussnail
Sonorella compar Pilsbry, 1919	T	Oak Creek talussnail
Sonorella dalli Bartsch, 1904	T	Garden Canyon talussnail
Sonorella danielsi Pilsbry and Ferriss, 1910	T	Bear Canyon talussnail
Sonorella delicata Pilsbry, 1919	T	Tollhouse Canyon talussnail
Sonorella dragoonensis Pilsbry and Ferriss, 1915	T	Stronghold Canyon talussnail
Sonorella eremita Pilsbry and Ferriss, 1915	T	San Xavier talussnail
Sonorella ferrissi Pilsbry, 1915	T	Dragoon talussnail
Sonorella franciscana Pilsbry and Ferriss, 1919	T	St. Francis talussnail
Sonorella galiurensis Pilsbry and Ferriss, 1919	T	Galiuro talussnail
Sonorella grahamensis Pilsbry and Ferriss, 1919	T	Pinaleno talussnail
Sonorella granulatissima Pilsbry, 1905	T	Ramsey Canyon talussnail
Sonorella hachitana (Dall, 1896)	T	New Mexico talussnail
Sonorella huachucana Pilsbry, 1905	T	Huachuca talussnail
Sonorella imitator Gregg and W. B. Miller, 1972	T	mimic talussnail
Sonorella imperatrix Pilsbry, 1939	T	Total Wreck talussnail
Sonorella imperialis Pilsbry and Ferriss, 1923	T	Empire Mountain talussnail
Sonorella insignis Pilsbry and Ferriss, 1919	T	Whetstone talussnail
Sonorella macrophallus Fairbanks and Reeder, 1980	T	Wet Canyon talussnail
Sonorella magdalenensis (Stearns, 1890)	T	Sonoran talussnail
Sonorella meadi W. B. Miller, 1966	T	Agua Dulce talussnail
Sonorella metcalfi W. B. Miller, 1976	T	Franklin Mountain talussnail
Sonorella micra Pilsbry and Ferriss, 1910	T	pygmy talussnail
Sonorella micromphala Pilsbry, 1939	T	Mik Ranch talussnail

SCIENTIFIC NAME	OCCURRENCE	COMMON NAME
Sonorella milleri Christensen and Reeder, 1981	T	Table Top talussnail
Sonorella mustang Pilsbry and Ferriss, 1919	T	Mustang talussnail
Sonorella neglecta Gregg, 1951	T	Portal talussnail
Sonorella odorata Pilsbry and Ferriss, 1919	T	pungent talussnail
Sonorella optata Pilsbry and Ferris, 1910	T	Big Emigrant talussnail
Sonorella orientis Pilsbry, 1936	T	Organ Mountain talussnail
Sonorella papagorum Pilsbry and Ferriss, 1915	T	Black Mountain talussnail
Sonorella parva Pilsbry, 1905	T	little talussnail
Sonorella rinconensis Pilsbry and Ferriss, 1910	T	Posta Quemada talussnail
Sonorella rooseveltiana S. S. Berry, 1917	T	Roosevelt talussnail
Sonorella rosemontensis Pilsbry, 1939	T	Rosemont talussnail
Sonorella sabinoensis Pilsbry and Ferriss, 1919	T	Santa Catalina talussnail
Sonorella santaritana Pilsbry and Ferriss, 1910	T	Agua Caliente talussnail
Sonorella simmonsi W. B. Miller, 1966	T	Picacho talussnail
Sonorella sitiens Pilsbry and Ferriss, 1915	T	Las Guijas talussnail
Sonorella superstitionis Pilsbry, 1939	T	Superstition Mountain talussnail
Sonorella todseni W. B. Miller, 1976	T	Dona Ana talussnail
Sonorella tortillita Pilsbry and Ferriss, 1919	T	Tortolita talussnail
Sonorella tryoniana Pilsbry and Ferriss, 1923	T	Sanford talussnail
Sonorella vespertina Pilsbry and Ferriss, 1915	T	evening talussnail
Sonorella virilis Pilsbry, 1905	T	Chiricahua talussnail
Sonorella walkeri Pilsbry and Ferriss, 1915	T	Santa Rita talussnail
Sonorella waltoni W. B. Miller, 1968	T	Doubtful Canyon talussnail
Sonorella xanthenes Pilsbry and Ferriss, 1923	T	Kitt Peak talussnail
Xerarionta intercisa (W. G. Binney, 1851)	T	plain cactussnail
Xerarionta kelletti (Forbes, 1850)	T	Catalina cactussnail
Xerarionta redimita (W. G. Binney, 1857)	T	wreathed cactussnail
Xerarionta tryoni (Newcomb, 1864)	T	bicolor cactussnail

Helicellidae

SCIENTIFIC NAME	OCCURRENCE	COMMON NAME
Candidula intersecta (Poiret, 1801)	T [I]	wrinkled helicellid
Cochlicella barbara (Linnaeus, 1758)	T [I]	potbellied helicellid
Helicella obvia (Menke, 1828)	T [I]	heath helicellid
Trichia hispida (Linnaeus, 1758)	T [I]	hairy helicellid
Trichia striolata (Pfeiffer, 1828)	T [I]	furrowed helicellid
Trochoidea elegans (Gmelin, 1791)	T [I]	elegant helicellid

Helicidae

SCIENTIFIC NAME	OCCURRENCE	COMMON NAME
Helix aperta Born, 1778	T [I]	green gardensnail
Helix aspersa Müller, 1774	T [I]	brown gardensnail
Helix pomatia Linnaeus, 1758	T [I]	escargot
Cepaea hortensis (Müller, 1774)	T	white-lip gardensnail
Cepaea nemoralis (Linnaeus, 1798)	T [I]	grovesnail
Eobania vermiculata (Müller, 1774)	T [I]	chocolate-band snail
Monacha cantiana (Montagu, 1803)	T [I]	Kentish gardensnail
Otala lactea Müller, 1774	T [I]	milk snail
Theba pisana (Müller, 1774)	T [I]	white gardensnail

SCIENTIFIC NAME	OCCURRENCE	COMMON NAME
ORDER SYSTELLOMMATOPHORA		
Onchidiidae		
Onchidella borealis Dall, 1871...............	P	...........northwest onchidella
Onchidella carpenteri (W. G. Binney, 1860).......	P	
Onchidella floridana (Dall, 1885)..............	A	.............Florida onchidella
Veronicellidae		
Angustipes ameghini (Gambetta, 1923)...........	T [I]	.........black-velvet leatherleaf
Diplosolenodes occidentalis (Guilding, 1825)......	T [I]	...............spotted leatherleaf
Laevicaulus alte (Férussac, 1821)...............	T [I]	.... tropical leatherleaf
Leidyula floridana (Leidy, 1851)..............	T	.............Florida leatherleaf
Leidyula kraussi (Férussac, 1823)..............	T [I]	.... dappled leatherleaf
Leidyula moreleti (Crosse and P. Fischer, 1872) ...	T [I]	.... tan leatherleaf
Leidyula sloani (Cuvier, 1816).................	T [I]	.... Sloan leatherleaf
Sarasomia plebeia (P. Fischer, 1898)	T [I]	.... Caribbean leatherleaf

SCIENTIFIC NAME	OCCURRENCE	COMMON NAME

CLASS POLYPLACOPHORA—CHITONS

Order Neoloricata

Lepidopleuridae

Hanleyella oldroydi (Dall, 1919)	P	
Leptochiton alveolus (M. Sars in Lovén, 1846)	A, P, Ac	
Leptochiton nexus Carpenter, 1864	P	
Leptochiton rugatus (Pilsbry, 1892)	P	
Oldroydia percrassa (Dall, 1894)	P	

Hanleyidae

Hanleya hanleyi (Bean, 1844)	A	eastern hanleya

Ischnochitonidae

Callistochiton asthenes (S. S. Berry, 1919)	P	
Callistochiton crassicostatus Pilsbry, 1892	P	
Callistochiton decoratus Pilsbry, 1892	P	
Callistochiton palmulatus Carpenter in Dall, 1879	P	
Callistochiton portobelensis Ferreira, 1976	A	Portobelo chiton
Callistochiton shuttleworthianus Pilsbry, 1892	A	eastern orange chiton
Dendrochiton gothicus (Carpenter, 1863)	P	
Dendrochiton psaltes S. S. Berry, 1963	P	
Dendrochiton semiliratus S. S. Berry, 1927	P	
Dendrochiton thamnoporus S. S. Berry, 1911	P	
Ischnochiton allyni Ferreira, 1977	P	
Ischnochiton boogi Haddon, 1886	A	Atlantic rose chiton
Ischnochiton dilatosculptus Kaas, 1982	A	
Ischnochiton erythronotus (C. B. Adams, 1845)	A	multihue chiton
Ischnochiton hartmeyeri Thiele, 1910	A	multiring chiton
Ischnochiton interstinctus (Gould, 1846)	P	
Ischnochiton lividus Middendorff, 1847	P	
Ischnochiton newcombi Pilsbry, 1892	P	
Ischnochiton pseudovirgatus Kaas, 1972	A	blue-spot chiton
Ischnochiton regularis (Carpenter, 1855)	P	
Ischnochiton scrobiculatus Middendorff, 1847	P	
Ischnochiton striolatus (J. E. Gray, 1828)	A	
Ischnochiton trifidus Carpenter, 1864	P	
Lepidochitona aleutica (Dall, 1879)	P	
Lepidochitona beanii (Carpenter, 1864)	P	
Lepidochitona dentiens cryptica Kues, 1974	P	
Lepidochitona dentiens dentiens (Gould, 1846)	P	
Lepidochitona flectens (Carpenter, 1864)	P	
Lepidochitona hartwegii (Carpenter, 1855)	P	
Lepidochitona heathii (Pilsbry, 1898)	P	
Lepidochitona keepiana (S. S. Berry, 1948)	P	
Lepidochitona liozonis (Dall and Simpson, 1901)	A	Caribbean red chiton
Lepidochitona lobium (S. S. Berry, 1925)	P	

SCIENTIFIC NAME	OCCURRENCE	COMMON NAME
Lepidochitona lowei fackenthallae (S. S Berry, 1919)	P	
Lepidochitona lowei lowei (Pilsbry, 1918)	P	
Lepidochitona sharpii (Pilsbry, 1896)	P	
Lepidozona cooperi (Dall, 1879)	P	
Lepidozona mertensii (Middendorff, 1847)	P	
Lepidozona pectinulata (Carpenter in Pilsbry, 1893)	P	
Lepidozona retiporosa (Carpenter, 1864)	P	
Lepidozona scabricostata (Carpenter, 1864)	P	
Lepidozona serrata (Carpenter, 1864)	P	
Lepidozona sinudentata (Carpenter in Pilsbry, 1892)	P	
Lepidozona willetti (S. S. Berry, 1917)	P	
Nuttallina californica (Reeve, 1847)	P	
Nuttallina thomasi Pilsbry, 1898	P	
Schizoplax brandtii (Middendorff, 1846)	P	
Schizoplax multicolor Dall, 1920	P	
Stenoplax conspicua (Dall, 1879)	P	
Stenoplax corrugata (Pilsbry, 1892)	P	
Stenoplax fallax (Pilsbry, 1892)	P	
Stenoplax floridana (Pilsbry, 1892)	A	
Stenoplax heathiana S. S. Berry, 1946	P	
Stenoplax producta (Reeve, 1847)	A	
Stenosemus albus (Linnaeus, 1767)	A, P	northern white chiton
Stenosemus exaratus (G. O. Sars, 1878)	A	
Tonicella beringensis Yakovleva, 1952	P	
Tonicella blaneyi Dall, 1905	A	
Tonicella granulata Yakovleva, 1952	P	
Tonicella insignis (Reeve, 1847)	P	
Tonicella lineata (W. Wood, 1815)	P	
Tonicella marmorea (Fabricius, 1780)	A, P	
Tonicella rubra (Linnaeus, 1767)	A, P, Ac	northern red chiton
Tonicella saccharina Dall, 1879	P	
Tonicella sitkensis (Middendorff, 1846)	P	
Tonicella submarmorea (Middendorff, 1846)	P	
Tonicella undocarulea Sirenko, 1973	P	

Chaetopleuridae

SCIENTIFIC NAME	OCCURRENCE	COMMON NAME
Chaetopleura apiculata (Say, 1834)	A	eastern beaded chiton
Chaetopleura gemma Carpenter in Dall, 1879	P	
Chaetopleura janeirensis (J. E. Gray, 1828)	A	West Indian ribbed chiton
Chaetopleura lanuginosa (Carpenter in Dall, 1879)	P	
Chaetopleura scabricula (Sowerby, 1832)	P	
Chaetopleura staphylophera Lyons, 1985	a	

Mopaliidae

SCIENTIFIC NAME	OCCURRENCE	COMMON NAME
Amicula amiculata (Pallas, 1786)	P	
Amicula pallasii (Middendorff, 1846)	P	
Amicula vestita (Broderip and Sowerby, 1829)	P	
Ceratozona squalida (C.B. Adams, 1845)	A	eastern surf chiton
Mopalia acuta (Carpenter, 1855)	P	

SCIENTIFIC NAME	OCCURRENCE	COMMON NAME
Mopalia ciliata (Sowerby, 1840)	P	
Mopalia cirrata S. S. Berry, 1919	P	
Mopalia cithara S. S. Berry, 1951	P	
Mopalia egretta S. S. Berry, 1919	P	
Mopalia hindsi (Reeve, 1847)	P	
Mopalia imporcata Carpenter, 1864	P	
Mopalia laevior Pilsbry, 1918	P	
Mopalia lignosa (Gould, 1846)	P	
Mopalia lionotus Pilsbry, 1918	P	
Mopalia lowei Pilsbry, 1918	P	
Mopalia muscosa (Gould, 1846)	P	
Mopalia phorminx S. S. Berry, 1919	P	
Mopalia porifera Pilsbry, 1892	P	
Mopalia recurvans Barnawell, 1960	P	
Mopalia sinuata Carpenter, 1864	P	
Mopalia spectabilis G. I. M. Cowan and I. M. Cowan, 1977	P	
Mopalia swanii (Carpenter, 1864)	P	
Placiphorella borealis Pilsbry, 1892	P	boreal veiled chiton
Placiphorella pacifica S. S. Berry, 1919	P	Pacific veiled chiton
Placiphorella rufa S. S. Berry, 1917	P	red veiled chiton
Placiphorella stimpsoni (Gould, 1859)	P	
Placiphorella velata Dall, 1879	P	veiled chiton

Katharinidae

SCIENTIFIC NAME	OCCURRENCE	COMMON NAME
Katharina tunicata (W. Wood, 1815)	P	black katy

Chitonidae

SCIENTIFIC NAME	OCCURRENCE	COMMON NAME
Acanthopleura granulata (Gmelin, 1791)	A	West Indian fuzzy chiton
Chiton tuberculatus Linnaeus, 1758	A	West Indian green chiton
Tonicia pustulifera Dall, 1919	P	
Tonicia schrammi (Shuttleworth, 1856)	A	gold-flecked chiton

Acanthochitonidae

SCIENTIFIC NAME	OCCURRENCE	COMMON NAME
Acanthochitona andersoni Watters, 1981	A	smooth glass-hair chiton
Acanthochitona angelica Dall, 1919	P	
Acanthochitona avicula (Carpenter, 1864)	P	
Acanthochitona balesae Abbott, 1954	A	slender glass-hair chiton
Acanthochitona hemphilli (Pilsbry, 1893)	A	red glass-hair chiton
Acanthochitona imperatrix Watters, 1981	P	
Acanthochitona pygmaea (Pilsbry, 1893)	A	striate glass-hair chiton
Cryptochiton stelleri (Middendorff, 1847)	P	giant Pacific chiton
Cryptoconchus floridanus (Dall, 1889)	A	white-barred chiton

SCIENTIFIC NAME	OCCURRENCE	COMMON NAME

CLASS CEPHALOPODA—SQUIDS AND OCTOPUSES

ORDER SEPIOIDEA

Spirulidae

Spirula spirula (Linnaeus, 1758)	A, P	ram's horn squid

Sepiolidae

Rossia antillensis Voss, 1955	A	Antilles bobtail
Rossia bullisi Voss, 1956	A	Gulf bobtail
Rossia pacifica S. S. Berry, 1911	A	eastern Pacific bobtail
Rossia tortugaensis Voss, 1956	A	Tortugas bobtail
Semirossia equalis (Voss, 1950)	A	greater shining bobtail
Semirossia tenera (A. E. Verrill, 1880)	A	lesser shining bobtail

ORDER TEUTHOIDEA

Loliginidae

Loligo ocula Cohen, 1976	A	bigeye squid
Loligo opalescens S. S. Berry, 1911	P	California market squid
Loligo pealeii Lesueur, 1821	A	longfin squid
Loligo pleii de Blainville, 1823	A	arrow squid
Loligo roperi Cohen, 1976	A	Roper squid
Loliolopsis diomedeae (Hoyle, 1904)	P	dart squid
Lolliguncula brevis (de Blainville, 1823)	A	Atlantic brief squid
Lolliguncula panamensis S. S. Berry, 1911	P	Panama brief squid
Sepioteuthis sepioidea (de Blainville, 1823)	A	Caribbean reef squid

Gonatidae

Berryteuthis magister (S. S. Berry, 1913)	P	magistrate armhook squid
Gonatus berryi Naef, 1923	P	Berry armhook squid
Gonatus fabrici (Lichtenstein, 1818)	A, P	boreal armhook squid
Gonatus onyx Young, 1972	P	clawed armhook squid
Gonatus oregonensis Jefferts, 1985	P	Oregon armhook squid
Gonatus ursabrunae Jefferts, 1985	P	Brown Bear armhook squid

Onychoteuthididae

Moroteuthis robusta (A. E. Verrill, 1876)	P	robust clubhook squid
Onychoteuthis banksi (Leach, 1817)	A, P	common clubhook squid
Onychoteuthis borealijaponica Okada, 1927	P	boreal clubhook squid

Ommastrephidae

Dosidicus gigas (d'Orbigny, 1835)	P	jumbo squid

SCIENTIFIC NAME	OCCURRENCE	COMMON NAME
Illex coindeti (Verany, 1837)...............	A	southern shortfin squid
Illex illecebrosus (Lesueur, 1821).............	A	northern shortfin squid
Illex oxygonius C. F. E. Roper, Lu, and Mangold, 1969........	A	sharptail shortfin squid
Ommastrephes bartrami (Lesueur, 1821).........	A, P	flying squid
Ommastrephes caroli (Furtado, 1887)............	A	webbed squid
Ommastrephes pteropus (Steenstrup, 1855).......	A	orangeback squid
Ornithoteuthis antillarum Adam, 1957...........	A	Atlantic bird squid
Symplectoteuthis luminosa Sasaki, 1915..........	A, P	luminous flying squid
Symplectoteuthis oualaniensis (Lesson, 1830).....	P	purpleback squid

Thysanoteuthidae

Thysanoteuthis rhombus Troschel, 1857..........	A, P	diamondback squid

ORDER OCTOPODA

Octopodidae

Bathypolypus arcticus (Prosch, 1849)............	A	spoonarm octopus
Octopus bimaculatus A. E. Verrill, 1883; *O. bimaculoides* Pickford and McConnaughey, 1949[5].....................	P	California two-spot octopus
Octopus briareus Robson, 1929................	A	Caribbean reef octopus
Octopus burryi Voss, 1953	A	brownstripe octopus
Octopus californicus (S. S. Berry, 1911)	P	orange bigeye octopus
Octopus defilippi Verany, 1851................	A	longarm octopus
Octopus digueti Perrier and Rochebrune, 1894	P	Pacific pygmy octopus
Octopus dofleini (Wulker, 1910)	P	giant octopus
Octopus fitchi S. S. Berry, 1953	P	lilliput octopus
Octopus hummelincki Adam, 1936	A	Caribbean two-spot octopus
Octopus joubini Robson, 1929.................	A	Atlantic pygmy octopus
Octopus leioderma (S. S. Berry, 1911)...........	P	smoothskin octopus
Octopus macropus Risso, 1826.................	A, P	white-spotted octopus
Octopus rubescens S. S. Berry, 1953	P	red octopus
Octopus vulgaris Lamarck, 1798................	A, P	common octopus
Scaeurgus unicirrhus (d'Orbigny, 1840)...........	A	unicorn octopus

Argonautidae

Argonauta argo Linnaeus, 1758	A, P	greater argonaut
Argonauta hians Solander, 1786................	A	winged argonaut
Argonauta nodosus Solander, 1786..............	A	knobby argonaut

Tremoctopodidae

Tremoctopus violaceus delle Chiaje, 1830	A, P	blanket octopus

---

[5]These two species appear to be morphologically indistinguishable; they are separated only by differences in egg size and behavior.

APPENDIX 1

Endangered and Threatened Mollusks of North America

In the USA, the Endangered Species Act of 1973 (16 U.S. Code 1531 *et seq.*) authorized an official federal list of endangered and threatened wildlife. The list need not be geographically restricted. Responsibility for adding a species to the list, for changing a species' status within the list, or for removing a species from the list is divided between the Secretaries of Interior and Commerce. Mollusks that occur on land, in fresh water, or in estuaries are the responsibility of the Department of the Interior. Marine mollusks and harvested estuarine species are under the jurisdiction of the Department of Commerce.

At present, 40 taxa of mollusks are on the federal list of endangered and threatened wildlife. Of these 40 taxa, 36 occur in North America north of Mexico. Two taxa occur in Mexico (Nicklin pearlymussel *Megalonaias nicklineana* and a subspecies of Tampico pearlymussel *Cyrtonaias tampicoensis tecomatensis*), one on the island of Manus, Papua New Guinea (Manus Island treesnail *Papustyla pulcherrima*), and one on the island of Ohau, Hawaii (tree snail genus *Achatinella*, all species of which are endangered).

Of the 36 North American mollusk taxa listed officially and on the following pages, 29 are freshwater bivalves and 7 are land snails.

The U.S. Fish and Wildlife Service has published a "Review of Invertebrate Wildlife for Listing as Endangered or Threatened Species" (*Federal Register* 49:21664–21675, 22 May 1984). This review includes 207 species of terrestrial and freshwater mollusks in the USA. The Service expects to publish a revised invertebrate review in 1988; copies will be available from the U.S. Fish and Wildlife Service, Office of Endangered Species, Washington D.C. 20240, USA.

SCIENTIFIC NAME	OCCURRENCE BY STATE	COMMON NAME

CLASS BIVALVIA—BIVALVE MOLLUSKS

Order Unionoida

Endangered Species (all freshwater)

Unionidae

Dromus dromas (I. Lea, 1834)	TN, VA	dromedary pearlymussel
Elliptio steinstansana R. I. Johnson and Clarke, 1983	NC	Tar spinymussel
Epioblasma florentina curtisi (Utterback, 1916)	MO	Curtis pearlymussel
Epioblasma florentina florentina (I. Lea, 1857)	AL, TN	yellow blossom
Epioblasma florentina walkeri (Wilson and H. W. Clark, 1914)	KY, TN, VA	tan riffleshell
Epioblasma obliquata perobliqua (Conrad, 1836) = *E. sulcata delicata* (Simpson, 1914)	IN, MI, OH	white catspaw
Epioblasma penita (Conrad, 1834)	AL, MS	southern combshell
Epioblasma torulosa gubernaculum (Reeve, 1865)	TN, VA	green blossom
Epioblasma torulosa torulosa (Rafinesque, 1820)	IL, IN, KY, TN, WV	tubercled blossom
Epioblasma turgidula (I. Lea, 1858)	AL, TN	turgid blossom
Fusconaia cor (Conrad, 1834)	AL, TN, VA	shiny pigtoe
Fusconaia cuneolus (I. Lea, 1840)	AL, TN, VA	fine-rayed pigtoe
Lampsilis abrupta (Say, 1831) = *L. orbiculata* (Hildreth, 1828)	AL, IL, IN, KY, MO, OH, PA, TN, WV	pink mucket
Lampsilis higginsi (I. Lea, 1857)	IL, IA, MN, MO, NE, WI	Higgins eye
Lampsilis virescens (I. Lea, 1858)	AL, TN	Alabama lampmussel
Lemiox rimosus (Rafinesque, 1820)	TN, VA	birdwing pearlymussel
Margartifera hembeli (Conrad, 1838)	AL	Louisiana pearlshell
Plethobasus cicatricosus (Say, 1829)	AL, IN, TN	white wartyback
Plethobasus cooperianus (I. Lea, 1834)	AL, IN, IA, KY, OH, PA, TN	orange-foot pimpleback
Pleurobema collina (Conrad, 1837)	VA, WV	James spinymussel
Pleurobema curtum (I. Lea, 1859)	AL, MS	black clubshell
Pleurobema marshalli Frierson, 1927	AL, MS	flat pigtoe
Pleurobema plenum (I. Lea, 1840)	KY, TN, VA	rough pigtoe
Pleurobema taitianum (I. Lea, 1834)	AL, MS	heavy pigtoe
Potamilus capax (Green, 1832)	AR, IN, MO, OH	fat pocketbook
Quadrula intermedia (Conrad, 1836)	AL, TN, VA	Cumberland monkeyface
Quadrula sparsa (I. Lea, 1841)	TN, VA	Appalachian monkeyface
Quadrula stapes (I. Lea, 1831)	AL, MS	stirrupshell
Toxolasma cylindrellus (I. Lea, 1868)	AL, TN	pale lilliput
Villosa trabalis (Conrad, 1834)	KY, TN	Cumberland bean

SCIENTIFIC NAME	OCCURRENCE BY STATE	COMMON NAME
CLASS GASTROPODA—GASTROPODS		
ORDER STYLOMMATOPHORA		
Endangered Species (all terrestrial)		
Helicodiscidae		
Polygyriscus virginicus P. R. Burch, 1947	VA	Virginia coil
Discidae		
Discus macclintocki (F. C. Baker, 1928)	IA	Pleistocene disc
Threatened Species (all terrestrial)		
Bulimulidae		
Orthalicus reses reses (Say, 1830)	FL	Stock Island treesnail
Discidae		
Anguispira picta (G. H. Clapp, 1920)	TN	painted disc
Succineidae		
Succinea chittenangoensis Pilsbry, 1908	NY	Appalachian ambersnail
Polygyridae		
Mesodon clarki nantahala Clench and Banks, 1932	NC	noonday globe
Triodopsis platysayoides (Brooks, 1933)	WV	Cheat threetooth

APPENDIX 2

Possibly Extinct Mollusks of the United States

There is a finality to the word "extinct," which implies the absolute end of a species or subspecies. For this reason, biologists exercise considerable caution in applying the label *extinct* to a particular taxon. In recent years, several mollusks have been written off, as "extinct," only to be rediscovered a few years later. Such species generally have been poorly known, and often live in habitats that are difficult to sample.

The mollusks listed below, all freshwater species, either have not been collected alive for a long time or their known habitat has been so drastically altered that their survival is unlikely.

SCIENTIFIC NAME	OCCURRENCE BY STATE	COMMON NAME
CLASS BIVALVIA—BIVALVE MOLLUSKS		
ORDER UNIONOIDA		
Unionidae		
Alasmidonta maccordi Athearn, 1964	AL	Coosa elktoe
Alasmidonta wrightiana (Walker, 1901)	FL	Ochlacknee arc-mussel
Epioblasma arcaeformis (I. Lea, 1831)	AL, KY, TN	sugarspoon
Epioblasma biemarginata (I. Lea, 1857)	AL, KY, TN	angled riffleshell
Epioblasma flexuosa (Rafinesque, 1820)	AL, IL, IN, KY, OH, TN	leafshell
Epioblasma haysiana (I. Lea, 1833)	AL, KY, TN, VA	acornshell
Epioblasma lenior (I. Lea, 1843)	AL, TN, VA	narrow catspaw
Epioblasma lewisii (Walker, 1910)	AL, KY, TN	forkshell
Epioblasma personata (Say, 1829)	AL, IL, IN, KY, OH, TN	round combshell
Epioblasma propinqua (I. Lea, 1857)	AL, IL, IN, KY, OH, TN	Tennessee riffleshell
Epioblasma sampsonii (I. Lea, 1861)	IL, IN, OH	Wabash riffleshell
Epioblasma stewardsoni (I. Lea, 1852)	AL, KY, TN	Cumberland leafshell
Medionidus macglameriae van der Schalie, 1939	AL	Tombigbee moccasinshell

SCIENTIFIC NAME	OCCURRENCE BY STATE	COMMON NAME

CLASS GASTROPODA—SNAILS

ORDER MESOGASTROPODA

Hydrobiidae

Clappia umbilicata (Walker, 1904)	AL	umbilicate pebblesnail

Pleuroceridae

Elimia clausa (I. Lea, 1861)	AL	closed elimia
Elimia fusiformis (I. Lea, 1861)	AL	fusiform elimia
Elimia hartmaniana (I. Lea, 1861)	AL	high-spired elimia
Elimia impressa (I. Lea, 1841)	AL	constricted elimia
Elimia jonesi (Goodrich, 1936)	AL	hearty elimia
Elimia laeta (Jay, 1839)	AL	ribbed elimia
Elimia pilsbryi (Goodrich, 1927)	AL	rough-lined elimia
Elimia pupaeformis (I. Lea, 1864)	AL	pupa elimia
Elimia pygmaea (H. H. Smith, 1936)	AL	pygmy elimia
Elimia varians (I. Lea, 1861)	AL	puzzle elimia
Gyrotoma excisa (I. Lea, 1843)	AL	excised slitshell
Gyrotoma lewisii (I. Lea, 1869)	AL	striate slitshell
Gyrotoma pagoda (I. Lea, 1845)	AL	pagoda slitshell
Gyrotoma pumila (I. Lea, 1860)	AL	ribbed slitshell
Gyrotoma pyramidata (Shuttleworth, 1845)	AL	pyramid slitshell
Gyrotoma walkeri (H. H. Smith, 1924)	AL	round slitshell
Leptoxis clipeata (H. H. Smith, 1922)	AL	agate rocksnail
Leptoxis formanii (I. Lea, 1843)	AL	interrupted rocksnail
Leptoxis ligata (Anthony, 1860)	AL	rotund rocksnail
Leptoxis lirata (H. H. Smith, 1922)	AL	lirate rocksnail
Leptoxis occultata (H. H. Smith, 1922)	AL	bigmouth rocksnail
Leptoxis showalterii (I. Lea, 1860)	AL	Coosa rocksnail
Leptoxis vittata (I. Lea, 1860)	AL	striped rocksnail

INDEX

A

abaconis, Tivela	45
abalone	
black	52
flat	52
green	52
pink	52
pinto	52
red	52
threaded	52
white	52
abarbareus, Antiplanes	95
abbotti	
Aperiovula	80
Opalia	76
abbreviata, Gastrocopta	126
aberrans	
Coralliophila	86
Pulsellum	51
Stosicia	68
aberti, Cyprogenia	29
abietum, Radiodiscus	130
Abra	42
abra	
Atlantic	42
long	43
smooth	43
abronia, Cuthona	120
abrupta	
Lampsilis	30, 154
Panopea	46
abyssorum, Buccinum	87
Acanthina	84
Acanthochitona	150
Acanthochitonidae	150
acanthodes, Aequipecten	26
Acanthodoris	115
Acanthopleura	150
Acanthotrophon	84
accincta, Kurtziella	98
Acella	122
acerra, Ventridens	136
Achatina	129
Achatinidae	129
acicula	
Cecilioides	128
Creseis	110
Turritellopsis	72
acicularis, Cymbovula	80
Acila	21
acinula, Neilonela	21
Acirsa	75
Aclididae	76
Aclis	76
aclis	
angular	76
Carolina	76
Pacific	76
striate	76
Tampa	77
thin	76
turreted	76
wide	76
Aclistothyra	37
Acmaea	53
Acmaeidae	53
acolasta, Macoma	41
acornshell	30, 156
southern	30
acova, Inodrillia	97
acra, Turbonilla	104
acrior, Terebra	95
Acroloxidae	122
Acroloxus	122
acropora, Turritella	71
Acteocina	107
Acteon	107
Acteonidae	107
Actinocyclidae	117
Actinonaias	28
aculea, Mangelia	98
aculeata, Crepidula	79
aculeus, Onoba	67
acuminata, Proneomenia	20
acurugata, Drillia	97
acuta	
Barleeia	68
Elimia	63
Mopalia	149
Nuculana	21
Physella	123
Pleurocera	65
acutelirata, Alvania	67
acuticostata	
Lirularia	55
Verticordia	49
acuticostatus, Parviturbo	57
acutifilosa, Juga	65
acutissimus, Medionidus	31
acutus, Nassarius	91
Adalaria	115

adamsi
- *Arcopsis* 24
- *Finella* 73
- *Pisidium* 34
- *Pyramidella* 103
- *Seila* 74
- *Thracia* 48

adamsianus, Brachidontes 23
adamsii, Macromphalina 78
adansoni, Lasaea 37
adansonianus, Perotrochus 52
Addisonia 54
Addisoniidae 54

adelae
- *Calliostoma* 54
- *Olivella* 92

Admete 94
admete
- Aleutian 94
- California 94
- corded 94
- graceful 94
- modest 94
- noble 94
- northern 94
- slender 94
- stubby 94
- wrinkled 94

Adontorhina 35
Adrana 21
adria, Glyphostoma 97
adspersa, Tenellia 121
Adula 23
adunca, Crepidula 79
adusta, Turbonilla 104
advena, Spiculata 81
aegeensis, Nucula 21
Aegires 116
Aegiretidae 116
aegis, Lucapina 53
aeglees, Niso 77
aeneus, Strobilops 128
aenigmaticus, Navanax 108
Aeolidia 121
Aeolidiella 121
Aeolidiidae 121
aeolis
- blue-patch 120
- dwarf balloon 120
- green balloon 120
- long-tail 120
- miniature 121
- Misaki balloon 120
- orange-blotch 121
- painted balloon 120
- pellucid 120
- polyp 120
- predaceous 120
- pugnaceous 121
- purple 120
- red-finger 120
- salmon 120
- scarlet-tip 121
- shag-rug 121
- single-stalk 121
- smooth-tooth 120
- threeline 120
- vermillion Japanese 121
- white-tentacle Japanese 121

aepynota
- *Inodrillia* 97
- *Odostomia* 101
- *Turbonilla* 104

aepynotus, Fusinus 91
aequacostata, Haliris 49
aequalis
- *Abra* 42
- *Delonovolva* 81
- *Turbonilla* 104
- *vidleri, Delonovolva*

aequicostatus, Hyalopyrgus 61
Aequipecten 26
aequisculpa, Manzonia 67
aequisculpta
- *Odostomia* 101
- *Rimula* 53

aequistriata, Tellina 41
aequizonatum, Lucinoma 35
aeruginosum, Drepanotrema 124
Aesopus 86
affine, Periploma 48
affinis
- *Natica* 82
- *Strobilops* 128
- *Tagelus* 43
- *Tricolia* 58

Aforia 95
aforia, keeled 95
agarhecta, Marstonia 61
agassizi, Dentalium 50
agassizii,
- *Agatrix* 94
- *Berthella* 114
- *Gymnobela* 97
- *Meiocardia* 43
- *Polyschides* 51

Agathotoma 95
Agatrix 94

agavophila, Humboldtiana 144
agilis, Tellina 41
Aglaja 108
aglaja
   albatross 108
   California. 108
   eyespot 108
   leech 108
   mysterious. 108
   pusa 108
Aglajidae 108
agria, Peristichia 103
Agriopoma 44
aguayoi, Diodora 52
ahenea, Elliptio. 29
aila, Chromodoris 116
aivica timia, Taringa 118
Akera 111
Akeridae 111
Akiodoris 115
akutanica, Limopsis 24
Alaba 72
alabama, Anguispira 131
alabamensis
   Amphigyra 124
   Elimia 63
   Micromenetus 124
   Triodopsis 140
   Vertigo 127
alachuana, Hawaiia 134
alameda, Haplotrema. 129
alaskana
   Cingula 67
   Onoba 67
   Skeneopsis 69
   Turbonilla 104
alaskensis
   Cyclopecten 27
   Oenopota 98
   Strombiformis 77
   Tridonta 39
   Vitrinella 70
Alasmidonta 28, 156
alata, Mactrellona 40
alatus
   Isognomon 25
   Potamilus 32
   Strombus 78
alba
   Anodontia 34
   Atagema 117
   Conualevia 117
   Cylichna 108
   Discodoris 118

Lepeta 54
Philine 108
albanyensis, Elimia. 63
alberta, Fossaria 122
albescens, Erato 80
albicoma, Drilla 97
albida
   Chlamys. 26
   Cythnia 77
   Polystira 100
   Poromya 49
albidum
   Epitonium 75
   Homalopoma. 57
albidus, Pitar 44
albilabris, Pupoides 127
albocrusta, Cuthona 120
albolabris, Triodopsis. 140
albolineata
   Dirona 119
   Volvarina. 94
albomaculata, Pilsbryspira. 100
albopapillosa, Adalaria 115
albopunctata, Doriopsilla. 117
albopunctatus
   Aegires 116
   Dendronotus 118
albospinosus, Boreotrophon 84
albrechti, Oenopota 98
albula, Vallonia 128
albus
   Dendronotus 119
   Nassarius 91
   Stenosemus 149
Alcadia 58
alcima, Cerithiopsis 73
alcimus, Fusinus. 91
alcoviensis, Somatogyrus 62
alderi, Barleeia 68
Alderia 113
alderia, modest 113
Aldisa 117
Aldisidae. 117
aldrichi, Amnicola 59
aldrichianus, Helicodiscus 130
alesidota, Hindisclava 97
alethia, Paravitrea 135
aleutica
   Diplodonta. 36
   Falsicingula. 69
   Lepidochitona 148
   Mysella 37
   Odostomia. 101

INDEX

aleuticum
    Buccinum 87
    Periploma 48
aleuticus, Oenopota 98
Alexania 75
Algamorda 66
Alia 86
Aligena 36
aligena
    eastern 36
    San Diego 36
    Texas 36
alitakensis, Oenopota 98
alleneae, Cyphoma 80
alleni, Triodopsis 140
alleryi, Heliacus 74
alliarius, Oxychilus 135
Allogona 137
allyni, Ischnochiton 148
allyniana
    Helminthoglypta 143
    Vertigo 127
allynsmithi
    Helminthoglypta 143
    Sonorella 145
almo
    Manzonia 67
    Strombiformis 77
    Turbonilla 104
alpestris, Vertigo 127
alpina, Oreohelix 141
alta
    Lanx 123
    Siliqua 40
alte, Laevicaulus 147
alternata
    Anguispira 131
    Cuspidaria 49
    Siphonaria 126
    Tellina 41
alternatum, Bittium 72
alternatus, Rabdotus 130
althorpensis
    Margarites 55
    Oenopota 99
althorpi, Oenopota 99
altilis
    Gillia 61
    Lampsilis 30
altina, Odostomia 101
altispira, Stenotrema 139
altum, Pleurobema 32
altus, Polinices 82
Alvania 67

alvania
    carinate 68
    cosmic 67
    fine-cut 67
    Jan Mayen 67
    purple 67
    Santa Rosa 67
    sharp-rib 67
    West Indian 67
alveare, Pleurocera 66
alveata, Favartia 84
alveolus, Leptochiton 148
alveus, Collisella 53
amabilis
    Marginella 93
    Perotrochus 52
Amaea 75
amariradix, Oreohelix 142
amatula, Mangelia 98
Amauropsis 82
ambersnail
    Appalachian 133, 155
    blunt 133
    boundary 133
    Chesapeake 133
    chrome 133
    coastal plain 133
    crinkled 133
    depressed 133
    detritis 132
    diurnal 132
    dryland 133
    European 133
    golden 133
    Humboldt 133
    Kanab 133
    Lone Star 133
    Louisiana 133
    Marshall 133
    Mexico 133
    mudbank 133
    Ninepipes 133
    Niobrara 133
    oblique 133
    Oregon 133
    oval 133
    riblet 133
    ruddy 133
    rustic 133
    saltmarsh 133
    San Tomas 133
    Sanibel 133
    Santa Rita 133
    Sierra 133

slope 133
Snowhill........................... 132
spotted............................. 133
squatty............................. 133
striate.............................. 133
suboval 132
urban.............................. 133
weedpatch 132
xeric............................... 133
ambigua
Simpsonaias 33
Sonorella.......................... 145
Amblema.......................... 28
amblya, Ashmunella............ 137
ameghini, Angustipes........... 147
americana
Glycymeris....................... 24
Odostomia....................... 101
Tellina 41
Tylodina.......................... 114
americanum
Dentalium 50
Lithopoma....................... 57
americanus
Modiolus 23
Spondylus 27
Thyonicola....................... 78
Americardia 39
Amiantis 44
amiantis, Mitrella................ 87
amiantus
Linga.............................. 35
Neptunea......................... 89
Odostomia....................... 101
Oenopota........................ 99
amiata, Nuculana................ 21
amicalola, Paravitrea........... 135
Amicula............................ 149
amiculata, Amicula 149
amilda, Odostomia.............. 101
ammon, Planorbella............. 125
ammonia, Torellia 79
Ammonitella 141
Ammonitellidae.................. 141
Amnicola......................... 59
amnicola
Hoosier 59
Missouri.......................... 60
mud 60
peninsula........................ 59
squaremouth 60
stygian........................... 60
amnicoloides, Somatogyrus 62
amnicum, Pisidium............. 34

amphichaenus, Potamilus 32
Amphigyra 124
Amphissa 86
amphissa
Atlantic 86
reticulate 86
two-tone 86
variegate 86
wavy 86
wrinkled......................... 86
Amphithalamus................... 68
amphiurgus
Conus 94
Fusinus 91
ampla
Elimia 63
Leptoxis.......................... 65
Marsenina 80
Panomya 46
ampullacea, Volutharpa........ 90
Amusium.......................... 26
amygdala, Villosa............... 33
amygdalea, Yoldia 22
Amygdalum....................... 23
amyra, Doto..................... 119
Anachis............................ 86
Anadara 24
Anadenulus....................... 131
Anatina 40
anatina, Anatina................ 40
Anatoma 52
anceps, Helisoma................ 124
anchana
Oreohelix........................ 142
Sonorella......................... 145
ancillaria, Physella.............. 123
Ancistrobasis 56
Ancula............................. 114
ancula
Atlantic 114
freckled 114
Pacific 114
ancylid............................. 125
blackwater...................... 125
cloche 125
creeping......................... 125
cymbal........................... 125
domed 125
dusky............................ 125
excentric 125
fragile............................ 125
hood.............................. 125
knobby........................... 125
oblong 125

peninsula	125
wicker	125
Ancylidae	125
andersoni	
Acanthochitona	150
Prophysaon	132
andrewsae	
Mesodon	138
Mesomphix	134
Paravitrea	135
andrewsi, Calotrophon	84
andrewsii, Opalia	76
andrusiana, Vertigo	127
angelensis, Okenia	115
angelica, Acanthochitona	150
angelicae, Vitrina	136
angellum, Stenotrema	139
angelus, Sonorelix	145
angelwing	46
Campeche	47
false	45
Anguispira	131, 155
angularis, Odostomia	101
angulata	
Ashmunella	137
Gonidea	30
Philine	108
angulatum, Epitonium	75
angulatus	
Latirus	91
Pedipes	122
angulifera, Littorina	66
anguliferum, Periploma	48
angulosa, Tellina	41
angulosum	
angulosum, Buccinum	87
subcostatum, Buccinum	88
transliratum, Buccinum	88
angulosus, Oenopota	99
angusta, Fissurella	53
angustata, Elliptio	29
angustior, Littorina	66
Angustipes	147
angustipes, Platydoris	118
animasensis	
Ashmunella	137
Sonorella	145
Anisodoris	118
annettae, Elimia	63
anniae, Murex	84
Annulariidae	66
annulatum	
Calliostoma	55
Lucinoma	35
annulatus, Spiroglyptus	72
annulifera, Pleurocera	66
Anodonta	28
Anodontia	34
Anodontoides	29
anomala	
Bathyarca	24
Iselica	101
Anomalocardia	44
Anomia	27
Anomiidae	27
Antalis	50
antefilosa, Cerithiopsis	73
antemunda, Cerithiopsis	73
antestriata, Turbonilla	104
Anticlimax	69
antillarum	
Antalis	50
Bractechlamys	26
Caecum	71
Haminoea	109
Lithophaga	23
Lyropecten	27
Nassarius	91
Ornithoteuthis	152
Oxynoe	113
Trigoniocardia	40
Trivia	80
antillensis	
Aphelodoris	117
Cyerce	112
Limopsis	24
Rossia	151
Sphenia	45
Antillophos	87
Antiopella	120
Antiplanes	95
antiquatus, Hipponix	78
Antrobia	60
antroecetes, Fontigens	61
Antroselatus	60
anurus, Mesomphix	134
Aorotrema	69
apache, Sonorella	145
Aperiovula	80
aperta, Helix	146
apertus, Cecilioides	128
Aphaostracon	60
Aphelodoris	117
apicifulva, Peracle	110
apicina	
Marginella	93
Problacmaea	54
Stagnicola	123

apiculata
   Chaetopleura. 149
   Quadrula 32
apiculatum, Epitonium 75
Aplexa 123
aplexa
   lance 123
   marbled 124
   tawny. 124
Aplysia 112
Aplysiidae. 112
Aplysiopsis 113
Apodopsis. 122
Aporrhaididae. 78
Aporrhais 78
applesnail
   Florida. 59
   spiketop. 59
appressus, Mesodon. 138
approximans, Mesodon 138
approximata, Parvilucina. 35
aprica, Catinella. 132
aquaealbae, Eremarionta. 143
aquatile, Cymatium 83
arachnoidea, Elimia 63
aragoni, Turbonilla. 104
arata, Panacca. 47
arboreus, Zonitoides. 136
Arca 24
arca, Elliptio. 29
arcaeformis, Epioblasma 29, 156
arcana, Chama. 36
arcas, Terebra 95
arcellus, Ventridens 136
archeri, Mesodon 138
Archidorididae 117
Archidoris 117
archimedes, Pyrgulopsis. 62
architae, Heliacus 74
Architectonica. 74
Architectonicidae 74
Arcidae 24
Arcidens 29
Arcinella 36
arcinella, Arcinella. 36
arc-mussel
   Altamaha. 28
   Ochlocknee 28, 156
Arcopsis 24
arctata, Elliptio. 29
arctatum, Mesodesma 40
Arctica 43
arctica
   Hiatella 46

   Obestoma 98
   Odostomia. 101
   Panomya 46
   Portlandia 22
   Tridonta. 39
Arcticidae 43
arcticum, Pristiloma. 135
arcticus
   Bathypolypus. 152
   Plicifusus. 90
   Stagnicola 123
Arctinula. 26
Arctomelon 93
arcuata
   Diodora 52
   Melanella. 77
arcula, Alasmidonta. 28
arenaria, Mya. 45
Arene. 57
arenosa
   Lyonsia 47
   Pandora. 48
areolata
   Alvania 67
   Doriopsilla. 117
areolatus, Pleurobranchus 114
aresta, Turbonilla 104
arestus, Pitar 44
arga, Marstonia 61
argentea, Dimya. 27
argenteum, Chaetoderma. 20
argo, Argonauta 152
argonaut
   greater 152
   knobby. 152
   winged. 152
Argonauta. 152
Argonautidae. 152
Argopecten 26
argus
   Eremariontoides 143
   Mitrella 87
Ariolimax 131
ariomus, Pterochelus 85
Arion 131
arion
   black 131
   brown-banded 131
   chocolate. 131
   darkface. 131
   dusky. 131
   forest 131
   garden 131
   hedgehog. 131

orange-banded	131
warty	131
Arionidae	131
aristata, Lithophaga	23
arizonensis, Holospira	129
ark	24
Adams	24
Baily miniature	24
Baughman	24
blood	24
Chemnitz	24
cut-ribbed	24
Doc Bales	24
eared	24
incongruous	24
many-ribbed	24
mossy	24
ponderous	24
red-brown	24
transverse	24
white miniature	24
white-beard	24
arkansasensis, Villosa	33
arkansensis, Leptoxis	65
Arkansia	29
armata, Acanthodoris	115
armifera, Gastrocopta	126
armiger	
Conus	94
Vespericola	141
armigera	
Lithasia	65
Planorbella	125
armilla	
Cyclocardia	38
Pleuromeris	38
armillatum, Bittium	73
Armina	119
armina	
California	119
tiger	119
Arminidae	119
arnoldi	
Cerithiopsis	73
Diodora	52
Solariorbis	70
arrosa, Helminthoglypta	143
arteaga, Kurtzia	98
arthuri, Vertigo	127
Asaphis	42
asarca, Bactrocythara	96
ascidicola, Okenia	115
Ascobulla	112
Ashmunella	137
ashmuni	
Ashmunella	137
Gastrocopta	126
Sonorella	145
askewi, Fusconaia	30
asmi, Collisella	54
Aspella	84
aspella	
beggar	84
graybeard	84
aspera	
Cymakra	96
Diodora	52
asperata, Quadrula	32
Asperiscala	75
asperrima, Dentistyla	55
aspersa, Helix	146
asperum, Bittium	73
assimilis, Haliotis	52
Assiminea	68
assiminea	
Atlantic	68
California	68
Assimineidae	68
Astarte	38
astarte	
compact	38
crenulate	38
dwarf	38
Eskimo	38
polar	39
Smith	39
smooth	38
squarish	39
strange	38
wavy	39
Willett	39
Astartidae	38
asteriscus, Planogyra	128
Asteronotidae	117
Asterophila	77
asthenes	
Aphaostracon	60
Callistochiton	148
Asthenothaerus	48
Astralium	57
astricta	
Mangelia	98
Odostomia	101
asturiana, Puncturella	53
atacellana, Nucula	21
Atagema	117
ater, Arion	131
aterina, Elimia	63

athearni, Elimia 63
athleenae, Anticlimax 69
atkaensis, Lymnaea 123
Atlanta 81
atlantica
   Aclistothyra 37
   Cooperella 45
   Laevinesta 53
   Phylliroe 119
   Xylophaga 47
atlanticum
   atlanticum, Pneumoderma 111
   pacificum, Pneumoderma 111
atlanticus, Glaucus 122
Atlantidae 81
atlantis, Nystiella 76
atomus, Omalogyra 69
atossa, Odostomia 101
atra, Polycera 116
atrata, Acteocina 107
atratum, Cerithium 73
Atrina 25
atrogriseata, Acanthodoris 115
atropurpurea
   Alasmidonta 28
   Ocenebra 85
atrostyla, Kurtziella 98
attenuata, Limatula 26
attenuatum, Bittium 73
attenuatus
   Conus 94
   Volutopsius 90
Attiliosa 84
attonsa, Cylichna 108
attrita, Turbonilla 104
Atyidae 109
Atys 109
auberiana
   Alvania 67
   Anomalocardia 44
auger
   Arcas 95
   concave 95
   eastern 95
   fine-ribbed 95
   flame 95
   gray 95
   lilac 95
   Onslow 95
   porcelain 95
   San Pedro 95
   shiny 95
   Texas 95
   tongue 95

   western crenate 95
   woven 95
   yellow 95
aulacogyra, Paravitrea 135
aurantia
   Cuthona 120
   Dirona 119
   Turbonilla 104
aurantiaca, Mitrella 87
aurea
   Pilsbryna 135
   Quadrula 32
aureocinctum, Cyphoma 80
aureocinctus, Dentimargo 93
aureotincta, Tegula 56
aureus, Somatogyrus 62
auricinctus, Strombiformis 77
auricoma
   Turbonilla 104
   Zachrysia 141
auricula, Crucibulum 79
auricularia, Radix 123
auriculata
   Ashmunella 137
   Polygyra 139
aurifilia, Opalia 76
auriformis, Polygyra 139
aurisula, Polycera 116
aurita, Limopsis 24
auritula, Pisania 90
auritulus, Favorinus 121
aurivillii, Onoba 67
Austraeolis 121
australis, Lampsilis 30
Austrotrophon 84
avalonense, Radiocentrum 142
avalonensis, Boreotrophon 84
avara
   Anachis 86
   Catinella 132
   Polygyra 139
avawatzica, Sonorelix 145
avellanum, Pleurobema 32
avena
   ruthturnerae, Neosimnia 81
   Volvarina 94
avenacea, Volvarina 94
avernalis, Fluminicola 60
avicula, Acanthochitona 150
avira, Inodrillia 97
awlsnail
   blind 128
   dwarf 128
   graceful 128

Mauritian	128
miniature	128
obtuse	128
sharp	128
spike	128
tiny	128
awningclam	
Atlantic	22
boreal	22
grand	22
gutless	22
Pacific	22
West Indian	22
axinopsid	
Carey	35
circle	35
green	35
heart	35
inflated	35
minute	35
orbicular	35
Redondo	35
silky	35
Axinopsida	35
Axinulus	35
ayresiana, Helminthoglypta	143
azaria, Hiatella	46
azuropunctata, Oxynoe	113

B

Babakina	121
Babakinidae	121
Babelomurex	86
baboquivariensis, Sonorella	145
babybody	
Gulf	127
yam	127
baby-bubble	
estuarine	107
miniature	107
pitted	107
short	107
baby-ear	
brown	82
dwarf	82
fat	82
slight	82
white	82
babylonia, Scalenostoma	77
babylonium, Epitonium	75
babylonius, Oenopota	99
bachia, Odostomia	101
Bactrocythara	96

baculum, Homalopoma	57
Baeolidia	121
baerii, Buccinum	88
baetica, Olivella	92
bagnarai, Sonorella	145
baileyi, Sonorelix	145
Bailya	87
bailyi	
Barbatia	24
Cyclocardia	38
bairdii	
Arene	57
Bathybembix	54
Calliostoma	55
bakeri	
Bernardina	43
Melanella	77
Onoba	67
Orobitella	37
Philine	108
Praticolella	139
Schwartziella	68
Turbonilla	104
bakerianus, Amnicola	59
balboae, Cardiomya	49
Balcis	77
baldridgae, Odostomia	101
balesae, Acanthochitona	150
balesi, Pleuromalaxis	70
balliana, Erycina	36
balteata, Pyrgocythara	100
balthica, Macoma	41
bananaslug	
California	131
Pacific	132
slender	132
bandella, Mangelia	98
bankclimber	31
purple	29
Bankia	47
banksi, Onychoteuthis	151
barbadensis	
Fissurella	53
Mathilda	74
Mitra	92
barbara, Cochlicella	146
barbarae, Brachycythara	96
barbarensis	
Antiopella	120
Cyclocardia	38
Cyclopecten	26
Finella	73
Fusinus	91
Ocenebra	85

Spiculata	81
Thyasira	35
barbarinus, *Colus*	88
barbata, *Oreohelix*	142
Barbatia	24
barbatum, *Stenotrema*	139
barberi, *Succinea*	133
barbigerum, *Stenotrema*	139
barbouri, *Calliostoma*	55
barcleyensis, *Turbonilla*	104
barkleyensis, *Odostomia*	101
Barleeia	68
Barleeiidae	68
barleysnail	
abalone	68
acute	68
California	68
Caribbean	68
fragile	68
Barnea	46
barnesiana, *Fusconaia*	30
barrattiana, *Corbula*	46
barrel-bubble	
Arctic	109
blackback	107
channeled	107
engraved	109
grain	107
harp	109
intermediate	107
ivory	107
keeled	109
Monterey	109
ovate	109
pillow	107
polished	109
rice	107
rude	107
straight	107
striate	107
sulcate	109
two-tooth	107
barri, *Helicodiscus*	130
bartletti, *Tenaturris*	100
bartonensis, *Stygopyrgus*	63
bartrami, *Ommastrephes*	152
bartschi	
Borsonella	96
Fargoa	101
Oocorys	83
Sonorella	145
Teredo	47
Bartschia	87
Basiliomya	27
basilissa, depressed	56
Basterotia	38
Bathyarca	24
bathyark	
anomalous	24
comb	24
Friele	24
glacial	24
little-ball	24
Bathybembix	54
Bathypolypus	152
Bathytoma	96
Batillaria	72
baughmani, *Anadara*	24
baxteri, *Anatoma*	52
bayeri	
bayeri, *Tritonia*	118
misa, *Tritonia*	118
beadleanum, *Pleurobema*	32
baeliana, *Cerodrillia*	96
bean	
Choctaw	33
Cumberland	33, 154
purple	33
rayed	33
beana, *Entodesma*	47
beanclam	
California	42
Gould	42
beanii	
Cocculina	54
Lepidochitona	148
beauii	
Cyclostremiscus	69
Cylindrobulla	112
Siratus	85
beaumonti, *Cumanotus*	120
beckii, *Oenopota*	99
behringi, *Volutopsius*	90
behringiana, *Chlamys*	26
belcheri, *Forreria*	84
bella	
Marginella	93
Scelidotoma	53
Tricolia	58
bellacrenata, *Elimia*	63
bellamaris, *Subsimnia*	81
Bellaspira	95
bellastriata	
Asperiscala	75
Semele	43
bellegladeensis, *Murex*	84
bellerophon, *Scintillona*	37
bellona, *Paravitrea*	135

bellotii, Nucula	21
bellula, Elimia	63
bellus, Fossarus	78
bellytooth	
Appalachia	134
brown	134
belotheca, Turbonilla	104
benedicti, Chlamys	26
benitoensis, Helminthoglypta	143
bennettii, Tridonta	39
benteva, Baeolidia	121
benthalis	
Fusinus	91
Terebra	95
Benthonella	67
benthonella, regal	67
bentleyi	
Barleeia	68
Boreotrophon	84
bentoniensis, Elimia	63
bequaerti	
Ashmunella	137
Pteropurpura	85
Sonorella	145
Berghia	121
beringensis	
Mysella	37
Tonicella	149
beringi	
Boreotrophon	84
Odostomia	101
Thracia	48
beringiana	
Anodonta	28
Neptunea	89
Panomya	46
beringii, Beringius	87
Beringius	87
berlandieriana, Praticolell	139
Bermudaclis	77
bermudensis	
Diodora	52
Lucina	35
bermudezi	
Folinia	67
Physella	123
bernardclam	
Baker	43
salmon	43
subtrigonal	43
bernardi	
Calliostoma	55
Carditopsis	38
Bernardina	43
Bernardinidae	43
berryi	
Antalis	50
Cerithiopsis	73
Gonatus	151
Helminthoglypta	143
Melanella	77
Vertigo	127
Vitrinella	70
Berryteuthis	151
Berthelinia	113
Berthella	114
Berthellina	114
beta, Ocenebra	85
biangulata, Americardia	39
biangulatus, Somatogyrus	62
bibbianum, Teinostoma	70
bicarinata	
Trichotropis	79
Valvata	58
bicarinatus, Oenopota	99
bicaudata, Vitrinella	70
bicipitis, Sonorella	145
bicolor	
Amphissa	86
Cerithiopsis	73
Huttonella	129
Isognomon	25
biconica, Brachycythara	96
bicostata, Lirularia	55
bidens, Paravitrea	135
bidentata, Acteocina	107
bidentatus, Melampus	122
biemarginata, Epioblasma	29, 156
bifasciatus, Strombiformis	78
bifurcatus, Septifer	24
bilamellata	
Holospira	129
Onchidoris	115
bilineatus, Strombiformis	78
bilirata, Pandora	48
billsae, Puncturella	53
bimaculata	
Fissurellidea	53
Thordisa	118
bimaculatus	
Heterodonax	42
Octopus	152
bimaculoides, Octopus	152
Binneya	132
binneyana	
Fontigens	61
Nesovitrea	134
Vertigo	127

binneyanus, Mesodon................ 138
binneyi
   Ashmunella 137
   Planorbella 125
   Pomatiopsis...................... 63
   Sonorella........................ 145
binominata, Lampsilis 30
Biomphalaria 124
bipartitum, Caecum 71
biplicata
   Mitrolumna 98
   Olivella 92
birddrop
   California........................ 127
   Caribbean 127
   insular 127
   Mexican.......................... 127
   pitted 126
Birgella 60
biscaynense, Teinostoma 70
bisecta, Conchocele 35
bispinosa, Peracle 110
bisuculata, Lithophaga................ 23
bisulcatus, Heliacus 74
bisuturalis, Boonea................... 101
Bithynia............................ 59
bithynia, mud 59
Bithyniidae 59
bitruncata, Panopea.................. 46
bittersweet
   Bermuda 25
   California........................ 25
   comb 25
   Cortez 24
   decussate........................ 24
   giant............................ 24
   northern......................... 25
   spectral 25
   wavy 25
Bittium 72
bladderclam.......................... 48
bladetooth
   engraved 138
   flat.............................. 138
   grand 139
   Virginia 138
blainei, Epitonium 75
blakeana
   Microvoluta...................... 93
   Pleurotomella 100
blakeanum, Lucinoma 35
blakei, Solariorbis.................... 70
blanda, Parvilucina 35
blandi, Pupilla 127

blandianum
   Punctum 130
   Stenotrema 140
blandum, Clinocardium 39
blanesi, Jaspidella 92
blaneyi
   Oenopota........................ 99
   Tonicella 149
blarina, Paravitrea 135
Blauneria........................... 122
bleufer.............................. 32
blossom
   green 30, 154
   tubercled 30, 154
   turgid.......................... 30, 154
   yellow 29, 154
bobtail
   Antilles 151
   eastern Pacific................... 151
   greater shining................... 151
   Gulf............................ 151
   lesser shining.................... 151
   Tortugas 151
bodegensis, Tellina................... 41
bollesiana, Vertigo 127
bonamicus, Helicodiscus 130
bonnet
   Coronado........................ 82
   royal 82
   Scotch 82
bonnevillensis, Stagnicola 123
bonnieae, Pleurobranchaea 114
boogi, Ischnochiton 148
Boonea 101
booneae, Hydrobia................... 61
Boreacola 37
boreale, Prophysaon.................. 132
borealijaponica, Onychoteuthis 151
borealis
   Acirsa 75
   Collisella 54
   Cyclocardia...................... 38
   Heliacus 74
   Marginella....................... 93
   Odontogena..................... 37
   Onchidella...................... 147
   Opalia 76
   Placiphorella..................... 150
   Solemya........................ 22
   Trichotropis..................... 79
   Tridonta........................ 39
Boreocingula....................... 67
Boreoscala 75
Boreotrophon 84

Bornia	36
borniaclam	36
borregoensis, Sonorelix	145
Borsonella	95
Bosellia	112
Boselliidae	112
bostoniensis, Facelina	121
Bostrychocentrum	129
Bothriopupa	126
bottimeri	
Paludestrina	61
Physella	123
Botula	23
boucardi, Physella	123
bournianum, Pleurobema	32
bowiensis, Sonorella	145
boykiniana	
Elimia	63
Megalonaias	31
Brachidontes	23
Brachycythara	96
bracteata	
Lampsilis	30
Lyonsia	47
Bractechlamys	26
Bradybaena	142
Bradybaenidae	142
brandtii, Schizoplax	149
brannani, Siphonaria	126
brasiliana	
Anadara	24
Anomalocardia	44
Aplysia	112
Iphigenia	42
brevicula, Microglyphis	107
brevidens, Epioblasma	29
brevifrons, Macoma	41
brevipila, Stenotrema	140
brevis	
Elimia	63
Lolliguncula	151
Thyasira	35
brevispirum, Campeloma	59
brevissimus, Hyalopyrgus	61
briareus	
Arene	57
Octopus	152
bridgesi, Pomacea	59
briseis, Antiplanes	95
bristlesnail, Trinity	144
bristolensis, Colus	88
brittsi, Ventridens	136
brogniartianus, Micromenetu	124
broken-ray	
Arkansas	31
northern	31
ozark	31
bronniana, Limea	26
brophyi, Addisonia	54
brota, Macoma	41
browni, Amnicola	59
browniana, Zebina	68
brumbyi, Pleurocera	66
brunnea	
Acanthodoris	115
Diaphana	109
Eremarionta	143
Tegula	56
brunneus, Plicifusus	90
bryanti, Discus	131
brychia, Frigidoalvania	67
bryerea, Schwartziella	68
bubble	
California	109
exquisite	107
imperforate	109
jewel	109
solid	109
striate	109
buccata, Nuculana	21
Buccinidae	87
Buccinum	87
bucephalum, Phylliroe	119
buckleyi, Elliptio	29
bud	
ornate	135
prominent	135
bufo, Bufonaria	83
Bufonaria	83
bulbosa, Juga	65
Bulbus	82
Bulimnaea	122
bulimoides	
Antiplanes	95
Fossaria	122
Limacina	110
Bulimulidae	130, 155
Bulimulus	130
bulimulus, West Indian	130
Bulla	109
bullata, Acteocina	107
Bullidae	109
Bullina	107
Bullinidae	107
bullisi	
Fasciolaria	91
Rossia	151

bulloides, Detracia 122
bullula, Olivella 92
burchi
    Calyptraea........................ 79
    Doridella 114
    Triodopsis 140
    Turbonilla 104
burkei, Quincuncina 33
burringtoni, Hemphillia 132
burryi
    Octopus.......................... 152
    Opalia 76
Bursa............................... 83
Bursatella 112
Bursidae 83
bushclam, elegant..................... 48
Bushia 48
bushiae, Nassarina................... 87
bushiana
    Fargoa.......................... 101
    Pandora......................... 48
Busycon............................ 90
Busycotypus 90
butterfly............................ 29
button
    brittle............................. 134
    broad............................. 134
    common.......................... 134
    copper 134
    dusky............................ 134
    flat............................... 134
    fragile............................ 134
    frog 134
    globose 134
    mountain 134
    plain.............................. 134
    smooth........................... 134
    striate............................ 134
    wrinkled.......................... 134
buttonsnail 72
Bythinella 60

C

caamanoi, Nitidiscala 76
cabritii, Murex 84
cactussnail
    bicolor............................ 146
    Catalina.......................... 146
    plain.............................. 146
    speckled.......................... 144
    wreathed 146
Cadlina 116

cadlina
    dark-spot......................... 116
    white Atlantic 116
    yellow-rim........................ 116
    yellow-spot 116
Cadlinidae........................... 116
Caducifer 88
Cadulus............................ 51
caeca, Lepeta....................... 54
Caecidae 71
caecoides, Lepeta................... 54
Caecum............................ 71
caecum
    Antillean 71
    beautiful.......................... 71
    bone.............................. 71
    California......................... 71
    Carolina.......................... 71
    club 71
    deepwater 71
    fatlip 71
    fine-line 71
    Florida........................... 71
    horn-of-plenty 71
    imbricate 71
    little horn........................ 71
    many-named 71
    minute 71
    plicate 71
    smooth........................... 71
    textile............................ 71
    twisted........................... 71
    Vera Cruz 71
    western 71
    windpipe 71
caelatum
    Haplotrema...................... 129
    Lithopoma....................... 57
Caelatura 68
caelatura, Elimia 63
caelatus
    Colus............................ 88
    Pyrunculus...................... 109
caerulescens, Acanthodoris 115
caerulifluminis, Sonorella............ 145
caffea, Turcica 56
cahabensis, Clappia 60
cahawbensis, Elimia................ 63
cailleti, Siratus 85
cailletii, Turbo 58
calamitosa, Sterkia................. 127
calamus, Graptacme 50
calcarea, Macoma 41
calcarella, Odostomia 101

calceola, Corolla 110
calcicola
    Nucula........................... 21
    Paravitrea 135
calianus, Tornus 71
calidimaris, Parviturbo................ 57
californiana
    Dimya 27
    Rimula.......................... 53
    Trivia........................... 80
californianus
    Mytilus.......................... 24
    Tagelus 43
californica
    Admete 94
    Aplysia.......................... 112
    Armina.......................... 119
    Assiminea 68
    Barleeia......................... 68
    Berthella 114
    Cardiomya....................... 49
    Cerithidea 72
    Crossata 83
    Cryptomya....................... 45
    Cumingia........................ 43
    Cyclostremella.................... 101
    Diaphana........................ 109
    Dondersia 20
    Epilucina 35
    Finella 73
    Gari............................ 42
    Grippina 46
    Hancockia....................... 118
    Lyonsia 47
    Macromphalina................... 78
    Mactra.......................... 40
    Melanella........................ 77
    Metzgeria 93
    Nitidiscala....................... 76
    Nuttallina 149
    Oreohelix........................ 142
    Parapholas 47
    Pedicularia 81
    Philine.......................... 108
    Pleurobranchae................... 114
    Pomatiopsis...................... 63
    Rissoina......................... 68
    Skenea.......................... 57
    Succinea 133
    Truncatella 69
    Vertigo.......................... 127
    Volvulella 109
californicum
    Caecum......................... 71

    Prochaetoderm 20
    Punctum 130
californicus
    Ariolimax........................ 131
    Capulus......................... 78
    Conus 94
    Donax 42
    Megomphix 141
    Octopus......................... 152
    Polyschides 51
    Strombiformis 78
californiense, Clinocardium 39
californiensis
    Adula........................... 23
    Anodonta........................ 28
    Chione.......................... 44
    Fontelicella 61
    Helminthoglpta 143
    Hypselodoris 116
    Petricola 45
Calinaticina........................... 82
Caliphylla 112
Caliphyllidae........................... 112
callia, Turbonilla 104
calliglypta, Cosmioconcha.............. 87
callimene
    Nuculana........................ 21
    Odostomia....................... 101
    Turbonilla 104
callimorpha
    Globivenus....................... 44
    Mitrella 87
    Odostomia....................... 101
calliope, Odostomia 101
Calliostoma........................... 54
Calliotropis........................... 55
callipeplum, Laevidentalium 50
callipeplus, Monadenia 144
callipyrga, Triphora 75
Callista 44
Callistochiton 148
callistoderma, Helminthoglypta.......... 143
callithrix, Heteroschismoides............ 51
callomarginata, Lucapinella 53
callorhinus
    callorhinus, Volutopsius 90
    stejnegeri, Volutopsius 90
callosa, Amiantis 44
Calotrophon 84
calvescens, Stenotrema 140
Calycidorididae......................... 117
Calycidoris 117
Calyptraea 79
Calyptraeidae 79

Camaenidae.......................... 141
camelus, Hemphillia................. 132
campanulata, Planorbella............ 125
campechiensis
   Mercenaria...................... 44
   Pholas.......................... 47
Campeloma........................... 59
campeloma
   cylinder.......................... 59
   file.............................. 59
   maiden........................... 59
   ovate............................ 59
   pointed.......................... 59
   ponderous........................ 59
   purple-throat.................... 59
   slender.......................... 59
campestris
   Planorbella..................... 125
   Succinea........................ 133
canadensis, Turbonilla.............. 104
canaliculata
   Acteocina....................... 107
   Nucella......................... 85
   Pleurocera...................... 66
canaliculatum, Calliostoma.......... 55
canaliculatus
   Busycotypus..................... 90
   Turbo........................... 58
Cancellaria......................... 94
cancellaria, Barbatia............... 24
Cancellariidae....................... 94
cancellarius, Cantharus............. 88
cancellata
   Chione.......................... 44
   Odostomia....................... 101
   Ophiodermella................... 100
   Rissoina........................ 68
   Trichotropis.................... 79
cancellatum
   Cyclostrema..................... 57
   Propeamussium................... 27
   Rupellaria...................... 45
cancellatus
   Conus........................... 94
   Oenopota........................ 99
   Platyodon....................... 45
candeana
   Diplodonta...................... 36
   Tellina......................... 41
candeanum, Epitonium................ 75
candeanus, Malleus.................. 25
candei
   Acteocina....................... 107
   Antillophos..................... 87

candelabrum, Busycon................ 90
candens, Acteon..................... 107
candida
   Barbatia........................ 24
   Praticolella.................... 139
   Pyramidella..................... 103
candidissima, Pyrgocythara.......... 100
Candidula........................... 146
candidula, Trivia................... 80
canfieldi
   Clathurella..................... 96
   Odostomia....................... 101
   Turbonilla...................... 104
canguzua, Elysia.................... 113
canna, Drillia...................... 97
canoe-bubble, giant.................. 107
canrena, Natica..................... 82
Cantharus........................... 88
cantharus
   cancellate....................... 88
   gaudy............................ 90
   ribbed........................... 88
   tinted........................... 90
cantiana, Monacha................... 146
capaniola, Clathurella.............. 96
capax
   Modiolus........................ 23
   Potamilus................... 32, 154
   Tresus.......................... 40
caperatus, Stagnicola............... 123
capillaris, Elimia.................. 63
capitana, Odostomia................. 101
capnodes, Mesomphix................. 134
capponius, Colus.................... 88
capsaeformis, Epioblasma............ 29
capsella, Paravitrea................ 135
capshell, Rocky Mountain............. 122
capsnail
   California....................... 78
   fools............................ 78
   incurved......................... 78
Capulacmaea......................... 79
Capulidae............................ 78
Capulus............................. 78
cardiapod, flat...................... 81
Cardiapoda.......................... 81
Cardiidae............................ 39
Cardiomya........................... 49
cardiomya
   Balboa........................... 49
   California....................... 49
   carved........................... 49
   costate.......................... 49
   lirate........................... 49

little-ribbed	49
Oldroyd	49
ornate	49
pectinate	49
planet	49
rostrate	49
striate	49
Carditamera	38
carditid	
armlet	38
broad-ribbed	38
Carpenter	38
elongate	38
flattened	38
Santo Domingo	38
thick	38
threetooth	38
Carditidae	38
carditoides, Rupellaria	45
Carditopsis	38
cardium, Lampsilis	30
carduus, Fissidentalium	50
careyi, Axinulus	35
caribaea	
Coralliophila	86
Cosa	25
Xenophora	79
caribaeensis, Truncatella	69
caribaeus, Atys	109
caribbaea, Engina	89
caribbea, Berthelinia	113
caribea, Rissoella	69
carica, Busycon	90
Carinaria	81
carinaria	
harp	81
helmet	81
Carinariidae	81
carinata	
Alia	86
Bankia	47
Coleophysis	109
Lacuna	66
Leptoxis	65
Vitrinella	70
carinatus	
Neoplanorbis	125
Trimusculus	126
carinicallus, Teinostoma	70
carinicostata, Elimia	63
carinifer, Latirus	91
carinifera	
Elimia	63
Oreohelix	142
cariosa, Lampsilis	30
carlotta, Calliotropis	55
carlottensis, Macoma	41
carlsbadensis, Ashmunella	137
carmelensis, Skenea	57
carnaria, Strigilla	41
carnea	
Marginella	93
Pinna	25
carnegiei, Gastrocopta	126
carolae, Ficus	83
caroli, Ommastrephes	152
carolinensis	
Aclis	76
Glyphyalinia	134
Mohnia	89
Polyschides	51
Sigatica	82
caroliniana	
Polymesoda	44
Uniomerus	33
carolinianum, Caecum	71
carolinianus, Philomycus	132
caroliniensis, Triodopsis	140
carpenteri	
Barleeia	68
Caecum	71
Cerithiopsis	73
Glans	38
Helminthoglypta	143
Homalopoma	57
Inodrillia	97
Mitromorpha	98
Modiolus	23
Nuculana	21
Onchidella	147
Onoba	67
Tellina	41
Triphora	75
Turbonilla	104
carpenteriana, Megasurcula	98
carriersnail	
American	79
Caribbean	79
shingled	79
caruanae, Deroceras	136
caruthersi, Helminthoglypta	143
Carychiidae	125
Carychium	125
caryophylla, Puncturella	53
casablancae, Cerion	128
casanica, Cocculina	54
casertanum, Pisidium	34
Casmaria	82

casmaria, Atlantic................... 82
cassandra, Odostomia 101
Cassidae 82
Cassis 82
cassis, Marginella................... 93
casta, Mysella...................... 37
castanea
   Alvania 67
   Astarte.......................... 38
   Boreocingula..................... 67
   Pilsbryna 135
   Turbo........................... 58
   Turbonilla 104
castanella
   Onoba 67
   Turbonilla 104
castaneum
   castaneum, Buccinum.............. 88
   fluctuatum, Buccinum............. 88
   triplostephanum, Buccinum 88
castaneus
   Lioberus 23
   Volutopsius 90
castianira, Clathurella 96
castor, Marstonia.................... 61
castrensis
   Acila 21
   Natica 82
castus, Atys 109
catalinae
   Antiplanes 95
   Nitidiscala...................... 76
   Triopha 115
catalinense, Haplotrema 129
catalinensis
   Alaba........................... 72
   Cyclopecten..................... 26
   Melanella....................... 77
   Nitidiscala...................... 76
   Odostomia...................... 101
   Triphora........................ 75
cataracta
   cataracta, Anodonta............... 28
   fragilis, Anodonta................ 28
   marginata, Anodonta 28
catascopium, Stagnicola 123
catenaria, Elimia 63
catenata
   Anachis 86
   Persicula 94
catesbyana, Schwartziella 68
catilliformis, Spisula................. 40
catina, Austraeolis 121
Catinella.......................... 132

Catriona 120
catskillensis, Discus 131
catspaw........................... 30
   narrow...................... 30, 156
   white 30, 154
catulus, Elysia 113
caudata
   Eupleura 84
   Linatella 83
   Nuculana....................... 21
caurinus, Patinopecten............... 27
cavesnail
   Barton 63
   beaked......................... 62
   disc 62
   domed 62
   flattened........................ 62
   Foushee........................ 59
   Greenbrier 61
   high-hat 62
   Hueco 61
   mimic.......................... 61
   organ 61
   Proserpine 60
   shaggy 60
   tapered......................... 61
   Tumbling Creek................. 60
cavoline
   fourtooth 110
   gibbose 110
   inflexed 110
   longsnout...................... 110
   threespine 110
   threetooth 110
   uncinate....................... 110
Cavolinia.......................... 110
Cavoliniidae 110
cayenensis, Diodora 52
cayohuesonicus, Dolicholatirus 91
Cecilioides........................ 128
cedrosa, Kleinella................... 101
cellarius, Oxychilus 135
cellulita, Nuculana 21
cellulosa, Favartia 84
centifilosum, Nemocardium 39
centralis, Calyptraea 79
centrifuga, Spirolaxis 74
Cepaea 146
cepio, Pododesmus.................. 27
cepulus, Boreotrophon 84
Cerastoderma 39
Ceratostoma 84
Ceratozona 149
ceratum, Antalis 50

Cerberilla	121	*Chaetopleura*	149
cerealis, Acteocina	107	Chaetopleuridae	149
cereolus, Polygyra	139	*chalarogyrus, Aphaostracon*	60
ceres, Paravitrea	135	chalice-bubble	
Ceriidae	128	concealed	108
cerina		ivory	108
Fusconaia	30	kernel	108
Gouldia	44	lined	108
Kurtziella	98	San Diego	108
Macoma	41	white	108
cerinella		chalksnail, Florida	133
Cryoturris	96	*challisae, Bittium*	73
Onoba	67	*challisiana, Thracia*	48
cerinoideus, Ventridens	136	*Chama*	36
Cerion	128	Chamidae	36
cerith		*championi, Epitonium*	75
California	73	*chani, Hallaxa*	117
dark	73	chaparral	
fine-sculpted	73	Bridge Creek	141
flyspeck	73	Mormon Island	141
four-thread	73	scaley	141
grass	73	Shasta	141
Guinea	73	Tehama	141
ivory	73	*chaptalii, Clio*	110
Japanese false	72	*charlottensis, Cerithiopsis*	73
Santa Barbara	73	*Charonia*	83
slender	73	Charopidae	130
stocky	73	*charybdis, Crepipatella*	79
threaded	73	*chattanoogaense, Pleurobema*	32
variable	73	*chazaliei, Pecten*	27
varicose	72	*cheatumi*	
West Indian false	72	*Euchemotrema*	138
Cerithidea	72	*Humboldtiana*	144
Cerithiidae	72	*Pseudosubulina*	129
Cerithiopsidae	73	*Tryonia*	63
Cerithiopsis	73	*Cheilea*	79
Cerithium	73	*Chelidonura*	108
Cerodrillia	96	*chemnitzii, Anadara*	24
ceroplasta, Mangelia	98	*chenui, Gregariella*	23
cerrosensis catalinensis, Austrotrophon	84	*chersinellum, Pristiloma*	135
cervina		*chersinus, Euconulus*	133
Aplysia	112	*chesapeakea, Sayella*	104
Viridrillia	100	chestnut	
cervus, Cypraea	80	San Diego	141
cesta		San Gabriel	141
Cerithiopsis	73	*chiachianus, Oenopota*	99
Mangelia	98	*chica, Doto*	119
cestrota, Nuculana	21	*Chicoreus*	84
chacei, Pomatiopsis	63	*chilhoweensis, Mesodon*	138
Chaceia	46	*chilluna, Polycera*	116
chadwicki, Triodopsis	140	*chiltonensis, Elimia*	63
Chaenaxis	126	*chinensis malleata, Cipango*	59
Chaetoderma	20	Chinese-hat	
Chaetodermatidae	20	circular	79

Pacific ... 79
chinooki, Odostomia ... 101
Chione ... 44
chipolaensis, Elliptio ... 29
chiricahuana
    Ashmunella ... 137
    Holospira ... 129
chiricahuanum, Radiocentrum ... 142
chishimanum, Buccinum ... 88
chisosensis
    Humboldtiana ... 144
    Polygyra ... 139
Chiton ... 150
chiton
    Atlantic rose ... 148
    blue-spot ... 148
    boreal veiled ... 150
    Caribbean red ... 148
    eastern beaded ... 149
    eastern orange ... 148
    eastern surf ... 149
    giant Pacific ... 150
    gold-flecked ... 150
    multihue ... 148
    multiring ... 148
    northern white ... 149
    Pacific veiled ... 150
    Portobelo ... 148
    red glass-hair ... 150
    red veiled ... 150
    slender glass-hair ... 150
    smooth glass-hair ... 150
    striate glass-hair ... 150
    veiled ... 150
    West Indian fuzzy ... 150
    West Indian green ... 150
    West Indian ribbed ... 149
    white-barred ... 150
Chitonidae ... 150
chittenangoensis, Succinea ... 133, 155
chittyana, Corbula ... 46
Chlamydoconcha ... 37
Chlamydoconchidae ... 37
Chlamys ... 26
chlorotica, Elysia ... 113
chocolata, Turbonilla ... 104
choctawensis, Villosa ... 33
Chondropoma ... 66
christenseni, Sonorella ... 145
christyi, Mesodon ... 138
Chromodorididae ... 116
Chromodoris ... 116
chromosoma, Spurilla ... 121
chrysalis fastigatum, Cerion ... 128

chrysalloideus, Aesopus ... 86
churchi
    Monadenia ... 144
    Odostomia ... 101
cicatricosus, Plethobasus ... 31, 154
Cidarina ... 55
cidaris, Cidarina ... 55
ciliata, Mopalia ... 150
ciliatum
    Buccinum ... 88
    Clinocardium ... 39
cimex, Drepanotrema ... 124
Cincinnatia ... 60
cincinnatiensis, Pomatiopsi ... 63
cincta, Odostomia ... 101
cineracea, Marginella ... 93
cinerea
    Cypraea ... 80
    Hastula ... 95
    Urosalpinx ... 86
cinereum, Gastropteron ... 108
Cingula ... 67
cingula, pointed ... 67
cingulata
    Philine ... 108
    Scissurella ... 52
Cipangopaludina ... 59
circinata, Aforia ... 95
Circomphalus ... 44
circularis, Argopecten ... 26
Circulus ... 69
circumcarinata, Monadenia ... 144
circumcincta, Admete ... 94
circumcinctum, Antalis ... 50
circumscriptus, Arion ... 131
circumstriatus, Gyraulus ... 124
circumtexta, Ocenebra ... 85
cirrata, Mopalia ... 150
Cirsotrema ... 75
cistelliformis, Elliptio ... 29
cistronium, Aorotrema ... 69
cithara
    Carinaria ... 81
    Mopalia ... 150
citrina, Berthellina ... 114
citrinus, Stylocheilus ... 112
citronella, Cryoturris ... 96
Cittarium ... 55
civitella, Borsonella ... 96
claibornensis, Triodopsis ... 140
clam
    Asian ... 44
    butter ... 45
    calico ... 44

chiton............................. 37
  mahogany 42
  ox-heart........................... 43
  pismo.............................. 45
  sea-cucumber 37
  Washington........................ 45
clappi
  Carychium....................... 125
  Cerodrillia....................... 96
  Discus 131
  Gastrocopta..................... 126
  Helicina......................... 58
  Paravitrea 135
  Planogyra 128
  Radiocentrum 142
  Sonorella....................... 145
  Vertigo......................... 127
Clappia 60, 157
clara, Elimia..................... 63
clarinda, Turbonilla 104
clarki
  Amnicola........................ 59
  clarki, Mesodon 138
  Detracia........................ 122
  nantahala, Mesodon 138, 155
clathrata
  Daphnella 97
  Distorsio 83
  Tryonia 63
clathratus, Boreotrophon 84
Clathromangelia..................... 96
Clathurella 96
clausa
  Bulla 109
  Elimia 63, 157
  Natica 82
clausus, Mesodon.................. 138
clava
  Caecum......................... 71
  Pleurobema 32
clavaeformis, Elimia................. 63
claviculata, Leiomya 49
clavium, Teinostoma 70
clavulinus, Lamellaxis 128
Cleba............................ 110
cleftclam
  Atlantic 36
  cut-off 36
  disjunct 35
  equal 35
  flexuose 35
  giant............................. 35
  Gould............................ 36
  granulose......................... 36
  Mya.............................. 35
  reddish........................... 35
  rotund 36
  Santa Barbara 35
  short 35
  simple 36
  swan 35
clementensis
  Mitrella 87
  Odostomia...................... 101
clementina
  Odostomia...................... 101
  Sterkia......................... 127
  Turbonilla 104
clenchi
  Chione.......................... 44
  Chromodoris 116
  Elimia 64
  Mesodon 138
  Neritina......................... 58
cleo, Schwartziella 68
cleryana, Mulinia 40
clingmani, Glyphyalinia 134
Clinocardium 39
clinocnemus, Mirachelus 56
Clio 110
clio
  cuspidate 110
  keeled 110
  pyramid 110
  striate........................... 110
  two-keel......................... 110
  wavy 110
Clione............................ 111
Clionidae........................... 111
Cliopsidae.......................... 111
Cliopsis 111
clipeata, Leptoxis.................. 65, 157
clubshell 32
  Alabama 32
  black 32, 154
  ovate 32
  painted........................... 32
  southern.......................... 32
  Tennessee 32
coarctata, Cumingia................. 43
coccineum, Pleurobema.............. 32
Cocculina 54
cocculina
  Alaska 54
  reticulate 54
Cocculinidae........................ 54
cochilaris, Elimia 64
cochisensis, Gastrocopta 126

cochlear, Neopycnodonte............... 28
Cochlespira......................... 96
Cochlicella 146
Cochlicopa 126
Cochlicopidae 126
Cochliolepis 71
Cochliopa 60
Cochliopina......................... 60
Cochlodinella 129
cockerelli
   Anadenulus 131
   Ashmunella 137
   Fossaria........................ 122
   Holospira....................... 129
   Laila 116
cockle
   Antillean 40
   broad 39
   California....................... 39
   Greenland 39
   hairy 39
   hundred-line 39
   Nuttall 39
   smooth........................... 39
   strait 39
cocoachroma, Cuthona 120
cocolitoris, Teinostoma 70
Codakia............................. 34
coelaxis, Ventridens 136
coerulea, Ercolania 113
coeruleum, Prophysaon 132
coffeus, Melampus 122
cognata, Truncilla 33
cohuttense, Stenotrema 140
coil
   bluff........................... 130
   burrowing 130
   compound 131
   cricket 130
   crosstimbers 131
   fringed 130
   glass........................... 131
   Mexican......................... 130
   oldfield........................ 130
   punctate........................ 131
   raccoon 130
   rubble 130
   salmon.......................... 131
   shaggy 130
   smooth.......................... 131
   spiral 130
   tallus 131
   temperate....................... 131
   tight 131

   toothy 130
   twilight........................ 131
   Virginia 131, 155
   wax 131
coin
   humped 141
   tight........................... 141
   toothed 141
coindeti, Illex 152
Coleophysis........................ 109
collina, Pleurobema 32, 154
Collisella.......................... 53
collisella, Ventridens 136
coloradensis, Acroloxus 122
coloradoensis, Sonorella 145
colorata, Tellina................... 41
coltoniana, Sonorella 145
Colubraria.......................... 90
Colubrariidae 90
Columbella 87
columbella, Erato................... 80
Columbellidae 86
columbiana
   Amphissa 86
   Catriona 120
   Doto........................... 119
   Fluminicola 60
   Megacrenella.................... 23
   Melanella....................... 77
   Odostomia...................... 101
   Physella....................... 124
   Vertigo........................ 127
   Vitrinella 70
columbianus
   Ariolimax...................... 132
   Eubranchus 120
   Vespericola 141
columbiensis, Planorbella 125
Columella 126
columella, Pseudosuccinea 123
column
   crestless....................... 127
   high-spire...................... 126
   Rocky Mountain 127
   threetooth 127
   toothless 126
   top-heavy....................... 127
   widespread 127
columna, Cerithiopsis............... 73
columnella, Cuvierina.............. 110
Colus............................... 88
colymbus, Pteria.................... 25
comalensis
   Cincinnatia 60

INDEX

Elimia 64
combshell
  cumberlandian 29
  round 30, 156
  southern 30, 154
  upland 30
comma, Elimia 64
commoda, Capulacmaea 79
communis
  Ficus 83
  Neptunea 89
comoxensis, Melanella 77
compacta
  Astarte 38
  Leptoxis 65
  Melanella 77
compactus
  Fossarus 78
  Petaloconchus 72
compar, Sonorella 145
complanata
  alabamensis, Lasmigona 31
  complanata, Lasmigona 31
  Elliptio 29
  Geitodoris 118
  Triodopsis 140
compressa
  Lasmigona 31
  Myrtea 35
  Mysella 37
  Neaeromya 37
Compressidens 51
compressum, Pisidium 34
Compsodrillia 96
Compsomyax 44
compta, Tricolia 58
comptum, Cymatium 83
concava
  Nystiella 76
  Terebra 95
concavum, Haplotrema 129
concentrata, Oreohelix 142
concentrica
  Ervilia 40
  Lepeta 54
  Nuculana 21
conch
  crown 90
  Florida fighting 78
  hawkwing 78
  horse 91
  milk 78
  queen 78
  roostertail 78
  West Indian fighting 78
conchaphila, Ostreola 28
Conchocele 35
conchyliophora, Xenophora 79
concinna, Cuthona 120
concinnula, Vertigo 127
concinnulus, Oenopota 99
concordia, Skenea 57
condylclam
  Bernard 38
  Dall 38
  tiny 38
Condylocardiidae 38
condylum, Caecum 71
cone
  agate 94
  alphabet 95
  Bermuda 95
  California 94
  cancellate 94
  carrot 94
  crown 95
  dusky 95
  flame 94
  flamingo 94
  Florida 94
  glory-of-the-Atlantic 95
  jasper 95
  mace 94
  mouse 95
  slender 94
  speckled 95
  sunrise 95
  yellow 95
conecuhensis
  Paravitrea 135
  Vertigo 127
coneja, Crimora 115
confragosa, Oreohelix 142
confragosus, Arcidens 29
confusa
  Limatula 26
  Transennella 45
congaraea, Elliptio 29
congregata, Chama 36
conica
  Phreatodrobia 61
  Sabia 78
  Spilochlamys 63
  Velutina 80
conicla, Eubranchus 120
Conidae 94
connasaugaensis, Strophitus 33

conoidea
    Melanella 77
    Physella 124
conradi
    Penitella 47
    Thracia 48
    Turbonilla 104
conradia, Cyclostremella 101
conradiana, Clathurella 96
conradicus, Medionidus 31
conradina, Transennella 45
consensus, Nassarius 91
consobrina, Tellina 41
conspectum, Punctum 130
conspicua, Stenoplax 149
constricta
    macgintyi, Distorsio 83
    Macoma 41
    Mangelia 98
    Villosa 33
constrictus, Somatogyrus 62
consuela, Siratus 85
continentis, Haplotrema 129
contracosta, Helminthoglypta 143
contracta
    Corbula 46
    Gastrocopta 126
    Vitrea 136
contractus, Stagnicola 123
Conualevia 117
Conualeviidae 117
Conus 94
conus, Collisella 54
conventus, Pisidium 34
convexa
    Crepidula 79
    Metaxia 74
conyna, Polycerella 116
cookeana
    Asperiscala 75
    Macrarene 57
    Odostomia 102
cookei
    Leptadrillia 98
    Microphysula 141
cooperclam
    Atlantic 45
    shiny 45
Cooperella 45
Cooperellidae 45
cooperi
    Aldisa 117
    Caecum 71
    Cancellaria 94

Coryphella 120
    Lepidozona 149
    Physella 124
    Puncturella 53
    Turritella 72
cooperianus, Plethobasus 31, 154
cooperii, Yoldia 22
coosaensis, Somatogyrus 62
coquina
    giant 42
    Texas 42
    variable 42
cor, Fusconaia 30, 154
cora, Amnicola 59
coralclam 43
Coralliophaga 43
coralliophaga
    Coralliophaga 43
    Gregariella 23
Coralliophila 86
Coralliophilidae 86
coralliotis, Dimya 27
coralsnail
    California 86
    Caribbean 86
    globose 86
    helmet 86
    staircase 86
Corambe 114
corambe
    frost-spot 114
    obscure 114
Corambidae 114
Corbicula 44
corbicula
    Cancellaria 94
    Daphnella 97
Corbiculidae 44
corbis, Mirachelus 56
Corbula 46
corbula
    Barratt 46
    contracted 46
    Cuba 46
    Dietz 46
    Kelsey 46
    Krebs 46
    nut 46
    ribbed 46
    snubby 46
    snubnose 46
    Swift 46
    wavy 46
    western 46

Corbulidae	46
corbuloidea, Basterotia	38
corbuloides, Thracia	48
cordata	
Axinopsida	35
Cleba	110
cordatum, Pleurobema	32
cordatus, Pitar	44
corinnae, Engina	89
corinneae, Bosellia	112
corneum, Sphaerium	34
cornuarietis, Marisa	59
cornucopiae, Caecum	71
cornuta, Arcinella	36
Corolla	110
corolla	
Atlantic	110
spectacular	111
corona, Melongena	90
coronadoensis	
Schwartziella	68
Skenea	57
coronadoi	
Borsonella	96
Phalium	82
coronata	
Doto	119
Erycina	36
Pterotrachea	81
Trichotropis	79
corpulenta	
Planorbella	125
Pleurocera	66
corrugata	
Haliotis	52
ponderosa, Bursa	83
Stenoplax	149
corrugatum	
amictum, Cymatium	83
krebsii, Cymatium	83
corrugatus, Musculus	23
corteziana, Glycymeris	24
corticaria, Gastrocopta	126
corvunculus, Toxolasma	33
Coryphella	120
Coryphellidae	120
corys, Onchidiopsis	80
Cosa	25
cosmia	
Cerithiopsis	73
Manzonia	67
Cosmioconcha	87
cosmius, Circulus	69
costai, Ercolania	113
costalis, Margarites	55
Costasiella	112
costasiella, eyespot	112
Costasiellidae	112
costata	
Cardiomya	49
Cerithidea	72
Codakia	34
Coralliophila	86
Cyrtopleura	46
Lasmigona	31
Physella	124
Siliqua	40
Vallonia	128
costatum, Haplotrema	129
costatus, Strombus	78
Costellariidae	92
costellata, Cardiomya	49
costifera, Elimia	64
costulata	
Cerithiopsis	73
Moelleria	57
couchiana, Quadrula	32
couei, Fusinus	91
couperiana, Anodonta	28
Couthouyella	75
couthouyi, Admete	94
covert	
big-tooth	138
cinnamon	139
clifty	139
engraved	138
glossy	138
Ocoee	138
Smoky Mountain	138
velvet	139
cowrie	
Atlantic deer	80
Atlantic gray	80
Atlantic yellow	80
chestnut	80
measled	80
Surinam	80
cowrie-helmet, reticulate	82
cracherodii, Haliotis	52
cragini, Triodopsis	140
crassa, Leptoxis	65
crassatella	
beautiful	39
wavy	39
Crassatellidae	39
Crassedoma	26
Crassicardia	38
crassicornis, Hermissenda	121

crassicostatus, Callistochiton............ 148
crassidens
    Crassicardia 38
    Elliptio........................... 29
crassilabris, Somatogyrus 62
Crassinella 39
crassinella
    lunate........................... 39
    Martinique....................... 39
    Pacific 39
crassior, Lacuna..................... 66
Crassispira 96
Crassostrea......................... 28
crassula
    Macoma 41
    Pyramidella..................... 103
crassulum, Campeloma 59
crassus
    Onobops 61
    Somatogyrus.................... 62
Cratena............................ 121
crater
    queen........................... 138
    spike-lip........................ 139
crawfordiana, Cancellaria 94
crebricinctum, Caecum 71
crebricostata, Cyclocardia............. 38
crebricostatus, Beringius 87
creekmussel
    Alabama 33
    southern......................... 33
creekshell
    Carolina......................... 33
    Coosa........................... 33
    eastern.......................... 33
    Kentucky........................ 33
    mountain 33
    Ouachita 33
    painted.......................... 33
    rayed 29
crenata
    Astarte.......................... 38
    Opalia 76
crenatella, Elimia.................... 64
Crenella............................ 23
crenella
    bean............................ 23
    British Columbia.................. 23
    cross-sculpture 23
    fragile........................... 23
    glandular 23
    joking........................... 23
    Lea 23
    little-comb....................... 23
    partly-sculptured.................. 23
    spreading-sculpture................ 23
crenifera, Terebra.................... 95
crenulata
    Megathura...................... 53
    Nucula......................... 21
    Pyramidella..................... 104
Crepidula 79
Crepipatella 79
Creseis 110
cretaceus, Plicifusus.................. 90
Crimora............................ 115
crispata
    Anatoma 52
    Tridachia....................... 113
    Zirfaea......................... 47
crispatissima, Ocenebra............... 85
crista, Gyraulus 124
cristallina
    Cyerce 112
    Tellina 41
cristata
    Carinaria....................... 81
    Gastrocopta..................... 126
    Limopsis 24
    Tellidora 41
cristulata, Monadenia 144
cronkhitei, Discus.................... 131
Crossata 83
crosseana, Sayella 104
crosseanum, Eudolium............... 83
crossei, Holospira.................... 129
crotalina, Helminthoglypta 143
crownsnail
    serrate 62
    spiny 62
cruciata, Hermaea 113
cruciatum, Pisidium 34
Crucibulum 79
cruenta
    Sanguinolaria 42
    Tricolia 58
cruentata, Arene..................... 57
Cryoturris 96
cryptica, Fontigens................... 61
Cryptochiton....................... 150
Cryptoconchus 150
Cryptomastix....................... 138
cryptomphala, Glyphyalinia............ 134
Cryptomya 45
Cryptopecten....................... 26
cryptospira, Teinostoma............... 70
Cryptostrea......................... 28

crystallina
    Cerithiopsis 73
    Clathurella. 96
cubana
    Codakia. 34
    Crassispira 96
cubaniana
    Corbula 46
    Transennella 45
cubensis
    Eupera. 34
    Fossaria. 122
    Laemodonta 122
    Physella. 124
cubitatum, Caecum 71
cucullata, Puncturella 53
culcitella, Acteocina. 107
cultellus, Musculus. 23
culveri, Antrobia. 60
Cumanotidae. 120
Cumanotus 120
Cumberlandia 29
cumberlandiana
    Anguispira. 131
    Glyphyalinia 134
Cumingia. 43
cumingianus, Solecurtus 43
cumingii, Mysouffa. 107
cumshewaensis, Odostomia 102
Cuna 38
cuneata
    Pythenella 37
    Rangia. 40
cuneiformis, Martesia. 46
cuneolus, Fusconaia. 30, 154
cup-and-saucer
    false. 79
    spiny 79
    striate. 79
    West Indian. 79
cupella, Okenia. 115
cupreus, Mesomphix. 134
curreyana, Elimia. 64
currierianus, Somatogyrus 62
curta
    Lithasia 65
    Pleurocera. 66
    Thracia 48
    Turbonilla 104
curtum, Pleurobema. 32, 154
curvicostata, Elimia 64
Cuspidaria. 49
Cuspidariidae 49
cuspidata, Clio. 110
Cuthona 120
cuthona
    black-and-yellow. 120
    bornagain. 120
    brown 120
    colorful 120
    green 121
    Lake Merritt 120
    orange-face 120
    orange-tip 120
    red-tentacle 120
    rose-pink 121
    white-crust. 120
    yellowish 120
Cuvierina. 110
cuyama, Helminthoglypta 143
cuyamacensis, Helminthoglypta 143
cyaneum
    cyaneum, Buccinum 88
    patulum, Buccinum. 88
    perdix, Buccinum 88
Cyathodonta. 48
cyclia, Adontorhina 35
Cyclinella 44
cyclinella, thin 44
Cyclocardia. 38
cyclocardia
    bracelet 38
    bumpy 38
    cut 38
    many-rib 38
    New England. 38
    northern. 38
    ovate 38
    Rjabinina 38
    Santa Barbara 38
    stout. 38
    Umnak. 38
cycloferum, Caecum. 71
Cyclonaias 29
Cyclopecten 26
cyclophorella, Vallonia. 128
cyclostoma, Fossaria 122
cyclostomaformis, Lioplax 59
Cyclostrema 57
Cyclostrematidae 57
cyclostreme
    briar. 57
    California. 57
    cancellate. 57
    Farallon 57
    gem 57
    graceful 57
    Key West 57

sharp-rib	57
star	57
Tortugas	57
tropical	57
tuberculate	57
variable	57
warty	57
Cyclostremella	101
Cyclostremiscus	69
cydia, *Drillia*	97
Cyerce	112
cygnus, *Thyasira*	35
Cylichna	108
Cylichnidae	108
cylindracea, *Elimia*	64
cylindrellus, *Toxolasma*	33, 154
cylindrica	
cylindrica, *Quadrula*	32
strigillata, *Quadrula*	32
Volvulella	109
cylindricus, *Heliacus*	74
Cylindrobulla	112
Cylindrobullidae	112
cylleniformis, *Enaeta*	93
Cymakra	96
cymata	
Amphissa	86
Psephidia	45
Cymatium	83
Cymatoica	40
Cymatosyrinx	96
cymatus, *Boreotrophon*	84
Cymbovula	80
Cymbulia	111
Cymbuliidae	110
cymella, *Corbula*	46
cymodacea, *Phyllaplysia*	112
cymodoce, *Glyphostoma*	97
cynocephalum, *Cymatium*	83
Cyphoma	80
cyphoma	
bullroarer	81
fingerprint	81
gold-line	80
intermediate	81
plump	81
cyphyus, *Plethobasus*	31
Cypraea	80
Cypraecassis	82
Cypraeidae	80
cypreophila, *Helminthoglypta*	143
cypria, *Odostomia*	102
Cyprogenia	29
Cyrenoida	36
Cyrenoididae	36
Cyrtodaria	46
Cyrtonaias	29
Cyrtopleura	46
Cystiscus	93
Cythnia	77

D

Dacrydium	23
dacryon, *Rhapinema*	62
dactylomela, *Aplysia*	112
dagger	127
island	127
ribbed	127
Rocky Mountain	127
white-lip	127
dalli	
Amnicola	59
Babelomurex	86
Caecum	71
Chromodoris	116
Circulus	69
Cirsotrema	75
Cuna	38
Dendronotus	119
Fossaria	122
Inodrillia	97
Nodulotrophon	85
Onoba	67
Portlandia	22
Rhabdus	50
Schwartziella	68
Scissilabra	70
Sonorella	145
Turbonilla	104
dalliana	
Gastrocopta	126
Nesovitrea	134
Vertigo	127
dalva, *Trapania*	115
danai, *Terebra*	95
danielsi	
Ashmunella	137
Hemphillia	132
Holospira	129
Sonorella	145
Daphnella	97
daphnelle	
Morro	97
volute	97
dariensis, *Elliptio*	29
datemussel	
black	23

California	23
curved	23
giant	23
Kelsey	23
mahogany	23
Rogers	23
San Diego	23
scissor	23
dauca, Dryachloa	133
daucus, Conus	94
davenportii, Polycerella	116
dawsoni, Montacuta	37
dealbata, Olivella	92
dealbatus, Rabdotus	130
deauratum, Mesodesma	40
debile, Sinum	82
debilis, Diaphana	109
decampi	
Campeloma	59
Oxyloma	133
deceptum, Stenotrema	140
decipiens, Somatogyrus	62
decisa, Semele	43
decisum	
Campeloma	59
Pleurobema	32
decisus, Amnicola	59
declivis, Uniomerus	33
decollata, Rumina	128
decorata	
Lasmigona	31
Puncturella	53
Triphora	75
decoratus, Callistochiton	148
decussata	
Crenella	23
Glycymeris	24
Pedicularia	81
Rissoina	68
decussatus	
Oenopota	99
Serpulorbis	72
Ventridens	136
deertoe	33
defilippi, Octopus	152
deflectus, Gyraulus	124
deflorata, Asaphis	42
deformis, Volutopsius	90
delecta, Polygyra	139
Delectopecten	26
delessertii, Conus	94
delicata, Sonorella	145
delmontana, Turbonilla	104
delmontensis, Melanella	77
Delonovolva	81
delphinodonta, Nucula	21
deltoidea	
Polygyra	139
Thais	85
delumbis, Villosa	33
demarestia, Firoloida	81
demissa, Geukensia	23
demissus, Ventridens	136
Dendostrea	28
dendritica, Placida	113
Dendrochiton	148
Dendrodorididae	117
Dendrodoris	117
Dendronotidae	118
Dendronotus	118
Dendropoma	72
dennisoni, Morum	92
denotata, Triodopsis	140
densilineata, Mangelea	98
Dentaliidae	50
Dentalium	50
dentata, Divaricella	35
dentatum, Chondropoma	66
dentatus, Euconulus	133
denticulatum, Epitonium	75
dentiens	
cryptica, Lepidochitona	148
entiens, Lepidochitona	148
dentifera, Triodopsis	140
dentiferum, Glyphostoma	97
dentigera, Emarginula	53
dentilla, Paravitrea	135
Dentimargo	93
Dentistyla	55
depicta, Notoacmea	54
depilatum, Stenotrema	140
depressa, Ancistrobasis	56
Depressiscala	75
depressus	
Gyrodes	82
Somotogyrus	62
dermestinum, Vexillum	92
Dermomurex	84
Deroceras	136
deserta, Fontelicella	61
desertsnail	
Argus	143
Avawatz	145
Black Rock	145
Borrego	145
Chuckwalla Spring	143
Coachella	143
eastern	143

Joshua Tree............ 145
Morongo............ 143
Mountain Spring............ 145
Orocopia............ 143
Panamint............ 143
Resting Spring............ 145
Soledad............ 145
Thousand Palms............ 143
white............ 143
Whitewater............ 143
Desmopteridae............ 111
Desmopterus............ 111
despecta, Neptunea............ 89
Detracia............ 122
devexa
    Epicynia............ 70
    Thracia............ 48
devia, Cryptomastix............ 138
dexioptera, Macoma............ 41
diabloensis, Helminthoglypta............ 143
diaboli, Tryonia............ 63
Diacria............ 110
diadema, Helicodiscus............ 130
diademata, Onchidoris............ 115
diana, Paravitrea............ 135
dianthophila, Fargoa............ 101
Diaphana............ 109
diaphana
    Mitrella............ 87
    Vitrinella............ 70
Diaphanidae............ 109
diaphanus
    Laevapex............ 125
    Thliptodon............ 111
Diaphorodoris............ 115
diaulax, Antiplanes............ 95
Diaulula............ 118
dicella, Odostomia............ 102
dickinsoni, Elimia............ 64
Didianema............ 70
dido, Inodrillia............ 97
didyma, Trachypollia............ 86
didymus, Episiphon............ 50
diegensis
    Adula............ 23
    Borsonella............ 96
    Cerithiopsis............ 73
    Cylichna............ 108
    Limopsis............ 24
    Pecten............ 27
    Turbonilla............ 104
    Vitrinorbis............ 71
diegoana, Aligena............ 36

diegoensis
    Lamellaria............ 79
    Zonites............ 130
dietziana, Corbula............ 46
digitalis, Collisella............ 54
digueti
    Lamellaria............ 79
    Octopus............ 152
dilatata
    Elliptio............ 29
    Leptoxis............ 65
dilatatus, Micromenetus............ 124
dilatosculptus, Ischnochito............ 148
diminuta, Glyphoturris............ 97
Dimya............ 27
Dimyella............ 27
dimyid
    California............ 27
    coral............ 27
    Goreau............ 27
    silver............ 27
    Starck............ 27
Dimyidae............ 27
dinella, Odostomia............ 102
dingbat, flapping............ 108
Dinocardium............ 39
dinora
    Onoba............ 67
    Turbonilla............ 104
Diodora............ 52
diomedea
    Cerithiopsis............ 73
    Melanochlamys............ 108
    Kurtziella............ 98
    Tritonia............ 118
Loliolopsis............ 151
dioscoricola, Pupisoma............ 127
diplodon
    Aleutian............ 36
    Atlantic............ 36
    Cande............ 36
    equal............ 36
    marked............ 36
    nut............ 36
    orb............ 36
    pimpled............ 36
    rough............ 36
    sister............ 36
    subglobose............ 36
    Venezuela............ 36
    Verrill............ 36
Diplodonta............ 36
Diplosolenodes............ 147
Diplothyra............ 46

dipperclam
    alternate............................ 49
    California........................... 49
    glacial.............................. 49
    Jeffrey.............................. 49
    little-snout......................... 49
    median.............................. 49
    obese............................... 49
    rostrate............................ 49
    subglacial.......................... 49
    translucent......................... 49
dira, Searlesia........................ 90
directus, Ensis....................... 40
Dirona............................... 119
dirona
    golden.............................. 119
    painted............................. 119
    white-line.......................... 119
Dironidae.............................. 119
disc
    Alabama............................ 131
    angular............................. 131
    Appalachian........................ 131
    Black Mountain.................... 131
    channelled......................... 131
    coastal-plain...................... 131
    Cumberland........................ 131
    domed.............................. 131
    file................................ 131
    flamed............................. 131
    forest.............................. 131
    marbled............................ 131
    mountain........................... 131
    Nimapuna.......................... 131
    painted........................ 131, 155
    Pine Mountain..................... 131
    Pleistocene.................... 131, 155
    rotund............................. 131
    rustic.............................. 131
    saw-tooth.......................... 131
    southeastern....................... 131
    striate.............................. 131
Discidae........................... 131, 155
Discodorididae........................ 118
Discodoris............................ 118
discoidea, Triodopsis................. 140
Disconaias............................ 29
discors
    Lirularia........................... 55
    Musculus.......................... 23
Discotectonica........................ 74
Discus........................... 131, 155
discus
    Discotectonica..................... 74

Dosinia.............................. 44
    Periploma......................... 48
disjuncta, Conchocele................ 35
dislocata
    Elimia.............................. 64
    Terebra............................ 95
disparilis, Boreotrophon.............. 84
dissimilis, Yoldiella.................. 22
distinctus, Arion..................... 131
Distorsio............................. 83
distorsio
    Atlantic............................ 83
    hunchback......................... 83
distorta, Thracia..................... 48
divae
    Doto............................... 119
    Precuthona........................ 121
    Runcina........................... 109
divaricata, Crenella.................. 23
Divaricella............................ 35
diversa, Coryphella.................. 120
diversicolor, Dendronotus............ 119
divesta, Triodopsis................... 140
divisus, Tagelus...................... 43
docima, Heleobops................... 61
doerga, Doto......................... 119
dofleini, Octopus.................... 152
dogwinkle
    Atlantic............................ 85
    channeled.......................... 85
    emarginate......................... 85
    file................................ 85
    frilled.............................. 85
dolabelloides, Lexingtonia............ 31
dolabraeformis, Lampsilis............ 30
dolabrata, Pyramidella............... 104
Dolabrifera........................... 112
dolabrifera, Dolabrifera.............. 112
dolabriformis, Mactra................ 40
Dolicholatirus........................ 91
dolichophallus, Ariolimax............ 132
doliiformis, Paedoclione............. 111
dombeyanus, Plectomerus............ 31
dome
    bidentate.......................... 136
    blade............................... 136
    carinate............................ 136
    copper.............................. 136
    crossed............................ 136
    flat................................ 136
    globose............................ 136
    glossy............................. 136
    golden.............................. 136
    green............................... 136

hollow 136
perforate 136
pyramid 136
rounded 136
sculptured 136
Seminole 136
split-tooth 136
Tennessee 136
throaty.......................... 136
wax 136
western 136
yellow 136
domingensis
    Barbatia......................... 24
    Brachidontes..................... 23
dominguensis, Glans 38
dominicus, Drymaeus............... 130
Donacidae........................ 42
donaciformis, Truncilla 33
Donax 42
donca, Aplysia 112
Dondersia 20
Dondersiidae..................... 20
Dondice......................... 121
dondice, western................... 121
dora, Turbonilla 104
dorfueilliana, Polygyra.............. 139
Doridella 114
Dorididae 117
Doriopsilla 117
Doris 117
doris
    blood-spot...................... 117
    brown spiny..................... 115
    California blue.................. 116
    California flat................... 118
    feline 117
    Florida regal 116
    gritty 118
    harlequin blue 116
    hunchback...................... 117
    orange-peel 115
    rabbit.......................... 115
    red sponge..................... 117
    red-line blue 116
    ringed.......................... 118
    salt-and-pepper 116
    salted yellow 117
    tiny black-spot.................. 117
    three-stripe 116
    white night..................... 117
    white smooth-horn 117
    wine-plumed spiny 115
    yellow false 115

dormani, Drymaeus 130
dorriae, Rimula................... 53
dorsalis, Pallifera.................. 132
Dosidicus 151
Dosinia 44
dosinia
    disk 44
    elegant......................... 44
Doto 119
doto
    British Columbia................. 119
    crown.......................... 119
    dark........................... 119
    hammerhead 119
Dotoidae......................... 119
dovesnail
    ant-egg........................ 86
    argus 87
    banded......................... 87
    beautiful........................ 86
    brown-band 87
    carinate 86
    chain 86
    cocoon......................... 86
    deepwater 87
    engraved 87
    fat 86
    fenestrate...................... 87
    flame 87
    Florida......................... 86
    glossy.......................... 87
    golden 87
    graceful 86
    greedy 86
    Gulf 86
    lineate 86
    lunar 87
    many-spot 87
    muddy......................... 87
    penciled........................ 87
    rosy northern................... 87
    rusty 87
    San Clemente 87
    Santa Monica 86
    shaggy......................... 87
    shy............................ 87
    simple 87
    smooth......................... 87
    sparse 86
    translucent..................... 87
    variegate 87
    well-ribbed..................... 86
    West Indian..................... 87
    white-spot 87

downieanus, Mesodon	138	*dupetithoursi, Helminthogly*	143
downiei, Elliptio	29	*duplicata, Neverita*	82
dracona, Turbonilla	104	*duranti, Haplotrema*	129
draconis, Polinices	82	*duryi, Planorbella*	125
dragoonensis, Sonorella	145	duskysnail	
draparnaudi, Oxychilus	135	Canadian	60
draperi, Homalopoma	57	indented	60
Dreissenidae	43	lake	60
Drepanotrema	124	pupa	60
drill		Rocky Mountain	60
Atlantic oyster	86	slender	59
Gulf oyster	86	squat	59
mauve-mouth	84	*duttoniana, Lithasia*	65
rosy	84	dwarf-cockle	
sharp-rib	84	elegant	39
Tampa	86	northern	39
thick-lip	84	dwarf-tellin	
waxy	86	northern	41
Drillia	97	shiny	42
drillia		dwarf-turban	
orange-rib	96	berry	57
rough	97	dark	57
solid	100	few-rib	57
tea	96	northwest	57
white-band	100	rayed	57
white-knob	100	two-faced	57
Drilliola	97	white	57
dromas, Dromus	29, 154	dwarf-venus	
dromedarius, Hemphillia	132	Lord	45
Dromus	29, 154	oval	45
drop		Stephens	45
cherrystone	58	wavy	45
globular	58	*dysbatus, Floridiscrobs*	60
ochre	58	*dysoni, Diodora*	52
rainbow	58		
striate	58	**E**	
drupe			
blackberry	86	*ebena, Fusconaia*	30
dark	86	*ebenum, Elimia*	64
twin	86	ebonyshell	30
Dryachloa	133	round	31
Drymaeus	130	*eborea, Graptacme*	50
dubia		eburnea	
Finella	73	*Acteocina*	107
Polycera	116	*Cylichna*	108
Scaphella	93	*eburneolus, Dentimargo*	93
dubiosa, Cyathodonta	48	*eburneum, Cerithium*	73
dubium		*echinaticostum, Epitonium*	75
Pisidium	34	*edentatus, Mesodon*	138
Prophysaon	132	*edenticulata, Hypselodoris*	116
duckclam		*edentula, Columella*	126
channeled	40	*edgariana, Elimia*	64
Pacific	40	*edgarianum, Stenotrema*	140
smooth	40	*edithae, Humboldtiana*	144

edmondi, Odostomia 102
edulis
   Mytilus............................ 24
   Ostrea 28
edvardsi, Stenotrema 140
edwardensis, Turbonilla 104
edwardsi, Helminthoglypta 143
effusum, Oxyloma..................... 133
effusus, Vorticifex..................... 125
eggcockle 39
   delicate 39
   giant.............................. 39
   Morton............................ 39
   Pacific 39
   painted............................ 39
egmontianum, Trachycardium 39
egretta, Mopalia...................... 150
eigenmanni, Helicodiscus............... 130
eiseni
   Modiolus 23
   Tegula 56
Elachisina 67
Elachisinidae......................... 69
Elaeocyma 97
elata, Vitricythara 100
elatior
   Rhodacme........................ 125
   Vertigo........................... 127
elatum, Laevicardium.................. 39
eldorana, Odostomia 102
electrina, Nesovitrea 134
elegans
   Bushia 48
   Cochlespira....................... 96
   Dosinia 44
   Fossarus 78
   Haminoea 109
   Oenopota......................... 99
   Trochoidea 146
elegantula
   branfordensis, *Turbonilla* 104
   elegantula, *Turbonilla*............... 104
elegantulum
   Cerastoderma 39
Dacrydium 23
elephant-ear 29
   fluted 29
   Georgia 29
   Satilla............................. 29
elevata, Aligena 36
elevatus, Mesodon 138
elimata, Macoma..................... 41
Elimia 63, 157

elimia
   acute 63
   amber............................. 64
   ample............................. 63
   Balcones 64
   banded............................ 64
   black-crest......................... 63
   bot 64
   brook............................. 64
   Cahaba............................ 63
   caper 64
   carved 64
   closed 63, 157
   club 63
   coal 63
   cobble 64
   cockle 64
   coldwater.......................... 64
   compact........................... 64
   constricted..................... 64, 157
   corded 64
   Cumberland........................ 64
   cylinder 64
   dented 64
   ebony............................. 64
   elegant............................ 64
   engraved 64
   file 64
   fine-ridged 64
   flaxen............................. 63
   fluted 63
   fusiform....................... 64, 157
   gladiator........................... 64
   goblin............................. 64
   graphite 64
   gravel............................. 63
   hearty 64, 157
   high-spired.................... 64, 157
   knobby............................ 63
   knotty 64
   lacey 64
   lapped 64
   Lilyshoals 63
   liver.............................. 64
   mossy 64
   mud 63
   nymph 64
   oak............................... 64
   panel 64
   Piedmont.......................... 65
   princess........................... 63
   prune 63
   pupa.......................... 64, 157
   puzzle 65, 157

pygmy	64, 157
pyramid	64
rasp	64
ribbed	64, 157
riffle	63
rippled	63
rough-lined	64, 157
round-rib	64
rusty	63
sharp-crest	63
short-spire	63
silt	64
slackwater	64
slanted	64
slough	65
slowwater	64
smooth	64
sooty	64
spider	63
spindle	63
spring	64
sprite	64
squat	65
stately	64
symmetrical	64
walnut	63
yellow	64
elizabethae, *Dermomurex*	84
elktoe	28
Appalachian	28
Carolina	28
Coosa	28, 156
Cumberland	28
elliotti, *Zonitoides*	136
Ellipsaria	29
ellipse	33
ellipsiformis, *Venustaconcha*	33
elliptica, *Basterotia*	38
Elliptio	29, 154
elliptio, eastern	29
Elliptoideus	29
Ellobium	122
elodes, *Stagnicola*	123
elongata	
Aplexa	123
Melanella	77
Poromya	49
elongatus, *Polyschides*	51
elrodi	
Oreohelix	142
Stagnicola	123
elrodianus, *Stagnicola*	123
elsa, *Odostomia*	102
elsae, *Glyphostoma*	97

Elysia	113
elysia	
emerald	113
ornate	113
painted	113
papillose	113
seagrass	113
Elysiidae	113
Emarcusia	121
emarginata	
Hemitoma	53
Nucella	85
emarginatus, *Stagnicola*	123
Emarginula	53
emarginula	
dagger	53
eight-rib	53
elegant	53
emarginate	53
northern	53
pygmy	53
ruffled	53
toothed	53
tuberculate	53
emersonii, *Cerithiopsis*	73
emertoni, *Polycerella*	116
emipowlusi, *Ocinebrina*	85
emmonsi, *Erycina*	36
empyrosia, *Elaeocyma*	96
Enaeta	93
enbergi, *Odostomia*	102
encella, *Turbonilla*	104
engbergi	
Homalopoma	57
Turbonilla	105
engeli	
Berthellina	114
Phyllaplysia	112
Engina	89
engina	
Caribbean	89
white-spot	89
engonia, *Odostomia*	102
enna, *Turbonilla*	105
enneodon, *Helicodiscus*	130
enora, *Odostomia*	102
enosimensis, *Sakuraeolis*	121
Ensis	40
Ensitellops	38
entale occidentale, *Antalis*	50
Entalinidae	51
enteromorphae, *Aplysiopsis*	113
Enteroxenos	78
Entocolax	78

Entoconchidae	78
Entodesma	47
entodesma	
painted	47
pearly	47
rock	47
Entoliidae	26
Entovalva	37
Eobania	146
eolis	
Aclis	76
Lucapina	53
Opalia	76
Epicynia	70
Epilucina	35
Epioblasma	29, 154, 156
epiphaneum, Vexillum	92
Episiphon	50
Epitoniidae	75
Epitonium	75
equalis	
Semirossia	151
Thyasira	35
equestris	
Cheilea	79
Ostreola	28
equilaterale, Pisidium	34
Erato	80
erato	
appleseed	80
green	80
pigeon	80
whitish	80
Ercolania	113
erecta, Puncturella	53
erectus, Petaloconchus	72
Eremarionta	143
Eremariontoides	143
eremita, Sonorella	145
eriopsis, Oenopota	99
eriphyle, Mangelia	98
eritima, Glyphoturris	97
ermineus, Conus	94
erosus, Tachyrhynchus	71
erraticum, Aorotrema	69
errones, Colus	88
Ersilia	77
Ervilia	40
ervilia	
concentric	40
shining	40
subcancellate	40
Erycina	36
erycina	
Ball	36
crown	36
Emmons	36
lined	36
periscope	37
eryiesi rhoadsi, Sterkia	127
erythopoma, Fluminicola	60
erythrocoma, Pseudostomatella	56
erythronotus, Ischnochiton	148
escambia, Fusconaia	30
escargot	146
eschrichtii, Bittium	73
eschscholtzi, Turbonilla	105
escuritor, Ashmunella	137
eshnaurae, Vitrinella	70
esilda, Odostomia	102
esquimalti, Astarte	38
estuarina, Microglyphis	107
esychus, Colus	88
Eubela	97
Eubranchidae	120
Eubranchus	120
Euchelus	55
Euchemotrema	138
Euconulus	133
eucosmia	
Compsodrillia	96
Neptunea	89
Odostomia	102
eucosmius, Fusinus	91
eucosmobasis, Turbonilla	105
Eucrassatella	39
eucymata, Callista	44
eucymatus, Boreotrophon	84
Eudolium	83
eugena, Odostomia	102
Euglandina	129
euglypta, Odostomia	102
euglyptum, Calliostoma	55
eugrammatum, Buccinum	88
Eulima	77
eulima	
Alaska	77
auburn	77
Baja	77
brown	77
brown-line	77
brown-varice	77
California	78
Catalina	77
conoidal	77
gold-stripe	77
grooved	77

Jamaica	77
keeled	77
largemouth	78
Monterey	77
pencil-spine	77
shouldered	77
Tacoma	77
twisted	77
two-band	78
two-line	78
Eulimidae	77
eulimoides, Melanella	77
Eulimostraca	77
eumyaria, Thyasira	35
euomphalodes, Helminthoglypta	143
Eupera	34
Eupleura	84
eurekensis, Oreohelix	142
eurytoideus, Aesopus	86
Euspira	82
eutropis, Ventridens	136
euvitrea, Tellina	41
eva, Turbonilla	105
evadne, Mangelia	98
evelinae	
Ancula	114
Discodoris	118
Elysia	113
Miesea	119
evelynae, Marginella	93
exacuous, Promenetus	125
exara, Odostomia	102
exarata	
Alvania	67
Helminthoglypta	143
exaratus	
Oenopota	99
Stenosemus	149
excavata	
Lima	25
Tegula	56
excavatus, Somatogyrus	62
excentrica, Vallonia	128
excentricus, Hebetoncylus	125
excisa	
Gyrotoma	65, 157
Odostomia	102
excultus, Uniomerus	33
excurvatus, Oenopota	99
exerythra, Tellina	42
exigua	
Fossaria	122
Janthina	75
Nucula	21
Striatura	136
exiguum	
Carychium	125
Vexillum	92
exiguus, Eubranchus	120
exile	
Campeloma	59
Carychium	125
Exilioidea	89
exilis	
Stagnicola	123
Turbonilla	105
exodon, Stenotrema	140
exogyra, Pseudochama	36
exoleta, Turritella	72
expansa, Macoma	41
expansilabris, Helminthogly	143
exquisita, Bullina	107
extenuata, Macoma	41
exustus, Brachidontes	23
eye	
Higgins	30, 154
shark	82
eyerdami	
Beringius	87
Odostomia	102
Turbonilla	105

F

faba, Crenella	23
fabale, Sphaerium	34
fabalis, Villosa	33
fabrici, Gonatus	151
fabricii, Nipponotrophon	85
fabula, Pegias	31
Facelina	121
facelina, Boston	121
Facelinidae	121
fackenthallae, Turbonilla	105
facta, Micrarionta	144
falcata	
Adula	23
Margaritifera	31
Spisula	40
Falcidens	20
fallax	
Pisidium	34
Stenoplax	149
Triodopsis	140
false-bean	42
Pacific	42
false-dial	
exquisite	74

lamellate 74
noble 74
falsejingle
   Alaska 27
   Atlantic 28
   Pacific 27
   pedestal 27
falselimpet
   shield 126
   striped 126
falsemussel
   dark 43
   Santo Domingo 43
false-tun
   giant 83
   sulcate 83
Falsicingula 69
Falsicingulidae 69
fanshell 29
   western 29
farallonensis
   Arene 57
   Odostomia 102
Fargoa 101
fargoi
   Cryoturris 96
   Vermicularia 72
farma, Odostomia 102
Fartulum 71
fasciata, Tegula 56
fasciatum
   Euchemotrema 138
   Prophysaon 132
fasciatus
   Arion 131
   Liguus 130
fascicularis, Fissurella 53
fascinans
   Calliostoma 55
   Elimia 64
fasciola, Lampsilis 30
Fasciolaria 91
Fasciolariidae 91
fasciolaris, Ptychobranchus 32
fastigiata
   Calyptraea 79
   Polygyra 139
fastigiatum, Bittium 73
fatmucket 31
   Arkansas 30
   Carolina 30
   Louisiana 30
   rayed pink 31
   rough 31

   southern 31
   Texas 30
   Waccamaw 30
fat-tellin
   Atlantic 41
   California 41
fausta, Tellina 42
Favartia 84
Favorinus 121
fawnsfoot 33
   Mexican 33
   Texas 33
Felaniella 36
Felimare 116
femorale, Cymatium 83
fenestrata
   Liotia 57
   Notoacmea 54
Fenimorea 97
feralis, Micrarionta 144
fergusoni, Anguispira 131
ferrea, Striatura 136
ferrissi
   Ashmunella 137
   Helminthoglypta 143
   Holospira 129
   Mesodon 138
   Radiocentrum 142
   Sonorella 145
Ferrissia 125
ferrissiana, Humboldtiana 144
ferruginea, Thyasira 35
ferrugineum, Pisidium 34
ferussacianus, Anodontoides 29
Ferussaciidae 128
festiva
   Babakina 121
   Pteropurpura 85
   Tritonia 118
fetella, Odostomia 102
fetellum, Bittium 73
fewkesi, Vermicularia 72
Ficidae 83
Ficus 83
fidelis, Monadenia 144
fidicula, Oenopota 99
fieldi, Helminthoglypta 143
fieldslug
   evening 137
   gray 137
   longneck 137
   one-ridge 137
figsnail
   Atlantic 83

slender	83
filatovae, Tridonta	39
fileclam	
Antillean	26
attenuate	26
boreal	26
bristly	26
Bronn	26
confusing	26
excavated	25
Hemphill	26
Henderson	26
hyaline	26
Locklin	26
regular	26
rough	26
saturnine	26
small-ear	26
smooth	25
spiny	25
subovate	26
Vancouver	26
filifera	
Cryoturris	96
Vitrinella	70
filosa	
Alvania	67
Pandora	48
Pyrgocythara	100
Rhodacme	125
filosum, Lucinoma	35
filosus, Volutopsius	90
fimbriata, Torellia	79
fimbriatula, Bankia	47
fimbriatus, Helicodiscus	130
Finella	73
fingernailclam	
Arctic	34
European	34
grooved	34
Herrington	34
lake	34
long	34
mottled	34
pond	34
rhomboid	34
river	34
Rocky Mountain	34
striated	34
swamp	34
finlayi, Acteon	107
finmarchia, Philine	108
fiona	121
Fionidae	121
fiora, Oenopota	99
Firoloida	81
fischeriana, Verticordia	49
fischerianum, Buccinum	88
fisheri, Helminthoglypta	143
fisheriana, Elliptio	29
Fisherola	122
Fissidentalium	50
Fissurella	53
Fissurellidae	52
Fissurellidea	53
fitchi	
Octopus	152
Penitella	47
flabellatus, Modiolus	23
flamingo, Conus	94
flammea, Cassis	81
flatcoil	
Bahama	139
Florida	139
southern	139
flat-whorl	
eastern	128
western	128
flava	
Elimia	64
Fusconaia	30
flavescens, Conus	94
flavidulum, Pleurobema	32
flavomaculata, Cadlina	116
flavovulta, Cuthona	120
flavus, Limax	137
flectens, Lepidochitona	148
flexuolaris, Philomycus	132
flexuosa	
Epioblasma	29, 156
Halicardia	49
Rangia	40
Thyasira	35
floater	
alewife	28
barrel	28
brook	28
California	28
eastern	28
flat	28
Florida	28
Gaspe	28
giant	28
green	31
inflated	28
Newfoundland	28
Oregon	28
triangle	28

western 28
winged 28
Yukon 28
floralia, Olivella 92
florentina
  curtisi, Epioblasma 29, 154
  florentina, Epioblasma 29, 154
  walkeri, Epioblasma 29, 154
florida
  Chama 36
  Mitra 92
  Oreohelix 142
  Stenotrema 140
floridana
  Alexania 75
  Anachis 86
  Anadara 24
  Carditamera 38
  Cincinnatia 60
  Cyrenoida 36
  Detracia 122
  Elachisina 67
  Glyphyalinia 134
  Leidyula 147
  Lima 25
  Lyonsia 47
  Macromphalina 78
  Neaeromya 37
  Onchidella 147
  Pseudomiltha 35
  Stenoplax 149
  Succinea 133
  Terebra 95
  Thala 92
  Tivela 45
  Truncatella 69
  Vitrinella 70
floridanum, Caecum 71
floridanus
  Conus 94
  Cryptoconchus 150
  Microceramus 130
floridense
  Campeloma 59
  Fissidentalium 50
floridensis
  Conus 95
  Elimia 64
  Micromenetus 124
  Orthalicus 130
Floridiscrobs 60
florifer dilectus, Chicoreu 84
fluctifraga, Chione 44
fluctuata, Eucrassatella 39

fluctuosa, Liocyma 44
fluminea, Corbicula 44
Fluminicola 60
fluted-shell 31
fluvialis, Io 65
fluviana, Diodora 52
foliaceicostum, Epitonium 75
foliatum, Ceratostoma 84
Folinia 67
foliolatum, Prophysaon 132
folliculata, Elliptio 29
Fontelicella 61
fonticula, Gastrodonta 134
Fontigens 61
fontiphila, Helminthoglypta 143
fordi, Globivenus 44
foremani, Pleurocera 66
forestsnail
  broad-banded 137
  Idaho 137
  Oregon 137
  Selway 137
forkshell 30, 156
formanii, Leptoxis 65, 157
formosa
  Doto 119
  Leptoxis 65
formosus, Siratus 85
fornicata, Crepidula 79
Forreria 84
forreria, Catalina 84
forresterensis, Onoba 67
forsheyi, Succinea 133
fossa, Nuculana 21
Fossaria 122
fossaria
  attenuate 123
  boreal 122
  bugle 122
  Carib 122
  dusky 122
  glossy 123
  golden 122
  marsh 122
  prairie 122
  pygmy 122
  rock 122
  Sonoma 123
  Tazwell 123
Fossaridae 78
Fossarus 78
fossarus
  beautiful 78
  compact 78

elegant	78
fossatus, Nassarius	91
fosteri	
Pallifera	132
Triodopsis	140
foveata, Thala	92
foveolata, Ocenebra	85
foxsnail	
brown	129
Chisos	129
fractum, Epitonium	76
fradulenta, Triodopsis	140
fragile, Periploma	48
fragilis	
Bulbus	82
Crenella	23
Ferrissia	125
Leptodea	31
Mactra	40
Martesia	46
Philine	108
Volutopsius	90
fragosa, Quadrula	32
francesae, Parviturbo	57
franciscana	
Odostomia	102
Sonorella	145
Turbonilla	105
fraseri, Cerithiopsis	74
fraterculus, Nassarius	91
fraterna	
Cincinnatia	60
Elliptio	29
Yoldiella	22
fraternum, Euchemotrema	138
frenulata, Rimula	53
friabilis, Mesomphix	134
frielei	
Bathyarca	24
Beringius	87
Epitonium	76
frigida, Yoldiella	22
Frigidoalvania	67
frigidus, Margarites	55
fringed-snail	
lyrate	141
southwestern	141
fringillum, Buccinum	88
frogsnail	
California	83
chestnut	83
Cuba	83
elegant	83
fine-sculpted	83
gaudy	83
St. Thomas	83
frond-aeolis	119
giant	119
multicolor	119
red	119
robust	119
stubby	119
white	119
frondosus, Dendronotus	119
frons, Dendostrea	28
fucanum, Clinocardium	39
fucata	
Fenimorea	97
Gari	42
fulcidens, Triodopsis	140
fulgens	
Cuthona	120
Haliotis	52
fulgurans, Nerita	58
fulica, Achatina	129
fullerkati, Lampsilis	30
fullingtoni, Humboldtiana	144
fulminatus, Pitar	44
fultoni, Mitra	92
fulvescens, Muricanthus	85
fulvocincta, Eulima	77
fulvus, Euconulus	133
funebralis, Tegula	56
funerea, Ercolania	113
fungina, Tylodina	114
funiculata	
Acmaea	53
Opalia	76
furvum, Pleurobema	32
fusca	
Atlanta	81
Botula	23
Coryphella	120
Sayella	104
fuscata	
Ercolania	113
Tenellia	121
fuscocincta, Olivella	92
fuscoligata, Clathromangelia	96
Fusconaia	30, 154
fuscovittatus, Stiliger	113
fuscus	
Fluminicola	60
Laevapex	125
fusiformis	
Cerithiopsis	74
Elimia	64, 157
Gadila	51

Fusinus 91
Fusitriton 83

G

gabbi
   Glyphoturris 97
   Micrarionta 144
   Stagnicola 123
   Strigilla 41
   Succinea 133
gabbiana, Turbonilla 105
gabbii, Penitella 47
gabrielense, Glyptostoma 141
gabrielinum, Pristiloma 135
Gadila 51
Gadilidae............................ 51
Gadilinidae 50
gadinia
   carinate 126
   reticulate 126
gagates, Milax 137
galbana, Fossaria 122
galea
   Carinaria 81
   Coralliophila 86
   Tonna 83
galeata, Puncturella 53
galeommatid, Atlantic................. 37
Galeommatidae...................... 37
galgana, Oenopota 99
galiurensis, Sonorella 145
gallina, Tegula 56
gallus, Strombus 78
gaper
   fat 40
   lost.............................. 40
   Pacific 40
gardenslug
   giant............................ 137
   threeband....................... 137
   yellow 137
gardensnail
   brown 146
   green 146
   Kentish 146
   white 146
   white-lip......................... 146
Gari............................... 42
Gastrochaena 46
Gastrochaenidae..................... 46
Gastrocopta 126
Gastrodonta 134
Gastropteridae 108

Gastropteron....................... 108
gausapata, Nitidella.................. 87
Gaza 55
gaza, Benthonella.................... 67
gaza
   dwarf............................ 56
   superb 55
Geitodoris 118
gem
   minute 134
   southeastern 134
   striate........................... 134
gemclam
   amethyst 44
   brown 44
Gemma 44
gemma
   Bulla 109
   Chaetopleura................... 149
   Gemma 44
   Maxwellia 84
gemmatum, Vexillum 92
Gemmula........................... 97
gemmulatum, Calliostoma 55
gem-turris, Atlantic 97
geniculata, Lithasia 65
geniculum, Campeloma 59
geoduck
   Atlantic 46
   Pacific 46
geographica, Aplysia 112
georgianum, Pleurobema............. 32
georgianus
   Colus 88
   Somatogyrus 62
   Viviparus....................... 59
gerhardti, Elimia..................... 64
germaineae, Mourgona 112
germana, Cryptomastix 138
Geukensia 23
giant-cockle, Atlantic 39
giant-turris
   delicate 100
   twilled 100
   white 100
gibba
   Inodrillia 97
   Melanella....................... 77
gibberosum, Lithopoma 57
gibberum, Pleurobema............... 32
gibbosa
   Ancula 114
   Anodonta....................... 28
   Cavolinia....................... 110

Fargoa	101	harlequin	112
Plicatula	27	glass-snail	
gibbosum, Cyphoma	81	cellar	135
gibbulum, Pseudocyphoma	81	contracted	136
gibbum, Campeloma	59	dark-bodied	135
gibbus, Argopecten	26	eastern	136
gigantea		garlic	135
Lottia	54	Swiss	135
Pleuroploca	91	western	136
giganteum, Crassedoma	26	glassy-bubble	
Margarites	55	amber	109
Saxidomus	45	Antilles	109
gigas		blister	109
Crassostrea	28	Caribbean	109
Dosidicus	151	clean	109
Strombus	78	clean-slate	109
gilli		elegant	109
delmontensis, Turbonilla	105	green	109
gilli, Turbonilla	105	obscure	109
Gillia	61	solitary	109
giovia, Seguenzia	56	straight	109
girardi, Cerodrillia	96	Glaucidae	122
glabra, Marsenina	80	*Glaucus*	122
glabrata, Biomphalaria	124	glaucus, blue	122
glaciale, Buccinum	88	*Glebula*	30
glacialis		*Gliabates*	143
Bathyarca	24	globe	
Cuspidaria	49	balsam	138
Onchidiopsis	80	banded	131
Pandora	48	dwarf proud	138
glandula, Crenella	23	grand	138
glandulosa, Hemphillia	132	noonday	138, 155
Glans	38	proud	138
glass		toothed	139
amber	134	white-lip	139
blue	134	globelet	
carrot	133	brown	138
depressed	134	dwarf	138
live oak	135	proud	138
glass-scallop		sealed	138
Alaska	27	squat	139
cancellate	27	yellow	138
catalina	26	*Globivenus*	44
costate	26	*globosa*	
delicate	27	*Janthina*	75
dwarf	26	*Marsenina*	80
Holmes	27	*Physella*	124
netted	27	*globosus, Mesomphix*	134
Pourtales	27	*globula, Boreocingula*	67
Ringnes	27	*globuloides, Cingula*	67
Santa Barbara	26	*globulosus, Hydromyles*	111
shingle	27	*globulus, Ptychosalpinx*	90
glass-slug		*glomerula, Bathyarca*	24
Antilles	112		

gloriosa
  Cerithiopsis 74
  Odostomia 102
  Turbonilla 105
gloriosum, Calliostoma 55
gloss
  Appalachian 136
  black 136
  dull 136
  quick 136
  shadow 136
  striate 136
glossema, Terebra 95
Glossidae 43
Glycymerididae 24
Glycymeris 24
glyph
  blind 134
  blue-gray 134
  bright 134
  brilliant 134
  carved 134
  dark 134
  depressed 134
  fragile 134
  furrowed 134
  hill 134
  hollow 134
  honey 134
  Maryland 134
  Ocala 134
  pale 134
  pink 134
  pretty 134
  rust 134
  sculpted 134
  spiral mountain 134
  stone 134
  suborb 134
  Texas 134
  thin 134
  tongued 134
Glyphostoma 97
Glyphostomops 97
Glyphoturris 97
Glyphyalinia 134
glypta
  Cardiomya 49
  Murexiella 84
  Nassarina 87
Glyptostoma 141
glyptus, Aequipecten 26
Godiva 121
goforthi, Aesopus 86

goldfussi, Metastoma 129
golfoyaquense, Glyphostoma 97
Gonatidae 151
Gonatus 151
Gonidea 30
Goniodorididae 114
goniogyrus, Teinostoma 70
goreaui, Basiliomya 27
gothicus, Dendrochiton 148
gouldclam, waxy 44
gouldi
  Bankia 47
  Nitidella 87
  Vertigo 127
Gouldia 44
gouldiana
  Bulla 109
  Pandora 48
  Scaphella 93
gouldii
  Buccinum 88
  Cylichna 108
  Dentalium 50
  Donax 42
  Lyonsia 47
  Oenopota 99
  Tellina 42
  Thyasira 36
gracilentis, Odostomia 102
gracilicosta, Vallonia 128
gracilior
  Admete 94
  Cymakra 96
gracilis
  Lamellaxis 128
  Melanella 77
  Polygyra 139
gracillima
  Melaniella 129
  Ocenebra 85
gradata, Pleurocera 66
graduata, Oceanida 77
grahamensis
  Oreohelix 142
  Sonorella 145
grana, Amnicola 59
grandis
  Anodonta 28
  Crepidula 79
  Polyschides 51
  Solemya 22
granifera, Tarebia 63
graniticola, Helminthoglypta 143
Granoturris 97

granti, Pseudochama	36
granticus, Oenopota	99
granularis cubaniana, Bursa	83
granulata	
Acanthopleura	150
Poromya	49
Protothaca	44
Puncturella	53
Tonicella	149
Velutina	80
granulatissima, Sonorella	145
granulatum, Phalium	82
granulatus	
Conus	95
Plectodon	49
granule, glossy	133
granulifera, Lyonsia	47
Granulina	93
granulosa, Thyasira	36
grapeskin, glossy	136
Graphaeidae	28
Graptacme	50
gratiosa	
Nembrotha	116
Thala	92
gravida, Odostomia	102
gravis, Spilochlamys	63
grayana, Bursa	83
grayi, Nassarina	87
great-tellin	42
Alaska	42
greeleyi, Peltodoris	118
greeni	
Cerithiopsis	74
Ptychobranchus	32
greenlandica	
Arctinula	26
Boreoscala	75
greenlawi, Polyschides	51
greeri, Succinea	133
Gregariella	23
greggi	
Amnicola	60
Eremarionta	143
Helminthoglypta	143
grimmi, Paravitrea	135
grippclam, California	46
grippi	
Bellaspira	96
Cerithiopsis	74
Elachisina	69
Homalopoma	57
Melanella	77
Mysella	37
Ocenebra	85
Pseudomelatoma	100
Turbonilla	105
grippiana, Odostomia	102
Grippina	46
grisea, Onchidoris	115
griseola, Praticolella	139
griseus, Plicifusus	90
groenlandica	
Nucula	21
Volutomitra	93
groenlandicum, Oxyloma	133
groenlandicus	
Margarites	55
Serripes	39
grosvenori, Succinea	133
grovesnail	146
grus, Chione	44
guadalupensis	
Binneya	132
Bulimulus	130
guentheri, Calycidoris	117
guildingii, Tellina	42
guinaicum, Cerithium	73
gularis, Ventridens	136
gulella, two-tone	129
gundlachi, Guppya	133
Guppya	133
guppyi, Americardia	39
gurgulio, Caecum	71
guttarosea, Euchelus	55
guttata, Marginella	93
Gymnobela	97
Gymnodorididae	116
gymnota, Catriona	120
Gyraulus	124
gyrina, Physella	124
gyro	
ash	124
disc	124
flexed	124
star	124
tuba	124
Gyrodes	82
Gyrotoma	65, 157

H

habanensis, Diodora	52
hachetanum, Radiocentrum	142
hachitana, Sonorella	145
haddletoni, Lampsilis	30
hadenoecus, Helicodiscus	130
Hadoceras	61

hadria, Granulina	93	*hannai*	
haemastoma		*Onchidiopsis*	80
canaliculata, Thais	85	*Schwartziella*	68
floridana, Thais	85	*hansineensis, Olea*	113
hairysnail		*Haplotrema*	129
acorn	79	Haplotrematidae	129
cancellate	79	*haplus, Vespericola*	141
crowned	79	*harfordiana*	
gray	79	*Cryptomastix*	138
rams-horn	79	*Polygyroidea*	141
two-keel	79	*harfordii, Fusinus*	91
veiled	79	*harpa*	
wandering	79	*Acteocina*	107
halcyon, Marstonia	61	*Coleophysis*	109
haldemani, Acella	122	*Oenopota*	99
half-slippersnail, Pacific	79	*Volutopsius*	90
haliaeeti, Amphissa	86	*Zoogenetes*	128
halibrecta, Turbonilla	105	*harperi, Sonorelix*	145
Halicardia	49	Harpidae	92
halidonus, Colus	88	*harpularius, Oenopota*	99
halidorema, Fenimorea	97	*harrisi, Ashmunella*	137
halimeris, Colus	89	*hartfordensis, Odostomia*	102
haliostrephis, Compsodrillia	96	*hartleyana, Marginella*	93
Haliotididae	52	*hartmanae, Falcidens*	20
haliotidea, Testacella	137	*hartmaniana, Elimia*	64, 157
Haliotinella	82	*hartmeyeri, Ischnochiton*	148
haliotiphila, Barleeia	68	*hartwegii, Lepidochitona*	148
Haliotis	52	*hastata*	
Haliris	49	*Chlamys*	26
Halistrepta	48	*Hastula*	95
halistrepta, Turbonilla	105	*Hastula*	95
halistyle, pupa	55	*hatterasensis, Inodrillia*	97
Halistylus	55	*hausmani, Polygyra*	139
Hallaxa	117	*havanensis*	
halli, Colus	89	*Biomphalaria*	124
Haloconcha	66	*Miralda*	101
halocydne, Kylix	98	*Hawaiia*	134
Halodakra	43	*hawkinsi, Oxyloma*	133
hamata, Nuculana	21	*haydeni*	
hamillei, Phenacolepas	58	*Oreohelix*	142
hamiltoni, Metastoma	129	*Oxyloma*	133
Haminoea	109	*haysiana*	
hamlini, Finella	73	*Elimia*	64
hammer-oyster, Caribbean	25	*Epioblasma*	30, 156
hancocki, Chaetoderma	20	*healyi*	
Hancockia	118	*Margarites*	55
Hancockiidae	118	*Oenopota*	99
handi, Oreohelix	142	*heathi*	
Hanleya	148	*Discodoris*	118
hanleya, eastern	148	*Odostomia*	102
hanleyanum, Pleurobema	32	*heathiana, Stenoplax*	149
Hanleyella	148	*heathii, Lepidochitona*	148
hanleyi, Hanleya	148	*hebardi*	
Hanleyidae	148	*Ashmunella*	137

    Vertigo.......................... 127
hebes
    Oenopota........................ 99
    Pupilla.......................... 127
Hebetoncylus........................ 125
hecuba, Turbonilla................... 105
hedgpethi
    Elysia........................... 113
    Polycera 116
heelsplitter
    Alabama 31
    Carolina......................... 31
    creek 31
    inflated.......................... 32
    pink 32
    Tennessee 31
    Texas........................... 32
    white 31
Heilprinia 91
heladum, Caecum.................... 71
helena, Odostomia 102
helenae, Fusinus..................... 91
Heleobops.......................... 61
helga, Odostomia.................... 102
Heliacus 74
Helicarionidae....................... 133
Helicella 146
helicellid
    elegant.......................... 146
    furrowed 146
    hairy 146
    heath 146
    potbellied....................... 146
    wrinkled........................ 146
Helicellidae......................... 146
Helicidae........................... 146
Helicina............................ 58
helicina, Limacina 110
Helicinidae 58
helicinoides, Atlanta.................. 81
helicinus, Margarites 55
Helicodiscidae.................. 130, 155
Helicodiscus 130
helicogyra, Cincinnatia 60
helicoidea, Vitrinella 70
Helisoma........................... 124
Helix 146
helmet
    cameo 82
    Caribbean 81
    flame 81
helmet-bubble, striate................. 107
Helminthoglypta..................... 143
Helminthoglyptidae.................. 143

helveticus, Oxychilus 135
hematita, Marginella 93
hembeli, Margaritifera............. 31, 154
Hemistena.......................... 30
Hemitoma 53
Hemitrochus 144
hemphilli
    Acanthochitona.................. 150
    Asthenothaerus 48
    Bythinella 60
    Cymatosyrinx 96
    Eulimostraca.................... 77
    Hesperarion..................... 132
    Juga............................ 65
    Limaria 26
    Megomphix 141
    Odostomia...................... 102
    Oreohelix....................... 142
    Pallifera........................ 132
    Pyrgocythara................... 100
    Sayella......................... 104
    Sterkia......................... 127
    Terebra 95
    Turbonilla...................... 105
    Vitrinella 70
Hemphillia 132
hemphillii, Spisula 40
hendersoni
    Cryptomastix................... 138
    Ferrissia 125
    Fontelicella 61
    Fossaria........................ 122
    Glyphostoma 97
    Limatula 26
    Mathilda 74
    Niso............................ 77
    Odostomia...................... 102
    Oreohelix....................... 142
    Physella........................ 124
    Schwengelia 77
    Somatogyrus 62
    Vexillum 92
    Viridrillia....................... 100
Hendersonia 58
henriettae, Triodopsis................ 140
Henrya 77
henseni, Planktomya 38
henslowanum, Pisidium 34
hepaticus, Polinices 82
hepburni, Gadila..................... 51
hera, Paravitrea 135
Here 35
herendeenii, Colus 89
herilda, Odostomia................... 102

Hermaea	113	*hindsi*	
Hermaeidae	113	*Fluminicola*	60
herminea, Miraclathurella	98	*Mopalia*	150
hermissenda	121	*Hindsiclava*	97
heros		*hindsii, Nitidiscala*	76
Euspira	82	*hinkleyi*	
Neptunea	89	*Pomatiopsis*	63
hertleini		*Rhodacme*	125
Helminthoglypta	143	*Somatogyrus*	62
Rissoella	69	*Stagnicola*	123
hertzensteini, Buccinum	88	*Vertigo*	127
hesione, Glyphostoma	97	*hipolitensis, Niso*	77
Hesperarion	132	*hippocrepis, Polygyra*	139
hesperian		Hipponicidae	78
Butte Creek	141	*Hipponix*	78
Karok	141	*hirsutum, Stenotrema*	140
Monterey	141	*hirundinina, Chelidonura*	108
northwest	141	*hispida, Trichia*	146
redwood	141	*histrio, Vexillum*	92
Santa Cruz	141	hive	
Shasta	141	brown	133
Siskiyou	141	fat	133
hesperium, Deroceras	137	silk	133
heterocincta, Odostomia	102	toothed	133
heteroclita, Blauneria	122	wild	133
heterodon, Alasmidonta	28	*hoegiana praesidii, Humboldtiana*	144
Heterodonax	42	Hojeda	141
Heterodorididae	115	*holmesii, Propeamussium*	27
Heterodoris	115	*Holospira*	129
Heteroschismoides	51	holospira	
heterostropha, Physella	124	Arizona	129
heterura, Deroceras	137	Bartsch	129
hexagona, Mangelia	98	bilamellate	129
hexodon, Helicodiscus	130	Cave Creek	129
hians		Cockerell	129
Argonauta	152	Cross	129
Gastrochaena	46	Metcalf	129
Hiatella	46	mountain	129
hiatella		pinecone	129
Arctic	46	royal	129
dirt	46	stocky	129
hole-hugger	46	strongrib	129
striate	46	teasing	129
Hiatellidae	46	vagabond	129
hickmanae, Margarites	55	Whetstone	129
hickorynut		*holsingeri, Fontigens*	61
Alabama	31	*holstonia, Lasmigona*	31
round	31	*holzingeri, Gastrocopta*	126
southern	31	*Homalopoma*	57
hidalgoi, Murexiella	85	hoofsnail	
higginsi, Lampsilis	30, 154	orange	78
highnut	32	ribbed	78
hillebrandi, Monadenia	144	white	78
hiltoni, Phidiana	121	*hooveri, Mangelia*	98

hopetonensis
    Elliptio 29
    Triodopsis 140
Hopkinsia 114
hordacea, Physella 124
hordaceus, Pupoides 127
horn, crenulate 66
hornensis, Gyraulus 124
horni, Thysanophora 141
hornshell, Texas 32
hornsnail
    bottle 66
    broken 66
    brook 66
    California. 72
    corpulent 66
    costate 72
    dainty. 66
    ladder. 72
    noble 66
    pagoda. 66
    plicate 72
    ringed. 66
    rough 66
    rugged 66
    sharp 65
    shortspire. 66
    silty 66
    skirted 66
    smooth. 66
    spiral 66
    sulcate 66
    telescope 66
    upland 66
horsemussel
    American 23
    bag. 23
    California. 23
    Eisen 23
    fan 23
    fat 23
    Kurile 23
    neglected 23
    northern. 23
    straight. 23
hortensis
    Arion 131
    Cepaea 146
hotessieri, Nassarius 91
hotessieriana
    Anachis 86
    Opalia 76
    Tegula 56
houghi, Oreohelix 142

houstonensis, Quadrula 32
howardi, Oreohelix 142
Hoyia 61
huachucana, Sonorella. 145
hubbardi, Strobilops 128
hubrichti
    Catinella 132
    Euchemotrema 138
    Lithasia 65
hudsoni, Acanthodoris 115
huesonicum, Cyclostrema 57
Humboldtiana 144
humeralis, Valvata 58
humerosa, Physella 124
humerosus, Somatogyrus 62
Humilaria 44
humile, Prophysaon 132
humilis
    Cyclostremella. 101
    Fossaria. 122
hummelincki, Octopus 152
hummi, Polycera. 116
humphreysianum, Buccinum 88
humphreysii, Epitonium 76
Huttonella. 129
Huxleyia 25
Hyalina 93
hyalina
    Limatula 26
    Lyonsia 47
Hyalocylis. 110
Hyalopecten 27
Hyalopyrgus 61
Hydatina 107
Hydatinidae. 107
hydei, Elimia. 64
hydiana, Lampsilis 30
hydrobe
    bantum. 61
    Blue Spring 60
    Boone 61
    Clifton Spring 60
    cockscomb. 61
    delta. 62
    dense 60
    Fenney Spring. 60
    fine-lined 61
    freemouth 60
    henscomb 61
    minute 61
    narrowmouth 63
    oolite 61
    regal. 61
    saltmarsh 63

slough	60
smooth-rib	61
spartina	61
storm	61
Suwanee	60
thick-shell	60
Wekiwa	60
Hydrobia	61
Hydrobiidae	59, 157
Hydromyles	111
Hydromylidae	111
hydrophanum, Buccinum	88
hypatia, Odostomia	102
hyperborea	
Limatula	26
Yoldia	22
hypergonia, Aclis	76
hyperia, Antiplanes	95
hypocurta, Odostomia	102
hypodra, Mitrella	87
hypohyalinum, Aphaostracon	60
hypolispa, Turbonilla	105
hypolispus, Colus	89
Hypselodoris	116

I

ictericus, Spondylus	27
icterina, Elliptio	29
idahoense	
Pisidium	34
Pristiloma	135
idahoensis	
Fontelicella	61
Oreohelix	142
Stagnicola	123
Vertigo	127
Zacoleus	132
idalina, Mitrella	87
idiochila, Marginella	94
ilfa, Turbonilla	105
iliuliukensis, Odostomia	102
illecebrosus, Illex	152
Illex	152
Ilyanassa	91
imbecillis, Anodonta	28
imbricata	
Arca	24
Chlamys	26
Pinctada	25
imbricatum, Caecum	71
imbrifer, Cyclopecten	27
imitata, Phreatodrobia	61

imitator	
Sonorella	145
Tryonia	63
immaculata	
Eremarionta	143
Euspira	82
imperatrix	
Acanthochitona	150
Sonorella	145
imperialis	
Lischkeia	55
Sonorella	145
impexa, Okenia	115
implicata, Anodonta	28
impolita, Diplodonta	36
imporcata, Mopalia	150
impressa	
Boonea	101
Elimia	64, 157
Nodotoma	98
ina, Turbonilla	105
inaguensis, Hojeda	141
incanum, Cerion	128
incerta, Alaba	72
incisa	
constricta, Turbonilla	105
incisa, Turbonilla	105
Cyclocardia	38
incisulus, Oenopota	99
incisus, Plicifusus	90
inclinans, Elimia	64
inclinata, Atlanta	81
inclusus, Amphithalamus	68
incongrua, Semele	43
inconspicua	
Pleurobranchae	114
Yoldiella	22
inculta, Acteocina	107
incurvatus, Capulus	78
indentata	
Glyphyalinia	134
Macoma	41
indentatus, Beringius	87
indiana, Succinea	133
indianorum	
Mesodon	138
Nitidiscala	76
indioensis, Eremarionta	143
indutum, Homalopoma	57
inequita, Oenopota	99
inermis	
Helicodiscus	130
Navanax	108
Ophiodermella	100

inexhaustum, Buccinum 88
inezae, Pseudochama 36
infima, Assiminea 68
inflata
    Atlanta 81
    Limacina 110
    Odostomia 102
    Pandora 48
    Yoldiella 22
inflatus, Potamilus 32
inflectus, Mesodon 138
inflexa, Cavolinia 110
infracarinatus, Solariorbis 70
infrequens, Opalia 76
infucata, Quincuncina 33
infumata, Monadenia 144
infundibulum, Latirus 91
ingens, Cerithiopsis 74
ingersolli, Microphysula 141
inglei, Oceanida 77
inglesi, Helminthoglypta 143
ino, Inodrillia 97
Inodrillia 97
inornata
    Ocenebra 85
    Pandora 48
inornatus
    Mesomphix 134
    Pupoides 127
inquinata, Macoma 41
insculpta, Policordia 49
insculptus, Nassarius 91
insessa, Notoacmea 54
insigne, Pisidium 34
insignis
    Sonorella 145
    Tonicella 149
    Trichotropis 79
instabilis, Collisella 54
insularis, Neptunea 89
intapurpurea, Chione 44
intastriata, Leporimetis 41
integra
    Cincinnatia 60
    Physella 124
    Somatogyrus 62
intercisa, Xerarionta 146
interfossa
    Clathromangelia 96
    Ocenebra 85
interfossum, Bittium 73
interioris, Juga 65
intermedia
    Acteocina 107

    Corolla 111
    Fontelicella 61
    Quadrula 32, 154
    Solariella 56
    Triphora 75
    Yoldiella 22
intermedium, Pseudocyphoma 81
intermedius, Arion 131
interna, Gastrodonta 134
interrupta
    Elimia 64
    Turbonilla 105
interruptus, Parviturboides 70
intersecta, Candidula 146
interstinctus, Ischnochiton 148
intersum, Oreohelix 142
intertextus
    Ventridens 136
    Viviparus 59
interveniens, Elimia 64
intricata, Bailya 87
inversa, Graptacme 50
Io 65
io
    Cerithiopsis 74
    Finella 73
iodinea, Coryphella 120
iontha, Anachis 86
iornata, Epicynia 70
Iothia 54
Iphigenia 42
iris
    Dendronotus 119
    Tellina 42
    Villosa 33
    Yoldiella 22
irradians, Argopecten 26
irregularis, Spiroglyptus 72
irrorata, Littorina 66
Irusella 44
isabella, Helminthoglypta 143
Ischadium 23
Ischnochiton 148
Ischnochitonidae 148
Iselica 101
islandica
    Amauropsis 82
    Arctica 43
    Chlamys 26
islandicus, Colus 89
islandsnail
    pricklypear 144
    San Clemente 144
    San Nicolas 144

Santa Barbara 144
Santa Catalina 144
isocardia, Trachycardium............... 39
Isognomon 25
Isognomonidae 25
isolirata, Cardiomya.................. 49
Isorobitella 37
Issena 116
ista, Turbonilla..................... 105
Ithycythara 97

J

jackknife
    Atlantic 40
    California......................... 40
    green 40
    minor............................ 40
    oblique........................... 40
    rosy 40
    sickle 40
jacksoni
    Cingula 67
    Nuculana......................... 21
    Onobops 61
    Polygyra 139
jacksoniana, Obovaria 31
jaegeri
    Helminthoglypta 143
    Oreohelix......................... 142
jamaicensis
    Collisella 54
    Melanella......................... 77
    Nuculana......................... 21
janeirensis, Chaetopleura............... 149
janetae, Fenimorea................... 97
janira, Tenaturris 100
janmayeni, Frigidoalvania 67
Janolidae............................ 120
janthina
    brown 75
    dwarf............................ 75
    elongate.......................... 75
    pale 75
janthina, Janthina.................... 75
Janthinidae 75
janthostoma, Natica................... 82
janus, Isognomon.................... 25
Japonacteon 107
japonica
    Asterophila 77
    Cipangopaludina................... 59
japonicum, Netastoma................. 47
jaspidea, Jaspidella 92

Jaspidella 92
jaspideus, Conus..................... 95
jaumei, Diodora 52
javanicum, Calliostoma 55
jayana, Lithasia 65
jayensis, Elliptio..................... 29
jeanettae, Alaba 72
jeannae, Cyclostremiscus............... 69
jeffreysi, Cuspidaria 49
jennessi, Physa...................... 123
jessica, Anguispira 131
jewelbox
    Atlantic 36
    cherry 36
    corrugate 36
    Florida spiny...................... 36
    Grant 36
    Inez 36
    leafy............................. 36
    milky 36
    Pacific 36
    pretty............................ 36
    secret............................ 36
    smooth-edge 36
    spiny 36
jewetti
    Cystiscus 93
    Turbonilla 105
jingle
    common.......................... 27
    Peruvian 27
    prickly 27
johannis, Pleurobema.................. 32
johanseni, Plicifusus.................. 90
johnsoni
    Caecum 71
    Cymatosyrinx 97
    Physella.......................... 124
    Pristiloma 135
johnstonae, Bittium 73
jonesi
    Elimia 64, 157
    Ptychobranchus.................... 32
jonesianus, Mesodon 138
jordani
    Chlamys.......................... 26
    Colus 89
Jorunna............................ 118
jorunna, leopard 118
Jouannetia 46
joubini, Octopus..................... 152
juanense, Homalopoma 57
Juga 65

juga
- barren ... 65
- black ... 65
- bulb ... 65
- glassy ... 65
- oasis ... 65
- pleated ... 65
- scalloped ... 65
- smooth ... 65
- topaz ... 65

jugalis, Oreohelix ... 142
jujubinum, Calliostoma ... 55
jujuna, Praticolella ... 139
juliana, Aplysia ... 112
Juliidae ... 113
jumping-slug
- dromedary ... 132
- keeled ... 132
- Malone ... 132
- marbled ... 132
- pale ... 132
- panther ... 132
- warty ... 132

junaluskana, Glyphyalinia ... 134
junii, Oreohelix ... 142
juniperum, Pristiloma ... 135
junonia ... 93
junonia, Scaphella ... 93
juttingae, Tellina ... 42
juxtidens, Triodopsis ... 140

K

kachemakensis, Spiromoeller ... 57
kalmianus, Mesodon ... 138
kamakurensis, Penitella ... 47
kamchatkana, Trophonopsis ... 86
kamtschatkana, Haliotis ... 52
kanabense, Oxyloma ... 133
kaoruae, Cratena ... 121
karakorum, Vespericola ... 141
Katharina ... 150
Katherinidae ... 150
katy, black ... 150
keeledslug, American ... 131
keenae
- Calliostoma ... 55
- Littorina ... 66
- Lucina ... 35
- Rissoina ... 68

keepi
- Haplotrema ... 129
- Margarites ... 56
- Trophonopsis ... 86

keepiana, Lepidochitona ... 148
kelleti, Kelletia ... 89
Kelletia ... 89
kelletti, Xerarionta ... 146
Kellia ... 37
Kelliidae ... 36
kellyclam
- boreal ... 37
- La Perouse ... 37
- netted ... 37
- suborbicular ... 37

kelseyi
- Corbula ... 46
- Exilioidea ... 89
- Lirobarleeia ... 68
- Milneria ... 38
- Odostomia ... 102
- Pelycidion ... 69
- Turbonilla ... 105

kendeighi, Haplotrema ... 129
kennerleyi
- Humilaria ... 44
- Odostomia ... 102

kennerlyi, Anodonta ... 28
kennicottii
- Beringius ... 87
- Suavodrillia ... 100

Kentrodorididae ... 118
keraudrenii
- Oxygyrus ... 81
- Pterotrachea ... 81

kermatoides, Drepanotrema ... 124
kidneyshell ... 32
- fluted ... 32
- Ouachita ... 32
- southern ... 32
- triangular ... 32

killisnooensis, Odostomia ... 102
kincaidi
- Leucosyrinx ... 98
- Turbonilla ... 105

kingmaruensis, Onchidiopsis ... 80
kingstoni, Placida ... 113
kiowaensis, Mesodon ... 138
kirbyi, Zonitoides ... 136
kittenpaw, Atlantic ... 27
klamathensis, Lanx ... 123
Kleinella ... 101
knorrii, Vermicularia ... 72
knoxensis, Anguispira ... 131
knoxi, Nototeredo ... 47
kobelti, Fusinus ... 91
kochi
- Anguispira ... 131

Ashmunella 137
kodiakense, *Buccinum* 88
koto, *Lamellaria* 79
krausei
    Odostomia..................... 102
    Oenopota....................... 99
kraussi, *Leidyula* 147
krebsiana, *Corbula*................. 46
krebsii
    Dendrodoris.................... 117
    Epitonium 76
    Philippia 74
    Williamia...................... 126
krohni, *Cliopsis*................... 111
kroyeri
    Plicifusus...................... 90
    Trichotropis.................... 79
kurilensis, *Modiolus* 23
kurriana, *Cyrtodaria*............... 46
Kurtzia 98
Kurtziella 98
kurtzii, *Turbonilla*................. 105
kya, *Doto* 119
Kylix 98
kyolis, *Siraius*..................... 117
kyskanus, *Oenopota* 99
kyskensis, *Onoba* 67

L

labiosum, Cymatium................ 83
labrosum, Stenotrema 140
labyrinthicus, Strobilops............ 128
laciniata, Protothaca 44
lactea, Otala...................... 146
lacteodens, Paravitrea 135
Lacteoluna 141
lacteolus, Tachyrhynchus............ 71
lacteus, Polinices 82
lactuca, Chama 36
Lacuna 66
lacuna
    amber........................ 66
    carinate 66
    delightful 66
    lesser 66
    northern...................... 66
    one-band 66
    pale 66
    reflexed 66
    thick......................... 66
    tiny 66
    variegate 66
lacunatus, Amphithalamus........... 68

lacunella, *Solariella* 56
Lacunidae 66
lacustre, *Musculium*................ 34
lacustris, *Probythinella*.............. 62
Laemodonta 122
laeostomum, *Busycon*............... 90
laeta, *Elimia* 64, 157
Laevapex......................... 125
laeve, *Deroceras*................... 137
Laevicardium 39
Laevicaulus....................... 147
Laevidentaliidae 50
Laevidentalium.................... 50
laevigata
    Nitidella...................... 87
    Odostomia.................... 102
    Tellina 42
laevigatum, *Laevicardium* 39
laevigatus, *Oenopota* 99
Laevinesta........................ 53
laevior, *Mopalia* 150
laevis
    Cadlina 116
    Solariella..................... 56
    Turbonilla 105
lafresnayi, *Anachis*................. 86
lagunae, *Cuthona*.................. 120
Laila 116
lama, *Macoma* 41
lamarcki
    Carinaria..................... 81
    Morum....................... 92
Lamellaria........................ 79
lamellaria
    Baffin........................ 80
    bald 80
    conical....................... 80
    elongate..................... 80
    granular..................... 80
    great......................... 80
    icy 80
    oblique....................... 80
    red 80
    rhombic 80
    rotund 80
    San Diego 79
    smooth....................... 80
    transluscent.................. 80
    transparent 80
    wavy 80
    white-ball................... 80
    widemouth................... 79
    wooly........................ 80
Lamellariidae 79

Lamellaxis. 128
lamellidens, Paravitrea. 135
lamellifera
   Irusella 44
   Myonera 49
   Pseudomalaxis 74
lamellosa
   Nucella 85
   Solariella. 56
lamellosum, Epitonium. 76
laminata, Turbonilla. 105
lampmussel
   Alabama 31, 154
   eastern. 30
   Haddleton 30
   wavy-rayed 30
   yellow 30
Lampsilis. 30, 154
lananensis, Fusconaia 30
lance
   Altamaha. 29
   Carolina. 29
   Florida. 29
   northern. 29
   pod. 29
   southern. 29
   yellow 29
lanceolata
   Colubraria. 90
   Elliptio. 29
   Ithycythara 97
lancetooth
   Alameda 129
   beaded. 129
   blue-foot 129
   California. 129
   Catalina. 129
   costate. 129
   glassy. 129
   gray-foot 129
   grizzly 129
   hooded. 129
   ribbed 129
   robust 129
   slotted 129
   striate. 129
lanigera, Velutina. 80
lansingi, Pristiloma. 136
lanuginosa, Chaetopleura 149
Lanx 123
lanx
   highcap 123
   kneecap. 123
   rotund 123

   scale. 123
   shortface 122
laperousii
   Kellia. 37
   Serripes 39
lapicida, Petricola 45
lapidaria, Pomatiopsis 63
lapilla, Paravitrea. 135
lapillus, Nucella 85
laqueata, Elimia 64
laqueatum, Dentalium 50
larum, Bittium. 73
Lasaea 37
lasius, Nipponotrophon 85
Lasmigona 31
lasmodon, Ventridens. 136
lastra
   Melanella. 77
   Odostomia. 102
lata
   Aclis. 76
   Hemistena. 30
latebricola, Glyphyalinia 134
lateralis
   Mulinia 40
   Musculus. 24
laterculatum, Vexillum. 93
lateumbilicatus, Zonitoides 136
latiauratus, Leptopecten 27
laticordatus, Plicifusus. 90
latilirata, Chione. 44
latior
   Alvania 67
   Mesomphix 134
Latirus. 91
latirus
   brown-line 91
   chestnut. 91
   Key West 91
   short-tail 91
   slender. 91
   threaded. 91
   white-spot 91
   yellow 91
latissimus, Vitrizonites 136
laurae, Juga 65
lavalleeana, Marginella 94
lawae
   Praticolella 139
   Ventridens. 136
lawrenciana, Obestoma 98
laxa, Odostomia 102
leachii pleii, Bursatella. 112
leafshell. 29, 156

Cumberland 30, 156
leafslug
   emerald 112
   zebra 112
leai, Euchemotrema 138
leana, Crenella 23
leanum, Periploma 48
Learchis 121
leatherleaf
   black-velvet 147
   Caribbean 147
   dappled 147
   Florida 147
   Sloan 147
   spotted 147
   tan 147
   tropical 147
leatherwoodi, Mesodon 138
leia, Doriopsilla 117
Leidyula 147
leioderma, Octopus 152
Leiomya 49
Lemiox 31, 154
lenior, Epioblasma 30, 156
lens, Myrtea 35
lenticula, Yoldiella 22
lentiginosa
   Ancula 114
   Anisodoris 118
leonina, Melibe 119
Lepeta 54
Lepetidae 54
lepiderma, Ashmunella 137
Lepidochitona 148
Lepidopleuridae 148
Lepidozona 149
lepidum, Lepton 37
Leporimetis 41
leporina, Polygyra 139
Leptadrillia 98
leptaleus, Cyclopecten 27
Leptaxinus 35
Leptochiton 148
Leptodea 31
leptodon, Leptodea 31
Lepton 37
lepton
   Adanson 37
   broad 37
   graceful 37
   Pacific 37
leptonoidea, Macoma 41
Leptopecten 27
Leptoxis 65, 157

leptum, Campeloma 59
Lepyrium 61
lerema, Teinostoma 70
lesueuri, Atlanta 81
lesueurii, Limacina 110
letsoni, Pyrgulopsis 62
leucocyma
   Linga 35
   Pilsbryspira 100
leucophaeata, Mytilopsis 43
leucopleura, Collisella 54
leucosphaera, Lamellaria 80
Leucosyrinx 98
Leucozonia 91
leutarius, Megomphix 141
levettei, Ashmunella 137
levicula
   Euspira 82
   Murexiella 85
levidensis, Oenopota 99
lewisae, Solariella 56
lewisi
   Campeloma 59
   Valvata 58
lewisiana, Glyphyalinia 134
lewisii
   Epioblasma 30, 156
   Gyrotoma 65, 157
   Polinices 82
Lexingtonia 31
lienosa, Villosa 33
ligamentina, Actinonaias 28
ligata, Leptoxis 65, 157
ligatum, Calliostoma 55
ligatus, Ptychatractus 93
ligera, Ventridens 136
lignosa, Mopalia 150
Ligumia 31
Liguus 130
lilacina, Triphora 75
lilium
   branhamae, Fasciolaria 91
   hunteria, Fasciolaria 91
   lilium, Fasciolaria 91
   tortugana, Fasciolaria 91
lilliput 33
   iridescent 33
   pale 33, 154
   purple 33
   Savannah 33
   southern 33
   Texas 33
   western 33
lilljeborgi, Pisidium 34

Lima 25
lima
    Lima 25
    Lithasia 65
    Nucella 85
    Philine 108
Limacidae 136
Limacina 110
limacina, Clione 111
Limacinidae 110
Limapontia 113
Limaria 26
Limatula 26
limatula
    Collisella 54
    Lucapinella 53
    Yoldia 22
limatulus, Zonitoides 136
Limax 137
limbaughi, Cadlina 116
Limea 26
Limidae 25
Limifossor 20
Limifossoridae 20
limonitella, Kurtziella 98
limops
    Akutan 24
    Antillean 24
    crested 24
    eared 24
    minute 24
    San Diego 24
    sulcate 24
Limopsidae 24
Limopsis 24
limosus, Amnicola 60
limpet
    arcuate keyhole 52
    Barbados keyhole 53
    Bermuda keyhole 52
    black 54
    black-rib 54
    boreal 54
    bowl 53
    cancellate fleshy 53
    cap 54
    Cayenne keyhole 52
    corded white 53
    dead-end blind 54
    dwarf keyhole 52
    fenestrate 54
    file 54
    file fleshy 53
    giant keyhole 53
    green keyhole 53
    Jamaica 54
    knobby keyhole 53
    lush 54
    mask 54
    narrow keyhole 53
    neat-rib keyhole 52
    northern blind 54
    owl 54
    Pacific rosy 54
    painted 54
    pink 54
    plant 54
    plate 54
    punctate keyhole 53
    reddish 54
    ribbed 54
    rim fleshy 53
    ringed blind 54
    rosy keyhole 53
    rough 54
    rough keyhole 52
    seaweed 54
    shield 54
    shield fleshy 53
    spotted 54
    surfgrass 54
    triangular 54
    two-spot keyhole 53
    unstable 54
    variegate keyhole 52
    volcano keyhole 53
    white blind 54
    whitecap 53
    wobbly keyhole 53
    yellow 54
limpida, Montacuta 37
limula, Macoma 41
limum, Campeloma 59
Linatella 83
lindbergi, Iothia 54
linearis, Cylichna 108
lineata
    Tellina 42
    Tonicella 149
lineatus, Planaxis 72
linella, Erycina 36
lineolaris, Argopecten 26
lineolata
    Ellipsaria 29
    Littorina 66
Linga 35
lingulata, Crepipatella 79
linki, Nucula 21

Lioberus	23	sharp-rib	55
Liocyma	44	*lirulata, Lirularia*	55
liocyma		*lirulatocauda, Diaphorodori*	115
green	44	*Lischkeia*	55
wavy	44	*lissotropis, Splendrillia*	100
lioderma, Triodopsis	140	listeri	
liodoma, Helminthoglypta	143	*Periglypta*	44
lioica, Abra	43	*Tellina*	42
Liomesus	89	*Lithasia*	65
lionotus, Mopalia	150	*Lithophaga*	23
Lioplax	59	*Lithopoma*	57
lioplax		*Litiopa*	73
Choctaw	59	*litteratum, Cerithium*	73
cylindrical	59	littleneck	
furrowed	59	fringed	44
ridged	59	Japanese	45
Liotia	57	Pacific	45
liozonis, Lepidochitona	148	thin-shell	45
lipara, Macoma	41	*littoralis, Oreohelix*	142
liptooth		*littorea, Littorina*	66
bluegrass	139	*Littoridinops*	61
Chisos	139	*Littorina*	66
coastal	139	*Littorinidae*	66
Cumberland	139	*lituella, Dendropoma*	72
Dixie	139	*litus, Antiplanes*	95
Edwards Plateau	139	*lituspalmarum, Teinostoma*	70
Florida	139	*lituyana, Turbonilla*	105
grassland	139	*livescens, Elimia*	64
grooved	139	livida	
Gulf coast	139	*Natica*	82
Gulf Hammock	139	*Sayella*	104
horseshoe	139	*lividomaculata, Tegula*	56
Nashville	139	lividus	
oakwood	139	*Colus*	89
Ocala	139	*Ischnochiton*	148
Oklahoma	139	*Macron*	89
Ozark	139	*Toxolasma*	33
peninsula	139	*lobatum, Pulsellum*	51
rockpile	139	*Lobiger*	113
St. Johns	139	*Lobigeridae*	113
Suwannee	139	*lobium, Lepidochitona*	148
Tamaulipas	139	*locklini, Lima*	26
Texas	139	*loebbeckeana, Spiculata*	81
tiny	139	*Loliginidae*	151
White	139	*Loligo*	151
Wyandotte	139	*Loliolopsis*	151
lirata, Leptoxis	65, 157	*Lolliguncula*	151
lirellus, Helicodiscus	130	*lomana, Niso*	77
Lirobarleeia	68	*Lomanotidae*	119
Lirularia	55	*Lomanotus*	119
lirularia		*lombardii, Allogona*	137
few-spot	55	*longicallis americana, Abra*	43
girdled	55	*longicauda, Stylocheilus*	112
incongruous	55		

longicaudata
    Coryphella 120
    Paraclione 111
longifissa, Puncturella 53
longipes, Bornia 36
longirostris, Cavolinia 110
longisquamosa, Pteria 25
longissima, Triphora 75
longleyi, Xenophora 79
longnut 32
long-solid 30
loprestiana, Drilliola 97
lordi
    Physella 124
    Psephidia 45
    Turbonilla 105
loricata, Trilobopsis 141
lotta, Oenopota 99
Lottia 54
louisa, Mangelia 98
louiseae, Turbonilla 105
loveni, Macoma 41
lowei
    Asperiscala 75
    Chlamys 26
    fackenthallae, Lepido 149
    lowei, Lepidochitona 149
    Mopalia 150
lubrica
    Cochlicopa 126
    Theora 43
lubricella, Cochlicopa 126
Lucapina 53
Lucapinella 53
lucida
    Siliqua 40
    Yoldiella 22
Lucidella 58
Lucina 35
lucine
    approximate 35
    arrow 35
    Bermuda 35
    Blake 35
    buttercup 34
    California 35
    chalky buttercup 34
    compressed 35
    costate 34
    cross-hatched 35
    Cuba 34
    dentate 35
    Dosinia 35
    dwarf tiger 34

    equal-zone 35
    fine-lined 35
    Florida 35
    four-rib 35
    lamellate 35
    lens 35
    little-comb 35
    many-line 35
    miniature 35
    northeast 35
    Nuttall 35
    Pennsylvania 35
    pit 35
    ringed 35
    Sombrero 35
    spinose 35
    thick 35
    three-ridge 35
    tiger 34
    woven 35
Lucinidae 34
Lucinoma 35
ludwigii, Entocolax 78
luetkeni, Oenopota 99
lugubris
    Acanthina 84
    Trachypollia 86
luminosa, Symplectoteuthis 152
lunata, Mitrella 87
lunulata, Crassinella 39
lurida, Ocenebra 85
luridum, Homalopoma 57
lustrica, Marstonia 61
lutea
    Acanthodoris 115
    Tellina 42
luteola
    Corbula 46
    Succinea 133
luteomarginata, Cadlina 116
luteopictus, Fusinus 91
lutescens, Akiodoris 115
luticola
    Glyphyalinia 134
    Sphenia 45
lutkeanus, Oenopota 99
lutosum, Cerithium 73
lutulenta, Mitrella 87
lyalli, Turbonilla 105
Lymnaea 123
lymnaea
    frigid 123
    mammoth 122
    mimic 123

spindle 122
swamp 123
Lymnaeidae 122
lymneiformis, Daphnella 97
lynceus, Phidiana 121
Lyonsia 47
lyonsia
   California 47
   Florida 47
   giant 47
   glassy 47
   Gould 47
   island 47
   sandy 47
   scaly 47
   striate 48
Lyonsiidae 47
lyra, Scissurella 52
lyrata
   decemcostata, Neptunea 89
   lyrata, Neptunea 89
   turnerae, Neptunea 89
   Udosarx 142
lyria, sand 93
Lyrodus 47
lysonia, granulate 47

M

macclintocki, Discus 131, 155
maccordi, Alasmidonta 28, 156
macdonaldi, Notobranchaea 111
macerophylla, Chama 36
macfarlandi
   Chromodoris 116
   Platydoris 118
macgintyi
   Conus 95
   Cyphoma 81
   Eubola 98
   Murexiella 85
   Olivella 92
   Parahyotissa 28
macglameriae, Medionidus 31, 156
macmichaeli, Elliptio 29
macneilli, Pupisoma 127
Macoma 41
macoma
   Aleutian 41
   Baltic 41
   bent-nose 41
   chalky 41
   Charlotte 41
   cheating 41

   constricted 41
   elongate 41
   expanded 41
   file 41
   flat 41
   heavy 41
   indented 41
   Lepton 41
   little-file 41
   Loven 41
   Mera 41
   Middendorff 41
   Mitchell 41
   morsel 41
   oblique 41
   Pulley 41
   rightwing 41
   short 41
   sleek 41
   slender 41
   stained 41
   Tagelus 41
   thick 41
   waxy 41
   white-sand 41
   Yoldia 41
macouni
   Boreotrophon 84
   Turbonilla 105
macra
   Melanella 77
   Urosalpinx 86
Macrarene 57
Macrocallista 44
macrochira, Pneumodermopsis 111
macrodon, Truncilla 33
macromphala, Ashmunella 137
Macromphalina 78
macromphaline
   California 78
   Florida 78
   Palm Beach 78
Macron 89
macron, livid 89
macrophallus, Sonorella 145
macroptera, Pteropurpura 85
macropus, Octopus 152
macroschisma, Pododesmus 27
Mactra 40
Mactrellona 40
Mactridae 40
mactroides, Tivela 45
Mactromeris 40

maculata
    Macrocallista 44
    Triopha 115
maculatum, Sinum 82
maculosa
    Crepidula........................ 79
    Tonna 83
madagascariensis, Cassis.............. 81
magazinensis, Mesodon 138
magdalenae, Oreohelix................ 142
magdalenensis, Sonorella.............. 145
magellanicus, Placopecten............. 27
magister, Berryteuthis 151
magna
    Neptunea........................ 89
    Tellina 42
magnafumosum, Stenotrema 140
magnalacustris, Physella 124
magnidentata, Cryptomastix 138
magnifica
    Planorbella 125
    Tulotoma 59
Magnipelta 132
magnum, Trachycardium 39
major
    Antiplanes....................... 95
    Puncturella 53
    Triodopsis 140
malleatus, Beringius.................. 87
Mallcidac............................ 25
malleolus, Teredora 47
malletclam
    berry 21
    quadrangular 21
Malletiidae 21
Malleus 25
malonei, Hemphillia.................. 132
maltbiana, Trivia 80
Mangelia........................... 98
mangelia
    beaded 98
    brown-tip........................ 98
    filose 100
    plicate 100
    punctate......................... 98
    reddish.......................... 98
    spear 97
    tan.............................. 98
    violet-band....................... 98
    waxy 98
    yellow 98
mansfieldi, Babelomurex 86
mantleslug
    Alabama 132
    Arizona 132
    black 132
    blotchy.......................... 132
    brown-spotted 132
    Carolina......................... 132
    changeable...................... 132
    Foster 132
    lyre 142
    marbled 132
    Ozark........................... 132
    pale 132
    redfoot.......................... 132
    severed 132
    toga 132
    variable 132
    Virginia 132
    whiteface........................ 132
    winding 132
Manzanellidae........................ 25
Manzonia 67
mapleleaf
    ridged........................... 33
    southern......................... 32
    winged.......................... 32
maratima, Triodopsis 140
marcusi, Bosellia 112
marferula, Simnialena 81
margaretae, Daphnella................ 97
margaritaceum, Periploma............. 48
margarite
    beehive 56
    boreal rosy 55
    Charlotte Island spiny............. 55
    choice 55
    giant............................ 55
    Greenland 55
    imperial spiny 55
    lirulate.......................... 55
    little 56
    many-line....................... 56
    olive............................ 56
    Pacific rosy 56
    polar 55
    Port Althorp 55
    regal spiny...................... 55
    salmon.......................... 56
    spiral 55
    two-rib.......................... 55
    vortex 56
Margarites 55
Margaritifera 31, 154
margaritifera, Margaritifera............ 31
margaritula, Granulina................ 93

marginata
  Alasmidonta 28
  Cadlina 116
marginatus, Limax 137
Marginella. 93
marginella
  boreal. 93
  California. 94
  carmine 93
  common Atlantic. 93
  courtly. 94
  decorated. 94
  gaunt 94
  gold-line. 93
  gray 93
  helmet 93
  knave. 94
  la belle. 93
  little oat. 94
  orange 93
  orange-band. 94
  pallid 93
  pear 93
  polished. 93
  polite 93
  princess 94
  queen. 93
  seaboard 94
  snowflake. 94
  tan 93
  teardrop. 93
  threerib 94
  triangular. 93
  Virginia 94
  white-line. 94
  white-spot 93
Marginellidae 93
Marginellopsis 94
mariana, Turritella 72
marinelli, Chaetoderma 20
Marinula 122
marionae, Calliostoma 55
Marisa. 59
maritima, Polymesoda 44
marmorata
  Lacuna 66
  Stenophysa 124
marmoratis, Monadenia 144
marmoratus, Musculus. 24
marmorea
  Pallifera. 132
  Tonicella 149
marmorensis, Discus 131
marochiensis, Natica 82

marrianae, Margaritifera 31
Marsenina. 80
Marseniopsis. 80
marshalli
  Beringius 87
  Pleurobema. 32, 154
marshclam. 44
  Carolina. 44
  Florida. 36
  southern. 44
marshsnail
  Fish Springs 123
  mountain 123
  wrinkled. 123
Marstonia 61
marstonia
  armored. 61
  beaverpond 61
  boreal. 61
  ghost 61
  halcyon 61
  Ocmulgee. 61
  olive. 61
  royal 61
marsupiobesa, Elliptio 29
martensi
  Colus 89
  Odostomia. 102
Martesia 46
martha, Inodrillia 97
martinianum, Cymatium. 83
martinicensis
  Crassinella. 39
  Tellina 42
martyni, Boreocingula 67
martyria, Yoldia 22
maryleeae, Hastula 95
masoni, Fusconaia 30
Mathilda 74
mathilda
  Barbados 74
  Yucatan 74
Mathildidae. 74
matthewsae, Epitonium 76
maua, Catriona. 120
maugeriae
  Erato 80
  Stenophysa 124
maurellei, Oenopota 99
mauritianus, Lamellaxis 128
maxillatum, Stenotrema 140
maximus, Limax 137
Maxwellia 84
mayoi, Retusa. 109

mayori
    Gadila 51
    Rissoina......................... 68
mazatlanticus, Pyramidella 104
mcneili, Ferrissia 125
meadi, Sonorella..................... 145
mearnsi
    Ashmunella 137
    Toxolasma....................... 33
media
    Americardia...................... 39
    Cuspidaria....................... 49
Medionidus...................... 31, 156
mediterranea, Caliphylla 112
mediterraneum, Pneumoderma 111
Megacrenella 23
Megalonaias........................ 31
Megapallifera 132
megaradulatus, Scutopus............... 20
megasoma
    Bulimnaea....................... 122
    Vespericola 141
megastoma, Teinostoma 70
Megasurcula........................ 98
Megathura 53
megathyris, Fissidentalium 50
Megomphix......................... 141
megomphix
    Natural Bridge.................... 141
    Oregon.......................... 141
    Umatilla......................... 141
megotara, Psiloteredo 47
Meiocardia 43
Meiomenia 20
Meiomeniidae 20
Melampodidae...................... 122
Melampus 122
melampus
    bubble 122
    California........................ 122
    Caribbean 122
    coffee........................... 122
    eastern.......................... 122
    egg.............................. 122
    Florida.......................... 122
    left-hand 122
    white............................ 122
Melanella 77
melania
    fawn............................ 63
    quilted 63
    red-rim.......................... 63
Melaniella.......................... 129
melanitica, Nannodiella................ 98

Melanochlamys...................... 108
Melanoides 63
melanoidus, Leptoxis 65
melanophylon, Sonorelix 145
melanostoma, Litiopa................. 73
melanura, Triphora................... 75
meleagris, Littorina 66
Melibe............................. 119
melo, miniature....................... 107
Melongena 90
Melongenidae 90
mendax, Ashmunella 137
mendicus
    cooperi, Nassarius 91
    mendicus, Nassarius................ 91
Menetus 124
mera, Tellina........................ 42
meramecensis, Vertigo................ 127
mercatoria, Columbella 87
Mercenaria 44
mercenaria, Mercenaria............... 44
mergella, Valvata.................... 58
mergus, Chicoreus 84
meridionale, Fissidentalium 50
meridionalis, Striatura 136
merita, Mangelia 98
meroea, Platomysia 37
meropsis, Tellina 42
merriami, Fluminicola 60
mertensii, Lepidozona 149
Mesodesma........................ 40
Mesodesmatidae.................... 40
Mesodon 138, 155
mesolia, Metastoma................. 130
Mesomphix........................ 134
mespillum, Littorina.................. 66
messana, Triodopsis.................. 140
messanensis, Nuculana 21
meta, Diodora...................... 52
metallacta, Paravitrea 135
metanevra, Quadrula 32
Metastoma 129
metastoma
    distorted 130
    Hamilton 129
    New Braunfels 129
    Rio Grande 130
    robust 130
    widemouth...................... 130
metastriata, Epioblasma 30
metaxae, Metaxia.................... 74
Metaxia 74
metcalfei, Oreohelix 142

metcalfi
    Bostrychocentrum 129
    Sonorella 145
metria, Vitricythara 101
metschigmensis, Oenopota 99
Metzgeria 93
mexicana, Pyramidella 104
mexicanum, Carychium 125
Mexichromis 116
miamia, Inodrillia 97
miamiensis
    Guppya 133
    Polyschides 51
mica, Cincinnatia 60
micans, Melanella 77
micra
    Phreatodrobia 62
    Sonorella 145
Micrarionta 144
micraulax, Solariella 56
Microceramus 130
micrococcus, Fontelicella 61
micro-cockle
    dyed 39
    lovely 39
    transverse 39
Microgaza 56
Microglyphis 107
microglypta, Alvania 67
Microhedylidae 110
Micromelo 107
Micromenetus 124
micrometaleus, Mohavelix 144
micrometalleoides, Helminthoglypta 143
micromphala, Sonorella 146
Microphysula 141
micropoma, Buccinum 88
microrhina, Cuspidaria 49
Microstelma 67
microstriata, Physella 124
Microtralia 122
Microvoluta 93
micrus, Lamellaxis 128
middendorffi
    Macoma 41
    Turbonilla 105
middendorffiana, Neptunea 89
middendorffii, Volutopsius 90
Miesea 119
mighelsi
    Stagnicola 123
    Turbonilla 105
migrans, Trichotropis 79
Milacidae 137

Milax 137
mildredae, Chlamys 26
milesi, Campeloma 59
milium
    Pisidium 34
    Striatura 136
    Vertigo 127
millepalmarum, Eremarionta 143
milleri, Sonorella 146
millicostatus, Radiodiscus 130
Milneria 38
mimetica, Bosellia 112
mindanus, Conus 95
minima
    Batillaria 72
    Milneria 38
minimum
    Carychium 125
    Haplotrema 129
miniovula, robust 81
minirosea, Favartia 84
minusculum, Teinostoma 70
minor
    Ensis 40
    Haloconcha 66
    Leptoxis 65
    Nassarina 87
    Ocenebra 85
minus
    Pupisoma 127
    Sinum 82
minuscula, Hawaiia 134
minuta
    Diaphana 109
    Diodora 52
    Limopsis 24
    Nuculana 21
    Olivella 92
    Turtonia 38
minutissimum, Punctum 130
minutissimus, Fluminicola 60
minutus, Leptaxinus 35
Miodontiscus 38
mira, Paravitrea 135
mirabilis
    Astarte 38
    Pedipes 122
    Strigilla 41
Mirachelus 56
mirachelus
    basket 56
    stooped 56
Miraclathurella 98
Miralda 101

miris, Jaspidella	92
misakiensis, Eubranchus	120
missoula, Oxyloma	133
missouriensis, Amnicola	60
mitchelli	
Amaea	75
Macoma	41
Quincuncina	33
mitchellianus, Mesodon	138
miter	
Antillean	92
Barbados	92
beaded	92
beautiful	93
dwarf deepsea	93
esteemed	92
Florida	92
gem	92
Greenland	93
Gulf Stream	92
half-brown	92
half-pitted	92
harlequin	92
honeycomb	93
mottled	92
pitted	92
waxy	93
white-band	93
white-spot	93
Mitra	92
mitra, Acmaea	53
mitratus, Oenopota	99
Mitrella	87
Mitridae	92
Mitrolimna	98
Mitromorpha	98
mobiliana, Praticolella	139
moccasinshell	31
Alabama	31
Coosa	31
Cumberland	31
gulf	31
Ochlocknee	31
Suwannee	31
Tombigbee	31, 156
modesta	
Admete	94
Alderia	113
Cadlina	116
Okenia	115
Tellina	42
Vertigo	127
modicella, Fossaria	122
modicus, Pupoides	127
Modiolus	23
modiolus	
Brachidontes	23
Modiolus	23
modoci, Fluminicola	60
Modulidae	72
Modulus	72
modulus, Modulus	72
moelleri, Mysella	37
Moelleria	57
moelleria, ribbed	57
moerchi, Alvania	67
moesta, Macoma	41
mogollonensis, Ashmunella	137
mohaveana, Helminthoglypta	144
Mohavelix	144
Mohnia	89
Monacha	146
Monadenia	144
monas, Aphaostracon	60
monentolophus, Deroceras	137
monilifera, Nassarina	87
monilis	
Melampus	122
Pilsbryspira	100
monkeyface	
Appalachian	33, 154
Cumberland	32, 154
Rio Grande	32
monksae, Fusinus	91
monocingulata	
Saccharoturris	100
Seguenzia	56
monodon, Ventridens	136
monodonta, Cumberlandia	29
monroensis	
Cincinnatia	60
Littoridinops	61
Montacuta	37
montacutid	
Arctic	37
Baker	37
compressed	37
cuneate	37
Dawson	37
giant	37
limpid	37
much-compressed	37
Perrier	37
trigonal	37
wrinkled	37
Montacutidae	37
montagui, Tridonta	39
montanensis, Stagnicola	123

montereyana, Metzgeria................. 93
montereyensis
    Archidoris........................ 117
    Cerithiopsis...................... 74
    Cingula........................... 67
    Crassispira....................... 96
    Melanella......................... 77
    Opalia............................ 76
    Ophiodermella.................... 100
    Petaloconchus..................... 72
    Retusa............................ 109
    Seila............................. 74
    Triphora.......................... 75
montereyi, Tegula...................... 56
montivaga, Holospira................... 129
moonsnail
    Arctic............................ 82
    brown............................. 82
    Carolina.......................... 82
    colorful.......................... 82
    fingernail........................ 82
    fragile........................... 82
    grooved........................... 82
    Iceland........................... 82
    immaculate........................ 82
    lightweight....................... 82
    lined............................. 82
    livid............................. 82
    milk.............................. 82
    miniature......................... 82
    Morocco........................... 82
    netted............................ 82
    northern.......................... 82
    pale.............................. 82
    polished.......................... 82
    purplemouth....................... 82
    semisulcate....................... 82
    spotted........................... 82
    tall.............................. 82
    thin.............................. 82
    tiny.............................. 82
    white............................. 82
mooreana, Polygyra..................... 139
mooreanus
    Rabdotus.......................... 130
    Solariorbis....................... 70
Mopalia................................ 149
Mopaliidae.............................. 149
moratora, Odostomia.................... 102
morchi
    Oenopota.......................... 99
    Turbonilla........................ 105
mordax, Anguispira..................... 131
morditus, Colus........................ 89

moreleti, Leidyula..................... 147
morio, Aplysia......................... 112
mormonum, Monadenia.................... 144
morongoana, Eremarionta................ 143
Moroteuthis............................ 151
morra, Daphnella....................... 97
morrhuanus, Pitar...................... 44
morrisoni
    Henrya............................ 77
    Thracia........................... 48
morroensis, Emarcusia.................. 121
morseana, Cochlicopa................... 126
morsei, Vertigo........................ 127
mortoni, Laevicardium.................. 39
Morum.................................. 92
morum
    Atlantic.......................... 92
    Dennison.......................... 92
    rose-mouth........................ 92
moseri
    brunnescens, Splendrillia....... 100
    moseri, Splendrillia............ 100
moskalevi, Problacmaea................. 54
mosslandica, Cerberilla................ 121
mountainsnail
    alpine............................ 141
    Ancha............................. 142
    bearded........................... 142
    Bitter Root....................... 142
    Black Range....................... 142
    Boulder Pile...................... 142
    carinate.......................... 142
    Catalina.......................... 142
    Cave Creek........................ 142
    Chiricahuana...................... 142
    Clark............................. 142
    costate........................... 142
    Deep Slide........................ 142
    Deseret........................... 142
    Diablo............................ 142
    Eureka............................ 142
    Florida........................... 142
    fringed........................... 142
    Grand Coulee...................... 142
    Hacheta........................... 142
    Huachuca.......................... 142
    keeled............................ 142
    Kyle Canyon....................... 142
    lava rock......................... 142
    lyrate............................ 142
    Magdalena......................... 142
    Mill Creek........................ 142
    Mineral Creek..................... 142
    Morgan Creek...................... 142

pallid	142	multilineata	
Pinaleno	142	*Alvania*	67
Pinos Altos	142	*Mitrella*	87
pygmy	142	*Parvilucina*	35
Rocky	142	*Triodopsis*	140
San Augustin	142	multilineatus	
Schell Creek	142	*Drymaeus*	130
Socorro	142	*Margarites*	56
Spring	142	multisquamata, *Chlamys*	26
subalpine	142	multistriata, *Puncturella*	53
thin-ribbed	142	multistriatum	
Uinta	142	*Epitonium*	76
Whitepine	142	*Teinostoma*	70
whorled	142	multistriatus, *Circulus*	69
Yavapai	142	multivolvis, *Planorbella*	125
Mourgona	112	munita, *Huxleyia*	25
movilla, *Odostomia*	102	munitum, *Bittium*	73
mucket		murdochianus, *Oenopota*	99
Neosho	30	*Murex*	84
orange-nacre	30	murex	
pink	30, 154	Antilles	85
Salina	29	apple	85
tidewater	31	Belleglade	84
mudalia		carved	84
Arkansas	65	festive	85
black	65	frill-wing	85
broad	65	gem	84
crested	65	giant eastern	85
knob	65	hexagonal	85
seep	65	lace	84
smooth	65	lightweight	85
mudcloak		pitted	84
Keys	141	rose	84
moonlight	141	Santa Rosa	84
mudgei, *Ashmunella*	137	shining	85
mud-piddock		three-wing	85
Atlantic	46	wrinkle-wing	85
Pacific	46	*Murexiella*	84
mudsnail		*Muricanthus*	85
eastern	91	muricata	
threeline	91	*Lucina*	35
Mulinia	40	*Onchidoris*	115
mullani, *Cryptomastix*	138	muricatoides, *Turbonilla*	105
multangulus, *Cantharus*	88	muricatum	
multicarinata, *Epicynia*	70	*Trachycardium*	39
multicolor, *Schizoplax*	149	*Vasum*	93
multicostata		muricatus, *Tectarius*	67
Anadara	24	Muricidae	84
Rissoina	68	muricinum, *Cymatium*	83
Turbonilla	105	*Muricopsis*	85
multicostatus, *Boreotrophon*	84	muriei, *Onoba*	68
multidens, *Helicodiscus*	131	murrayense, *Pleurobema*	32
multidentata, *Paravitrea*	135	mus, *Conus*	95
		muscarum, *Cerithium*	73

muscorum, Pupilla 127
muscosa, Mopalia 150
muscosus, Aequipecten 26
Musculista......................... 23
Musculium 34
Musculus.......................... 23
mussel
    Adams...................... 23
    bifurcate 24
    black 24
    bleedingtooth................ 33
    California................... 24
    Chenu 23
    chestnut..................... 23
    cinnamon.................... 23
    coral-eating 23
    corrugate 23
    discordant 23
    blue 24
    elegant...................... 23
    glassy....................... 23
    hooked...................... 23
    knife........................ 23
    lateral...................... 24
    oyster....................... 29
    pygmy 24
    ribbed 23
    salamander 33
    Santo Domingo 23
    scorched 23
    Senhouse.................... 23
    slippershell 28
    spotted...................... 24
    Taylor 24
    varnished................... 24
    yellow 23
mustang, Sonorella................... 146
mutabilis
    Elimia 64
    Megapallifera 132
mutica, Olivella 92
Mya.............................. 45
myalis, Yoldia...................... 22
mycophaga, Magnipelta.............. 132
Myidae 45
Myonera 49
myoneria
    clavicle 49
    shingled 49
myopsis, Thracia 48
myosotis, Ovatella 122
myrae, Ensis........................ 40
myrmecoon, Aesopus 86
Myrtea............................ 35

Mysella 37
mysella
    Aleutian..................... 37
    Bering Sea................... 37
    compressed 37
    eliptical 37
    flat 37
    Moeller 37
    ovate 37
    plate........................ 37
    pure 37
    robust 37
    San Pedro 37
    scratched................... 37
    triandular................... 37
Mysouffa.......................... 107
mysterysnail
    banded...................... 59
    Chinese 59
    Japanese 59
    olive........................ 59
    rotund 59
Mytilidae........................... 23
Mytilimeria 48
Mytilopsis 43
Mytilus 24

N

nakedclam, Orcutt 37
nana
    Astarte...................... 38
    Cuthona..................... 120
    Neverita..................... 82
    Stiobia...................... 63
nanaimoensis, Acanthodoris 115
nannodes, Carychium................. 125
Nannodiella 98
nanulum, Chaetoderma 20
nanus
    Cyclopecten.................. 26
    Somatogyrus 62
napaea, Helminthoglypta.............. 144
nassa
    Antilles 91
    black-spot 91
    bruised...................... 91
    California................... 91
    carved 91
    channeled 91
    fat western.................. 91
    Japanese 91
    lean 91
    lean western 91

miniature	91
sharp	91
smooth western	91
striate	91
western mud	91
white	91
nassa, Leucozonia	91
Nassariidae	91
Nassarina	87
Nassarius	91
nassula	
Elimia	64
Liomesus	89
Lucina	35
Terebra	95
nasuta	
Corbula	46
Ligumia	31
Macoma	41
Mactra	40
Natica	82
Naticarius	82
naticarum, Crepidula	79
Naticidae	82
nautlae, Depressiscala	75
navalis, Teredo	47
Navanax	108
navisa, Odostomia	102
nazanensis, Oenopota	99
Neaeromya	37
neapolitana, Spurilla	121
nebulosa	
Littorina	66
Villosa	33
needle-pteropod	
curved	110
straight	110
neglecta	
Littorina	66
Sonorella	146
Triodopsis	140
neglectus, Modiolus	23
Neilonela	21
neislerii, Amblema	28
nematus, Latirus	91
Nembrotha	116
nemo, Odostomia	102
Nemocardium	39
nemoralis, Cepaea	146
neohexagonum, Dentalium	50
neomexicana	
Fontelicella	61
Hawaiia	134
neopalustris, Stagnicola	123
Neoplanorbis	125
Neopycnodonte	28
Neosimnia	81
Neptunea	89
nereia, Turbonilla	105
Nerita	58
nerite	
Antillean	58
checkered	58
emerald	58
four-tooth	58
olive	58
virgin	58
zebra	58
Neritidae	58
Neritina	58
nervosa, Megalonaias	31
nesaeum, Teinostoma	70
nesiotes, Lyonsia	47
Nesovitrea	134
nesta, Atlantic	53
Netastoma	47
nevadensis	
Fluminicola	60
Oreohelix	142
Pyrgulopsis	62
Neverita	82
newberryanum, Glyptostoma	141
newberryi, Helisoma	124
newcombei	
Kurtziella	98
Rissoina	68
Turbonilla	105
newcombi	
Eremarionta	143
Ischnochiton	148
newcombiana, Algamorda	66
newcombianus, Pitar	44
nexus, Leptochiton	148
nicholsoni, Pristiloma	136
nickliniana	
Fontigens	61
Helminthoglypta	144
nicobaricum, Cymatium	83
nicoli, Borsonella	96
nigella, Elliptio	29
niger	
Hesperarion	132
Musculus	24
nigra, Lithophaga	23
nigrimontanus, Discus	131
nigrina, Juga	65
nigrocincta, Triphora	75
nigromaculata, Doriopsilla	117

nigromontanus, Rabdotus	130	*Nodulotrophon*	85
nimapuna, Anguispira	131	*Noetia*	24
nimbosa, Macrocallista	44	*nonscriptus, Atys*	109
Nipponotrophon	85	*nordica, Emarginula*	53
Niso	77	*normalis, Mesodon*	138
niso		*norrisi, Norrisia*	56
brown-line	77	*Norrisia*	56
Hipolito	77	*norrisiarum, Crepidula*	79
nisonis, Benthonella	67	norrissnail	56
nitens		*norvagicus, Nototeredo*	47
Cochlicopa	126	*norvegicus, Volutopsius*	90
Ervilia	40	*nota, Odostomia*	102
Tellina	42	*notabilis*	
nitida		*Adrana*	21
Nitidella	87	*Anadara*	24
Ringicula	107	*Binneya*	132
Nitidella	87	*Notarchus*	112
nitidella, Depressiscala	75	*notata, Diplodonta*	36
Nitidiscala	76	*notius, Helicodiscus*	131
nitidulum, Chaetoderma	20	*Notoacmea*	54
nitidum		*Notobranchaea*	111
Caecum	71	Notobranchaeidae	111
Pisidium	34	*Notogillia*	61
Sphaerium	34	*Nototeredo*	47
nitidus, Zonitoides	136	*novajasemljensis, Oenopota*	99
nivalis, Typhlomangelia	100	*novangliae*	
nivea		*Cyclocardia*	38
Olivella	92	*Epitonium*	76
Turbonilla	105	*novimundi, Apodopsis*	122
niveus, Triptychus	104	*Nucella*	85
nix, Trivia	80	*nuciformis,*	
noachina, Puncturella	53	*Corbula*	46
nobiliana, Marginella	94	*Odostomia*	102
nobilis		nucinellid, minute	25
Anisodoris	118	*nucleiformis, Diplodonta*	36
Architectonica	74	*nucleolus, Cylichna*	108
Colus	89	*nucleopsis, Pleurobema*	32
Coryphella	120	nucleus	
Pleurocera	66	*Argopecten*	26
Pseudomalaxis	74	*Planaxis*	72
Nodilittorina	67	*Nucula*	21
Nodipecten	27	*Nuculana*	21
nodosa, Fissurella	53	Nuculanidae	21
nodosus		Nuculidae	21
Argonauta	152	nuculoides	
Nodipecten	27	*Semelina*	43
Nodotoma	98	*Tellina*	42
nodulata, Quadrula	32	*nuda, Solariella*	56
nodulosa		nudibranch	
Mitra	92	cockscomb	120
Trachypollia	86	lion	119
nodulosus		sand	110
Oenopota	99	sargassum	119
Tectininus	67	*nugax, Phreatodrobia*	62

nummaria, Crepidula 79
nummus, Helicodiscus 131
nunivakensis
   Odostomia....................... 102
   Oenopota......................... 99
nutclam
   Aegean........................... 21
   Arctic............................ 22
   Atlantic 21
   Belloti 21
   cancellate......................... 21
   Carpenter......................... 21
   concentric 21
   crenulate 21
   Dall 22
   divaricate......................... 21
   dolphintooth 21
   dwarf............................ 21
   fine-lined 21
   furrowed 22
   Greenland 21
   hammer 21
   hooked........................... 21
   Jackson 21
   Jamaica 21
   Link.............................. 21
   Messanean........................ 21
   minute 21
   moon 21
   northern.......................... 21
   notable........................... 21
   Pacific 21
   Pender 21
   pointed........................... 21
   rayed 22
   reef 21
   rostrate 21
   San Diego 22
   short 21
   skiff 21
   smooth........................... 21
   spear 21
   stout............................. 21
   tailed 21
   tellinoid 21
   thin 22
   trenched.......................... 21
   unequal 22
   Verrill 22
nutmeg
   basket 94
   rugose 94
nuttaliana, Anodonta 28
nuttalli
   Ceratostoma 84
   Fisherola 122
   Lucina 35
   Mytilimeria 48
   Saxidomus....................... 45
   Tresus 40
Nuttallia 42
nuttallianum. Oxyloma............... 133
nuttallianus, Fluminicola 61
nuttallii
   Clinocardium.................... 39
   Nuttallia 42
Nuttallina 149
nuttingi
   Paziella 85
   Turbonilla 105
nux, Liomesus...................... 89
nyalya, Chromodoris 116
nychia, Borsonella 96
nycteis, Mitrella 87
nyctelius, Limax 137
nylanderi, Vertigo................... 127
Nystiella 76

O

obeliscus, Turbonilla 105
obesa
   Anachis 86
   Cuspidaria....................... 49
   Leporimetis 41
   Turbonilla 105
Obestoma 98
obesus
   Colus............................ 89
   Uniomerus....................... 33
obliqua, Macoma 41
Obliquaria.......................... 31
obliquata
   obliquata, Epioblasma 30
   perobliqua, Epioblasma 30, 154
obliquus, Solen...................... 40
oblongus, Pristes 37
Obovaria........................... 31
obovata, Lithasia 65
obrussa, Fossaria.................... 122
obscura
   Colubraria....................... 90
   Doridella 114
   Solariella........................ 56
obscuratus, Atys..................... 109
obscurum, Prophysaon............... 132

obsoleta
    Ilyanassa 91
    Triodopsis 140
obstricta, Triodopsis 140
obtectum, Teinostoma 70
obtusa
    Iselica 101
    Retusa 109
obtusata, Littorina 66
obtusus, Somatogyrus 62
obvia, Helicella 146
occata, Juga 65
occidentale
    Calliostoma 55
    Carychium 125
    Cymatium 83
    Epitonium 76
    Fartulum 71
    Pulsellum 51
    Sphaerium 34
occidentalis
    Aclis 76
    Aporrhais 78
    Diplosolenodes 147
    Dondice 121
    Microtralia 122
    Planorbella 125
    Ptychatractus 93
    Ptychobranchus 32
    Solemya 22
    Triodopsis 140
    Triopha 115
    Turtonia 38
    Vertigo 127
occulta
    Cylichna 108
    Hendersonia 58
occultata, Leptoxis 65, 157
Oceanida 77
oceanodromae, Plicifusus 90
ocellata
    Leucozonia 91
    Mitrella 87
ocellifera, Costasiella 112
ocelligera, Aglaja 108
Ocenebra 85
ochotense, Buccinum 88
ochracea
    Collisella 54
    Leptodea 31
Ocinebrina 85
ocoae, Glyphyalinia 134
octona, Subulina 128
Octopodidae 152

Octopus 152
octopus
    Atlantic pygmy 152
    blanket 152
    brownstripe 152
    Caribbean reef 152
    Caribbean two-spot 152
    common 152
    lilliput 152
    longarm 152
    orange bigeye 152
    Pacific giant 152
    Pacific pygmy 152
    red 152
    smoothskin 152
    spoonarm 152
    unicorn 152
    white-spotted 152
octoradiata, Hemitoma 53
ocula, Loligo 151
odhneri
    Archidoris 117
    Polycera 116
odonoghuei, Doris 117
Odontogena 37
odorata, Sonorella 146
odostome
    half-smooth 101
    impressed 101
    serpulid 101
    two-groove 101
Odostomia 101
oedematum, Buccinum 88
Oenopota 98
ogmorhaphe, Marstonia 61
ohiensis, Potamilus 32
ohioensis, Pallifera 132
Okenia 115
okenia
    Angeles 115
    flat 115
    Vancouver 115
oklahomarum, Catinella 132
oldroydae
    Alvania 67
    Bittium 73
oldroydi
    Acteocina 107
    Babelomurex 86
    Barleeia 68
    Cardiomya 49
    Hanleyella 148
    Melanella 77
    Vitrinella 70

Oldroydia	148	*Onobops*	61
oldroydii		*onslowensis, Terebra*	95
Atrina	25	Onychoteuthididae	151
Calinaticina	82	*Onychoteuthis*	151
Olea	113	*onyx*	
Oligyra	58	*Crepidula*	79
Oliva	92	*Gonatus*	151
olivacea, Marstonia	61	*Oocorys*	83
olivaceus		Oocorythidae	83
Eubranchus	120	*ooides, Liomesus*	89
Margarites	56	*opalescens, Loligo*	151
Melampus	122	*Opalia*	76
olivae, Hermaea	113	*Opeas*	128
olivaria, Obovaria	31	*opercularis, Menetus*	124
olive		*operculata, Varicorbula*	46
beatic dwarf	92	*Ophiodermella*	100
bubble dwarf	92	*ophiodon, Compressidens*	51
jasper dwarf	92	*optabilis, Lirularia*	55
lettered	92	*optata, Sonorella*	146
minute dwarf	92	*opuntia, Micrarionta*	144
netted	92	*oralis, Vertigo*	127
purple dwarf	92	orb	
rice	92	Alabama	32
San Pedro	92	golden	32
snowy dwarf	92	*orbella, Diplodonta*	36
tiny dwarf	92	*orbicularis, Codakia*	34
variable dwarf	92	*orbiculata*	
whitened dwarf	92	*Axinopsida*	35
Olivella	92	*Codakia*	34
oliviae, Spurilla	121	*Crepipatella*	79
Olividae	92	*Lampsilis*	154
olivula, Elimia	64	*Oligyra*	58
Olssonella	94	*orcutti*	
Omalogyra	69	*Chlamydoconcha*	37
Omalogyridae	69	*Fartulum*	71
ombronius, Colus	89	*oregonensis*	
Ommastrephes	152	*Anodonta*	28
Ommastrephidae	151	*Fusitriton*	83
omphale, Borsonella	96	*Gonatus*	151
Onchidella	147	*Odostomia*	102
onchidella		*Planorbella*	125
Florida	147	*Succinea*	133
northwest	147	*Turbonilla*	105
Onchidiidae	147	oregonian	
Onchidiopsis	80	Coeur d'Alene	138
Onchidorididae	115	Columbia	138
Onchidoris	115	Kingston	138
onchidoris		Mission Creek	138
barnacle-eating	115	Puget	138
fuzzy	115	pygmy	138
onealensis, Cerithiopsis	74	*oregonius, Gliabates*	143
oniscus, Morum	92	Oreohelicidae	141
onismatopleura, Buccinum	88	*Oreohelix*	141
Onoba	67	*orestes, Mesodon*	138

organensis, Ashmunella 137
orientalis, Cymatoica 40
orientis, Sonorella 146
orina, Helminthoglypta 144
oritis, Holospira 129
ornata
    Chlamys...................... 26
    Elysia........................ 113
    Lampsilis..................... 30
    Triphora...................... 75
    Verticordia 49
ornatissima
    Cardiomya.................... 49
    Odostomia.................... 103
ornatus, Cyclostremiscus 70
Ornithoteuthis...................... 152
Orobitella 37
orocopia, Eremarionta 143
orolibas, Fontigens................. 61
oronoensis, Stagnicola.............. 123
orotis, Pristiloma 136
orpheus, Boreotrophon............... 84
Orthalicus 130, 155
orthosymmetrica, Turritella 72
ortmanni, Villosa 33
oryza, Acteocina................... 107
oscariana, Vertigo 127
osculans, Physella 124
Ostrea............................. 28
ostrearum
    Calotrophon 84
    Pyrgospira.................... 100
Ostreidae........................ 28
Ostreola 28
Otala............................. 146
othcaloogensis, Epioblasma.......... 30
oualaniensis, Symplectoteuthis 152
oval
    Calico Rock.................... 138
    drywoods...................... 138
    half-lidded.................... 138
    lidded......................... 138
    Pedernales..................... 138
    Texas.......................... 138
ovalis
    Anadara....................... 24
    Psephidia...................... 45
    Succinea 133
ovata
    Corolla........................ 111
    Cyclocardia.................... 38
    Gastrochaena 46
    Lampsilis...................... 30
    Mysella 37

Vertigo............................ 127
Ovatella........................... 122
ovatus, Pyrunculus................. 109
oviforme, Pleurobema 32
ovoidea
    Chaceia....................... 46
    Iselica 101
    Sphenia 45
ovula, Tralia 122
Ovulidae 80
ovuliformis, Granulina 93
owenii, Arion 131
oxia
    Nannodiella................... 98
    Nuculana..................... 21
Oxychilus 135
oxychone, Vanikoro 78
oxygonius, Illex................... 152
Oxygyrus......................... 81
Oxyloma 133
Oxynoe 113
oxynoe
    Antilles 113
    blue-spot 113
Oxynoidae........................ 113
oxytata, Muricopsis 85
oyster
    crested........................ 28
    deepsea 28
    eastern........................ 28
    edible......................... 28
    frond 28
    McGinty 28
    Olympia....................... 28
    Pacific 28
    sponge........................ 28
    Weber 28
ozarkensis
    Fusconaia 30
    Pyrgulopsis 62

P

pachyloma, Praticolella 139
pachynotum, Aphaostracon........... 60
pachyta, Marstonia 61
pacifica
    Adalaria...................... 115
    Ancula 114
    Cadlina 116
    Corambe 114
    Crassinella.................... 39
    Issena 116
    Placiphorella.................. 150

Rossia	151
Saxicavella	46
pacificum	
Chaetoderma	20
Gastropteron	108
pacificus	
Boreotrophon	84
Desmopterus	111
Heterodonax	42
packardii, *Pleurotomella*	100
Paedoclione	111
pagoda, *Gyrotoma*	65, 157
pagodula, *Cymatosyrinx*	97
painei	
Mangelia	98
Ocenebra	85
Rictaxis	107
pajaroana, *Tresus*	40
paleacea, *Notoacmea*	54
Palio	116
pallasii, *Amicula*	149
pallida	
Hyalina	93
Janthina	75
Palio	116
pallidula, *Lacuna*	66
pallidulum, *Amygdalum*	23
pallidus	
Eubranchus	120
Polinices	82
Pallifera	132
Palliolum	27
palmalitoris, *Macromphalina*	78
palmeri	
Humboldtiana	144
Onoba	67
Tritonia	118
palmulatus, *Callistochiton*	148
Paludestrina	61
paludosa, *Pomacea*	59
palustris	
Littoridinops	61
Triodopsis	140
Panacca	47
panamensis, *Lolliguncula*	151
panamica, *Volvulella*	109
pandionis, *Polyschides*	51
Pandora	48
pandora	
bilirate	48
Bush	48
glacial	48
Gould	48
inflated	48
inornate	48
punctate	48
sandy	48
threaded	48
threeline	48
Ward	48
Pandoridae	48
Panomya	46
Panopea	46
panselenus, *Mesodon*	138
pantherina, *Hemphillia*	132
papagorum, *Sonorella*	146
paperbubble	
angled	108
Arctic	109
axial	108
brown	109
brown-line	107
crenulate	108
file	108
Finmark	108
fragile	108
furrowed	109
girdled	108
quadrate	108
sinuate	108
tinted	108
weak	109
white	108
papercockle	
frilled	39
spiny	39
papermussel	
arrow	23
Atlantic	23
pallid	23
papershell	
cylindrical	29
fragile	31
pink	32
paphia, *Chione*	44
papilio, *Desmopterus*	111
papilligera, *Phyllidiopsis*	117
papillosa	
Aeolidia	121
Elysia	113
papyraceum, *Amusium*	26
papyratium, *Periploma*	48
Papyridea	39
papyrium, *Amygdalum*	23
Parabornia	37
Paraclione	111
paradigitalis, *Collisella*	54

paradoxa
   Addisonia 54
   Vertigo............................. 127
Parahyotissa............................ 28
paralia, Succinea 133
parallela, Ferrissia 125
parallelus, Helicodiscus 131
paramera, Tellina...................... 42
paramoea, Cerithiopsis 74
Paramya 45
Parapholas 47
parapodema, Cuspidaria 49
parasitica, Cochliolepis 71
parasitichopoli, Enteroxenos 78
Parastarte............................. 44
Paravitrea............................. 135
parcipicta, Lirularia 55
pardus, Jorunna 118
parilis, Felaniella 36
parkeri
   Ithycythara 98
   Physella.......................... 124
parthenopeum, Cymatium 83
parthenum, Campeloma............... 59
partumeium, Musculium............... 34
parva
   Cincinnatia 60
   Fossaria.......................... 122
   Lacuna 66
   Nitidella.......................... 87
   Nuculana......................... 21
   Olivella 92
   Pleurocera........................ 66
   Sonorella......................... 146
Parvamussium 27
parvicallum, Teinostoma 70
Parvilucina 35
Parviturbo............................. 57
Parviturboides......................... 70
parvula
   Aplysia........................... 112
   Vallonia.......................... 128
   Vertigo........................... 127
parvulus
   Medionidus 31
   Somatogyrus..................... 62
parvus
   Gyraulus 124
   Polyschides 51
   Toxolasma....................... 33
pasonis
   Ashmunella 137
   Metastoma 130
patae, Conus.......................... 95

patella, Sphaerium 34
Patelloida 54
patelloides, Lanx 123
patina, Elysia 113
patinaria, Haliotinella 82
Patinopecten........................... 27
patula
   Purpura 85
   Siliqua 40
   Strombiformis 78
patuloides, Zonitoides 136
patulus, Discus........................ 131
paucicostatum, Homalopoma 57
paucidens, Pneumodermopsis 111
pauciliarata, Acanthina 84
pauli
   Didianema....................... 70
   Dolicholatirus 91
   Turbonilla 105
paulus, Toxolasma 33
paupercula
   Elimia 64
   Volvulella 110
pauperculus, Dermomurex 84
pavlova, Oenopota 99
pazi, Paziella 85
Paziella............................... 85
Pazionotus 85
peaclam
   Adam............................. 34
   Alpine 34
   fat 34
   giant northern 34
   globular 34
   greater eastern.................... 34
   greater European.................. 34
   Henslow 34
   humpback 34
   Lilljeborg.......................... 34
   montane........................... 34
   ornamented 34
   perforated 34
   ridged-beak 34
   river............................... 34
   round............................. 34
   rusty 34
   shiny 34
   short-end......................... 34
   striate............................. 34
   tiny 34
   triangular......................... 34
   ubiquitous 34
   Walker............................ 34
pealeii, Loligo......................... 151

peanut-snail, gray 128
pearl-oyster, Atlantic 25
pearlshell
   Alabama 31
   eastern.......................... 31
   Louisiana..................... 31, 154
   round 30
   western 31
pearlymussel
   birdwing..................... 31, 154
   cracking........................ 30
   Curtis....................... 29, 154
   dromedary................... 29, 154
   little-wing...................... 31
   slabside 31
   Tampico........................ 29
pearwhelk 90
   shouldered..................... 90
pebblesnail
   angular......................... 62
   Ash Meadows 60
   ashy............................ 60
   Atlas 62
   buffalo 61
   Cahaba......................... 60
   channelled 62
   Cherokee....................... 62
   Choctaw 62
   coldwater....................... 62
   Columbia....................... 60
   compact........................ 62
   Coosa.......................... 62
   Cortez Hills.................... 60
   dusky.......................... 61
   dwarf 62
   flat 61
   flint 62
   fluted 62
   golden 62
   granite 62
   Gulf coast 62
   hidden 62
   knotty 62
   Laguna Mountain 61
   Moapa 60
   Modoc 60
   moon 62
   mud 62
   nugget 61
   Oachita 62
   Ohio........................... 62
   Olympia........................ 61
   opaque......................... 62
   ovate 62
Pahranagat........................... 60
panhandle 62
pixie................................. 60
pygmy 62
quadrate.............................. 62
reverse............................... 62
rolling 62
sandbar 62
Savannah............................. 62
shale 62
sparrow 62
spiral 60
stocky 62
Tallapoosa........................... 62
Tennessee 62
thick-lip 62
turban 61
umbilicate 60, 157
vagrant............................... 60
pecki, Glyphyalinia..................... 134
Pecten............................... 27
pectinata
   Cardiomya...................... 49
   Glycymeris...................... 25
   Lucina 35
   Siphonaria..................... 126
Pectinella 27
pectinella, Codakia..................... 35
Pectinidae............................ 26
pectinula, Crenella 23
pectinulata, Lepidozona................ 149
pectorosa, Actinonaias................. 28
pectunculoides, Bathyarca 24
pedicellatus, Lyrodus 47
Pedicularia 81
pedicularia
   California....................... 81
   hatched 81
pediculus, Trivia....................... 80
Pedipes............................. 122
pedipes
   angular......................... 122
   Cuba dwarf 122
   miraculous..................... 122
   one-groove..................... 122
pedroana
   Cerithiopsis 74
   Mysella........................ 37
   Olivella 92
   Peristichia 103
   Terebra 95
   Triphora....................... 75
peggyae, Anodonta.................... 28
Pegias 31

pelagica
  Onoba 68
  Scyllaea 119
pelicanfoot, American 78
pellucens, Ellobium 122
pellucida
  Coryphella 120
  Cuspidaria 49
  Gastrocopta 126
  Lamellaria 80
  Lima 26
  Vitrina 136
peloronta, Nerita 58
Pelseneeria 77
pelta, Collisella 54
Peltodoris 118
peltoides, Williamia 126
Pelycidiidae 69
Pelycidion 69
pemphigus major, Buccinum 88
penderi, Nuculana 21
pendula, Triodopsis 140
penicillata
  Nassarina 87
  Pseudomelatoma 100
penicillatus, Medionidus 31
peninsulae
  Fossaria 122
  Laevapex 125
  Polygyra 139
peninsularis
  Melanella 77
  Triphora 75
penita
  Epioblasma 30, 154
  Penitella 47
Penitella 47
penitens, Trilobopsis 141
pennsylvanicus
  Mesodon 138
  Somatogyrus 62
penshell
  amber 25
  half-naked 25
  Oldroyd 25
  rough 25
  sawtooth 25
  stiff 25
pensylvanica, Linga 35
pentadelphia, Glyphyalinia 134
pentagonalis
  Bellaspira 96
  Ithycythara 98
pentagonus, Cyclostremiscus 70

pentalopha, Turbonilla 105
pentodon, Gastrocopta 126
peoriense, Oxyloma 133
Peracle 110
Peraclididae 110
peracuta, Cincinnatia 60
peramabile, Nemocardium 39
peramabilis, Solariella 56
perattenuata, Mangelia 98
perca, Cuthona 120
percallosus, Ventridens 136
percompressa, Montacuta 37
percrassa, Oldroydia 148
percrassum, Buccinum 88
perdepressa, Valvata 58
perdistorta, Distorsio 83
peregrina, Polygyra 139
perexilis, Marginella 94
perforans, Crepidula 79
perfragilis, Mesomphix 134
Periglypta 44
perigraptus, Mesodon 138
peripherica, Oreohelix 142
Periploma 48
Periplomatidae 48
periscelida, Gemmula 97
periscelidus, Colus 89
periscopia, Solariella 56
periscopiana, Erycina 37
Peristichia 103
periwinkle
  beaded 67
  brown 66
  checkered 66
  cloudy 66
  common 66
  eroded 66
  lineolate 66
  mangrove 66
  marsh 66
  obscure 66
  rough 66
  Sitka 67
  slender 66
  white-spot 66
  yellow 66
  zebra 67
perlaevis, Mesomphix 134
perlepida, Turbonilla 105
perlonga, Graptacme 50
permodesta, Mitrella 87
permollis, Cryptostrea 28
pernobilis, Sthenorytis 76
pernoides, Pododesmus 27

pernula, Nuculana	21	*phaneus, Pterynotus*	85
peronii		*pharcida, Odostomia*	103
Atlanta	81	*pharpa, Doriopsilla*	117
Cymbulia	111	*phaseolina, Thracia*	48
Perotrochus	52	Phasianellidae	58
perovalis, Lampsilis	30	*phasma, Crassispira*	96
perovatum, Pleurobema	32	pheasant	
perpinguis, Nassarius	91	banded	58
perplana, Pleuromeris	38	checkered	58
perplexa		low-line	58
Fossaria	123	red-line	58
Olivella	92	rhodophyte	58
perpolita, Fossaria	123	shouldered	58
perpurpurea, Villosa	33	stained	58
perrieri		sullied	58
Entovalva	37	turtlegrass	58
Heliacus	74	pheasantshell	28
perrostrata, Cardiomya	49	Phenacolepadidae	58
perrugata, Urosalpinx	86	*Phenacolepas*	58
perryae		*Phenacovolva*	81
Cerodrillia	96	*phenax, Macoma*	41
Conus	95	*Phidiana*	121
Kurtziella	98	*Philine*	108
perryi, Vertigo	127	Philinidae	108
persica, Tellina	42	*Philinopsis*	108
Persicula	94	*Philippia*	74
persimilis, Volvulella	110	*philippiana*	
persona, Notoacmea	54	Anodontia	34
personata, Epioblasma	30, 156	Attiliosa	84
perspectiva, Vallonia	128	Lucapina	53
perspectivum, Sinum	82	*philippinarum, Tapes*	45
perspicua, Lamellaria	80	*Philobrya*	25
perstriata, Elimia	64	philobrya	
pertenuis, Retusa	109	Caribbean	25
peruviana, Anomia	27	hairy	25
perversus, Antiplanes	95	Philobryidae	25
pesa, Turbonilla	106	*philodice, Mangelia*	98
Petalifera	112	Philomycidae	132
petalifera, Petalifera	112	*Philomycu*s	132
Petaloconchus	72	*phoca, Discodoris*	118
petitii, Haminoea	109	*phoebium, Astralium*	57
petoskeyensis, Stagnicola	123	*phoenix, Cuthona*	120
Petricola	45	Pholadidae	46
petricola		*pholadiformis, Petricola*	45
boring	45	*pholadis, Hiatella*	46
California	45	Pholadomyidae	47
petricola, Helminthoglypta	144	*Pholas*	47
Petricolidae	45	*phorminx, Mopalia*	150
petrifons, Cincinnatia	60	phos	
petrina, Quadrula	33	banded	88
petrophila, Paravitrea	135	beaded	87
Phalium	82	intricate	87
phanea, Finella	73	*Phreatodrobia*	61
phanella, Odostomia	103	*phrixodes, Emarginula*	53

phrygium, Cryptopecten 26
Phyllaplysia 112
Phyllidiidae 117
Phyllidiopsis 117
Phylliroe 119
Phylliroidae 119
Phyllobranchillus 112
Phyllonotus 85
Physa 123
physa
  ashy 124
  banded 124
  bayou 124
  broadshoulder 124
  Carib 124
  cave 124
  cayuse 124
  Comanche 123
  corkscrew 124
  desert 123
  European 123
  Fish Lake 124
  glass 123
  globose 124
  grain 124
  Great Lakes 124
  hotwater 124
  lowdome 123
  obtuse 123
  olive 124
  ornate 124
  pewter 124
  protean 124
  pumpkin 123
  Rocky Mountain 124
  rotund 124
  sculpted 124
  squalid 124
  striate 124
  sunset 124
  tadpole 124
  Texas 124
  twisted 124
  Utah 124
  wet-rock 124
Physella 123
physematum, Buccinum 88
Physidae 123
physis, Hydatina 107
pica, Cittarium 55
picea
  Glyphyalinia 134
  Triodopsis 140

picta
  Anguispira 131, 155
  Dirona 119
  Elysia 113
  Leptoxis 65
  Synaptocochlea 56
pictum
  Entodesma 47
  Laevicardium 39
picturatum, Buccinum 88
piddock
  abalone 47
  Alaska 47
  Fitch 47
  flap-tip 47
  fragile 46
  Gabb 47
  great 47
  northern rostrate 47
  oyster 46
  Pilsbry 47
  Quilling 46
  rostrate 47
  scaleside 47
  striate 47
  Turner 47
  wartneck 46
  wedge 46
pierrot, Macromphalina 78
pigtoe
  Alabama 32
  Atlantic 30
  Coosa 32
  Cumberland 32
  dark 32
  fine-rayed 30, 154
  flat 32, 154
  fuzzy 32
  Georgia 32
  Gulf 30
  hazel 32
  heavy 32, 154
  Louisiana 32
  Mississippi 32
  narrow 30
  Ohio 32
  oval 32
  Ozark 30
  purple 30
  pyramid 32
  rough 32, 154
  round 32
  Scioto 32
  sculptured 33

shiny	30, 154	*Pilsbryna*	135
southern	32	*Pilsbryspira*	100
tapered	33	pilula	
Tennessee	30	Pitar	44
Texas	30	Stenotrema	149
triangle	30	pimpleback	33
true	32	orange-foot	31, 154
Virginia	31	purple	33
Wabash	30	smooth	32
Warrior	32	Texas	33
yellow	32	western	33
pilata, Cratena	121	*Pinctada*	25
Pilidae	59	pinecone	
pillar		bronze	128
Appalachian	126	eightfold	128
glossy	126	flattened	128
robust	126	maze	128
thin	126	southern	128
pillclam, quadrangular	34	*pingelii, Oenopota*	99
pillsnail		*pinicola, Vespericola*	141
carinate	138	pink, ring	31
lowland	138	*Pinna*	25
mountain	138	*pinnata, Fiona*	121
palmetto	138	Pinnidae	25
upland	138	*pinnulatum, Cerastoderma*	39
Wichita Mountain	138	*pinosensis, Borsonella*	96
pilosa, Acanthodoris	115	pinwheel	
pilsbry, Amnicola	60	fir	130
pilsbryana		ribbed	130
Ashmunella	137	*piragua, Phenacovolva*	81
Fontelicella	61	*pisana, Theba*	146
Gastrocopta	126	*Pisania*	90
Paravitrea	135	*piscinalis, Valvata*	58
pilsbryanus, Somatogyrus	62	*Pisidium*	34
pilsbryi		*pisiformis, Strigilla*	41
Antalis	50	pistolgrip	33
Anticlimax	69	*pita, Doto*	119
Cirsotrema	75	*Pitar*	44
Elimia	64, 157	pitar	
Ensitellops	38	banded	44
Glyphostoma	97	lightning	44
Lioplax	59	little-pill	44
Mesomphix	134	Newcomb	44
Oreohelix	142	pleasing	44
Pallifera	132	Schwengel	44
Planorbella	125	Simpson	44
Pristiloma	136	white	44
Rabdotus	130	*pityis, Holospira*	129
Scaphander	107	*placenta, Cardiapoda*	81
Stagnicola	123	*placentula, Paravitrea*	135
Stenotrema	140	*Placida*	113
Teinostoma	70	*Placiphorella*	150
Ventridens	136	*Placopecten*	27
Zirfaea	47	*plagioptycha, Thysanophora*	141

plagosus, Busycotypus 90
plana
   bahamensis, Polygyra 139
   Crepidula 79
   Okenia 115
   Phreatodrobia 62
planata, Mysella 37
Planaxidae 72
Planaxis 72
planaxis
   black 72
   dwarf 72
planetica, Cardiomya 49
planeticum, Buccinum 88
planiusculum, Periploma 48
Planktomya 38
Planogyra 128
planorb, bloodfluke 124
Planorbella 125
Planorbidae 124
planorbis, Skeneopsis 69
planulata
   Mysella 37
   Spisula 40
platinum, Calliostoma 55
Platomysia 37
Platydorididae 118
Platydoris 118
Platyodon 45
platyrachis, Pyrgophorus 62
platysayoides, Triodopsis 141, 155
pleasi, Venustaconcha 33
plebeia, Sarasomia 147
plebeius, Tagelus 43
Plectodon 49
plectodon
   grainy 49
   rough 49
Plectomerus 31
plectrum, Buccinum 88
pleii, Loligo 151
plenum, Pleurobema 32, 154
Plesarionta 144
Plethobasus 31, 154
Pleurobema 32, 154
Pleurobranchaea 114
Pleurobranchaeidae 114
Pleurobranchidae 114
Pleurobranchus 114
Pleurocera 65
Pleuroceridae 63, 157
Pleuromalaxis 70
Pleuromeris 38
Pleuroploca 91

Pleurotomariidae 52
pleurotomaris, Oenopota 99
Pleurotomella 100
plicata
   perplicata, Amblema 28
   plicata, Amblema 28
   Leptoxis 65
   Polygyra 139
plicatella, Raeta 40
plicatilis, Velutina 80
plicatostriata, Elimia 64
Plicatula 27
Plicatulidae 27
plicatum, Caecum 71
plicifera, Juga 65
Plicifusus 90
plicosa, Pyrgocythara 100
pliculosa, Cerithidea 72
plumbea, Kurtziella 98
plumula kelseyi, Lithophaga 23
pluto, Turbonilla 106
Pneumoderma 111
Pneumodermatidae 111
Pneumodermopsis 111
pocahontasae
   Odostomia 103
   Turbonilla 106
pocketbook
   Altamaha 30
   fat 32, 154
   fine-lined 30
   lined 30
   plain 30
   sandbank 31
   shiny-rayed 31
   southern 30
   speckled 31
Pododesmus 27
poeyana, Cochlodinella 129
poica, Learchis 121
pointed-venus 44
   Carib 44
polare, Buccinum 88
polaris
   Astarte 39
   Philine 108
Policordia 49
Polinices 82
polita
   Clio 110
   Depressiscala 75
   Polygyra 139
   Turbonilla 106
politiana, Neverita 82

politulus, Cystiscus	93	rustic	123
politus, Cystiscus	93	shortspire	123
Polycera	116	St. Lawrence	123
polycera		striate	123
eelgrass	116	thickshell	123
orange-spike	116	widelip	123
three-color	116	woodland	123
Polycerella	116	*pontificus, Microceramus*	130
Polyceridae	116	*pontis, Paravitrea*	135
polygonatus cinisculus, Nassarius	91	*pontogenes, Aorotrema*	69
Polygyra	139	*popei, Popenaias*	32
polygyratus, Euconulus	133	*Popenaias*	32
Polygyrella	141	*popovius, Oenopota*	99
polygyrella, Polygyrella	141	*porcella, Corbula*	46
Polygyridae	137, 155	*porifera, Mopalia*	150
Polygyriscus	131, 155	*Poromya*	49
Polygyroidea	141	poromya	
Polymesoda	44	elongate	49
polynyma, Mactromeris	40	granular	49
Polyschides	51	rostrate	49
Polystira	100	Trost	49
polystrigma, Woodbridgea	109	white	49
Pomacea	59	Poromyidae	49
pomara, Mangelia	98	*porrecta, Elimia*	64
pomatia, Helix	146	*porterae, Mexichromis*	116
Pomatiopsidae	63	*Portlandia*	22
Pomatiopsis	63	*portobelensis, Callistochiton*	148
pomum, Phyllonotus	85	*postelli, Pleurocera*	66
ponderosa		*postelliana, Polygyra*	139
atlantica, Casmaria	82	Potamididae	72
Cincinnatia	60	*Potamilus*	32, 154
Noetia	24	*potosiensis, Elimia*	64
pondhorn	33	pouchclam	
Florida	33	Kelsey	38
polished	33	tiny	38
southern	33	*poulsoni, Roperia*	85
tapered	33	*pourtalesianum, Propeamussium*	27
pondmussel, eastern	31	pourtalesii	
pondshell, paper	28	*Epitonium*	76
pondsnail		*Haliotis*	52
abbreviate	123	*powelli, Lampsilis*	30
arctic	123	*powhatani, Turbonilla*	106
bigmouth	123	praecox	
calabash	123	*Glyphyalinia*	134
coldwater	123	*Vitrinella*	70
deepwater	123	*praerosa, Leptoxis*	65
fat-whorled	123	*prasinata, Pleurocera*	66
flathead	123	*Praticolella*	139
flat-whorled	123	*pratoma, Odostomia*	103
longmouth	123	*pratomus, Tachyrhynchus*	71
marsh	123	prea	
obese	123	*Anisodoris*	118
Petoskey	123	*Atagema*	118
Piedmont	123	*precipitata, Alvania*	67

Precuthona 121
presleyi, Granoturris. 97
pressus, Compressidens 51
pretiosum, Antalis 50
pribiloffensis, Neptunea 89
pribilovus, Oenopota 99
pricei, Coryphella. 120
pricklycockle
    even. 39
    Florida. 39
    magnum. 39
    spiny 40
    yellow 39
pricklywinkle 67
    false. 67
Primovula 81
Prionoglossa. 111
Pristes. 37
Pristiloma 135
pristiphora, Myrtea. 35
probina, Tellina 42
Problacmaea. 54
Probythinella. 62
procera, Gastrocopta 126
Prochaetoderma. 20
Prochaetodermatidae 20
producta
    Elliptio. 29
    Odostomia. 103
    Stenoplax. 149
    Strigilla 41
productum, Chaetoderma. 20
proficua, Semele. 43
profunda, Allogona. 137
profundi, Mitrella. 87
profundicola
    Caecum. 71
    Odostomia. 103
profundior, Mya 45
proles, Helminthoglypta. 144
prolongata, Velutina. 80
prolongatus, Miodontiscus. 38
Promenetus. 125
Proneomenia. 20
Proneomeniidae 20
Propeamussiidae. 27
Propeamussium 27
propellerclam
    Kurr. 46
    northern. 46
Prophysaon. 132
propinqua
    Epioblasma 30, 156
    Physella. 124

proserpina, Amnicola. 60
Protatlanta 81
protea, Tryonia. 63
protexta
    Ensitellops. 38
    Terebra 95
Protothaca 44
prototypus, Gastrocopta. 126
protracta, Turbonilla 106
provisoria, Zachrysia 141
proxima
    Adalaria. 115
    Ashmunella. 137
    Elimia 64
    Nucula. 21
    Scissurella. 52
psaltes, Dendrochiton. 148
Psammobiidae. 42
Psephidia 45
Pseudantalis 50
pseudoarenaria, Mya. 45
pseudoareolata, Pusillina. 68
pseudocarnaria, Strigilla 41
Pseudochama 36
Pseudocyphoma 81
pseudodonta, Ashmunella 137
Pseudomalaxis 74
Pseudomelatoma 100
pseudomera, Macoma 41
Pseudomiltha 35
Pseudosimnia 81
Pseudostomatella 56
Pseudosubulina. 129
Pseudosuccinea 123
pseudovirgatus, Ischnochiton. 148
psila, Ithycythara 98
Psiloteredo 47
psyche, Calliostoma. 55
Pteria 25
Pteriidae 25
Pterochelus. 85
pterocladica, Tricolia 58
pteropod
    bulimoid. 110
    cigar. 110
    helicid 110
    planorbid 110
    reticulate 110
    retrovert 110
    trochiform 110
    two-spine. 110
Pteropurpura. 85
pteropus, Ommastrephes 152
Pterotrachea. 81

Pterotracheidae 81
Pterotyphis 85
Pterynotus. 85
Ptychatractus 93
Ptychobranchus 32
ptychophora, *Allogona*. 137
Ptychosalpinx 90
puber, *Chione*. 44
pubescens, *Colus* 89
puella, *Vexillum* 93
pugentensis, *Striatura* 136
pugetensis
    Lyonsia 47
    Turbonilla 106
pugilator, *Catinella*. 132
pugilis, *Strombus* 78
pulchella
    Anachis. 86
    Okenia. 115
    Triphora. 75
    Truncatella 69
    Vallonia. 128
pulchellum
    Caecum. 71
    Cerithiopsis 74
    Vexillum 93
pulcherrima
    Odostomia. 103
    Persicula 94
pulchra
    Rostanga. 117
    Semele. 43
pulchrum, *Calliostoma*. 55
pulcia, *Odostomia* 103
pulcius, *Colus*. 89
pulleyi
    Busycon. 90
    Macoma 41
pulligo, *Tegula* 56
pulloides, *Tricolia*. 58
pullus, *Toxolasma*. 33
Pulsellidae. 51
Pulsellum 51
pumila
    Emarginula 53
    Gyrotoma 65, 157
pumilio, *Opalia*. 76
pumilum, *Opeas* 128
pumilus, *Somatogyrus* 62
puncta, *Turbonilla* 106
punctata
    Diplodonta. 36
    Fissurella. 53
    Pandora. 48

Phreatodrobia 62
punctatellus, *Helicodiscus* 131
punctatum, *Pisidium*. 34
punctatus, *Notarchus*. 112
Punctidae 130
punctiferum, *Pisidium* 34
punctocaelatus, *Rictaxis* 107
punctocostata, *Puncturella* 53
punctostriatus
    Rictaxis 107
    Scaphander 107
punctulata, *Acanthina* 84
Punctum 130
Puncturella 53
puncturella
    Atlantic 53
    clove 53
    diluvian 53
    dot-rib 53
    erect. 53
    granulate 53
    great. 53
    helmet 53
    hood. 53
    long-slot. 53
    many-rib 53
    painted. 53
punicea
    Tellina 42
    Turbonilla 106
pupa, *Puperita* 58
pupaeformis, *Elimia*. 64, 157
Puperita 58
Pupilla. 127
Pupillidae 126
pupillus, *Margarites* 56
Pupisoma 127
pupoidea, *Elimia*. 64
Pupoides 127
pupoides, *Halistylus* 55
pupoideus, *Amnicola* 60
pura, *Mitrella* 87
purcina, *Discodoris* 118
Purpura. 85
purpurascens, *Semele*. 43
purpuratus, *Potamilus* 32
purpurea, *Manzonia*. 67
purpureum, *Bittium* 73
purse-oyster
    bicolor. 25
    Lister. 25
    thin 25
pusa, *Philinopsis*. 108
pusae, *Discodoris* 118

pusilla
    Olivella 92
    Tectonatica 82
    Turbonilla 106
Pusillina 68
pusillus, Japonacteon................ 107
pusio, Pisania...................... 90
pustula, Polygyra 139
pustulata
    Cuthona........................ 121
    Patelloida 54
pustulifera, Tonicia.................. 150
pustuloides, Polygyra................ 139
pustulosa
    mortoni, Quadrula 33
    pustulosa, Quadrula 33
putris, Succinea 133
pybasi, Elimia...................... 64
pycnonema, Rimula 53
pycnum, Aphaostracon................ 60
pygmaea
    Acanthochitona.................... 150
    Chione......................... 44
    Elimia 64, 157
    Oreohelix....................... 142
    Vertigo......................... 127
pygmaeus
    Colus 89
    Musculus 24
    Somatogyrus 62
pygmy-venus........................ 44
    gray 44
pyramidalis, Oenopota................ 99
pyramidata
    Clio 110
    Gyrotoma 65, 157
pyramidatum, Pleurobema............. 32
Pyramidella......................... 103
Pyramidellidae 101
pyrenella, Pleurocera 66
pyrg
    Archimedes 62
    corded 62
    gravel.......................... 62
    moss 62
    Ozark.......................... 62
Pyrgocythara 100
Pyrgophorus 62
Pyrgospira.......................... 100
pyrgula, Opeas 128
Pyrgulopsis 62
pyrifera, Pseudosimnia................ 81
pyriforme, Pleurobema................ 32
pyrrha, Triphora.................... 75

Pyrunculus 109
Pythenella.......................... 37

Q

quadrae
    Odostomia...................... 103
    Spiromoelleria................... 57
quadragenarium, Trachycardium........ 40
quadrangularis, Neilonela 21
quadrans, Astarte.................... 39
quadrata
    Basterotia 38
    Glyphoturris 97
    Philine 108
quadratus, Somatogyrus 62
quadridens, Gastrocopta 126
quadridentata, Diacria................ 110
quadridentatus, Polyschides............ 51
quadrifilatum, Bittium 73
quadrifissatum, Siphonodent 51
quadrilineata, Cryoturris 96
quadripunctata, Trivia 80
quadrisulcata, Divaricella 35
Quadrula....................... 32, 154
quadrula, Quadrula 33
quahog
    false........................... 44
    northern........................ 44
    ocean.......................... 43
    southern........................ 44
quillingi, Jouannetia.................. 46
Quincuncina 33
quirica, Nucula...................... 21
quoyanus, Perotrochus................ 52

R

rabbitsfoot........................... 32
    rough 32
Rabdotus........................... 130
rabdotus
    Black Mountain.................. 130
    elongate........................ 130
    prairie 130
    striped 130
    whitewashed 130
raderi, Glyphyalinia 134
radians, Pseudochama 36
radiata
    Cochlespira 96
    conspicua, Lampsilis 30
    Nuculana....................... 22
    radiata, Lampsilis.................. 30

Tellina	42
radiatum, *Homalopoma*	57
radiatus	
Anodontoides	29
Isognomon	25
radicula, *Vermicularia*	72
Radiocentrum	142
Radiodiscus	130
Radix	123
radix, big-ear	123
Raeta	40
rafinesqueana, *Lampsilis*	30
ragsdalei	
Megapallifera	132
Pallifera	132
rainbow	
Alabama	33
downy	34
Florida	33
notched	33
southern	33
rainesae, *Conus*	95
ramosa	
Issena	116
Petalifera	112
rams-horn	
acorn	125
Artemesian	125
bellmouth	125
caribou	125
coarse	125
corpulent	125
crested	124
druid	125
file	125
fine-lined	125
ghost	124
giant	59
Great Basin	124
Jupiter	125
keeled	125
lamb	125
magnificent	125
marsh	125
meadow	125
mesa	125
Mexican	125
ridged	124
rusty	124
Seminole	125
thicklip	125
two-ridge	124
randolphi	
Delectopecten	26
Melanella	77
Punctum	130
Ranellidae	83
ranelloides tenuisculpta, *Bursa*	83
Rangia	40
rangia	
Atlantic	40
brown	40
raninus, *Strombus*	78
rassinus, *Oenopota*	99
rastrus, *Spiroglyptus*	72
rathbuni	
Cocculina	54
Turbonilla	106
rava, *Clathurella*	96
raveneli	
Elliptio	29
Mitrella	87
Pecten	27
raveneliana, *Alasmidonta*	28
raymondi, *Turbonilla*	106
razor	
Alaska	40
Atlantic	40
northern	40
rough	40
Sloat	40
transparent	40
reclivata, *Neritina*	58
reclusiana, *Neverita*	82
reclusum, *Teinostoma*	70
Recluzia	75
recta	
Acteocina	107
Ligumia	31
Turbonilla	106
Volvulella	110
rectirostris, *Exilioidea*	89
rectius, *Rhabdus*	50
rectus, *Modiolus*	23
recurva, *Clio*	110
recurvans, *Mopalia*	150
recurvum, *Ischadium*	23
redimita, *Xerarionta*	146
redondoensis, *Axinulus*	35
reediana, *Helminthoglypta*	144
reesi	
Paravitrea	135
Pleurobranchus	114
reeviana	
brevicula, *Lampsilis*	31
brittsi, *Lampsilis*	31
reeviana, *Lampsilis*	31

reflexa
   Haloconcha 66
   Obliquaria 31
refulgens, Quadrula 33
regalis, Calliotropis 55
regina
   Admete 94
   Tegula 56
   Truncatella 69
regis, Holospira 129
regius, Conus 95
regulare, Campeloma 59
regularis
   Ischnochiton 148
   Limatula 26
   Volutopsius 90
   Yoldia 22
regulus, Oenopota 99
rehderi
   Catinella 133
   Cymatium 83
   Parviturbo 57
reidi, Solemya 22
remanei, Unela 110
remondii, Megasurcula 98
reses
   nesodryas, Orthalicus 130
   reses, Orthalicus 130, 155
resina, Odostomia 103
resticula, Pyramidella 104
reticularis, Oliva 92
reticulata
   adelae, Cancella 94
   reticulata, Cancella 94
   Amphissa 86
   Cocculina 54
   Peracle 110
   Turbonilla 106
reticulatum, Deroceras 137
reticulatus
   Oenopota 99
   Tachyrhynchus 71
   Trimusculus 126
reticulopoda, Aplysia 112
reticulosa, Daphnella 97
reticulum, Cyclopecten 27
retifera
   Amaea 75
   Daphnella 97
   Rhamphidonta 37
retiporosa, Lepidozona 149
retromargo, Amnicola 60
retroversa, Limacina 110
Retusa 109

retusa, Obovaria 31
Retusidae 109
retsum, Oxyloma 133
Rhabdus 50
rhadinum, Aphaostracon 60
Rhamphidonta 37
Rhapinema 62
rheophilas, Somatogyrus 62
rhinetes, Nassarius 91
rhoadsi, Glyphyalinia 134
Rhodacme 125
rhodia, Margarites 56
rhodoceras, Acanthodoris 115
rhodope, Rhodopetoma 100
Rhodopetala 54
Rhodopetoma 100
rhodostoma thomae, Bursa 83
rhomba, Cyphoma 81
rhombica, Marsenina 80
rhomboideum, Sphaerium 34
rhombostoma, Amnicola 60
rhombus, Thysanoteuthis 152
rhyssa
   Admete 94
   Ashmunella 138
   Solariella 56
rhythmica, Crassispira 96
richi, Odostomia 103
richthofeni, Here 35
rickettsi, Catriona 120
Rictaxis 107
riddelli, Pleurobema 32
ridgemussel, western 30
ridgwayi, Turbonilla 106
riffleshell
   angled 29, 156
   northern 30
   tan 29, 154
   Tennessee 30, 156
   Wabash 30, 156
rigida
   Atrina 25
   Clathurella 96
   Ventricolaria 45
riiseanus, Atys 109
rileyensis, Ashmunella 138
rimosus, Lemiox 31, 154
Rimula 53
rimula
   bridle 53
   California 53
   coarse 53
   threaded 53
   webbed 53

rimula, Glyphyalinia	134	curly	85
rimuloides, Sinezona	52	deltoid	85
rinconensis, Sonorella	146	dim.	85
rinella, Turbonilla	106	Florida	85
Ringicula	107	graceful	85
Ringiculidae	107	helmet	65
ringnesius, Cyclopecten	27	interrupted	65, 157
riograndense, Metastoma	130	Japanese	85
riograndensis		knobby	65
Cochliopina	60	lirate	65, 157
Gastrocopta	126	lurid	85
riparia, Gastrocopta	126	maiden	65
riparium, Carychium	125	minor	85
risso		muddy	65
Caribbean	69	oblong	65
Monterey	69	onyx	65
smooth	68	ornate	65
Rissoa	68	painted	65
Rissoella	69	plicate	65
Rissoellidae	69	purple	85
Rissoidae	67	ribbed	85
Rissoina	68	rotund	65, 157
ritteri		round	65
Odostomia	103	rugose	65
Trivia	80	rustic	86
riversnail, spiny	65	Santa Barbara	85
rivularis, Ferrissia	125	scaly	85
rixfordi, Sonorelix	145	Shawnee	65
rjabininae, Cyclocardia	38	spotted	65
roanokensis, Elliptio	29	striped	65, 157
robusta		umbilicate	65
Alasmidonta	28	varicose	65
Fontelicella	61	warty	65
Heterodoris	115	widemouth	85
Moroteuthis	151	*roemeri*	
scipio, Cingula	67	Glyphyalinia	134
robustum, Dinocardium	39	Mesodon	138
robustus, Dendronotus	119	Metastoma	130
rocellaria		*rogersi, Lithophaga*	23
Atlantic	46	*rollandi, Tridonta*	39
ovate	46	*rollandiana, Recluzia*	75
rostrate	46	*roosevelti, Marginella*	94
Stimpson	46	*rooseveltiana, Sonorella*	146
rock-pocketbook	29	*roperi*	
Ouchita	29	Loligo	151
rockshell, rough	33	Trilobopsis	141
rocksnail		*Roperia*	85
agate	65, 157	*rosacea*	
armored	65	Hopkinsia	114
beta	85	Mitrella	87
Big Black	65	Tectura	54
bigmouth	65, 157	*rosaceus, Solen*	40
circled	85	*rosana, Alvania*	67
Coosa	65, 157	*roscida, Marginella*	94

rose, Hopkins 114
rosea
   Euglandina 129
   Fissurella. 53
   Rhodopetala 54
rosemontensis, Sonorella 146
roseolum, Calliostoma 55
roseopicta, Chromodoris 116
roseus
   Colus 89
   Oenopota. 99, 100
rossellinus, Circulus 69
Rossia 151
Rostanga. 117
Rostangidae. 117
rostrata
   Cuspidaria. 49
   Poromya 49
   Spengleria 46
rostratum, Netastoma 47
rotella
   inornata, Microgaza 56
   rotella, Microgaza. 56
rotifera, Monadenia 144
rotulata, Obovaria 31
rotunda
   Olivella 92
   Phreatodrobia 62
   Thyasira. 36
rotundata, Glebula 30
rotundatum, Pisidium. 34
rotundatus
   Boreotrophon 84
   Discus 131
rotundus, Volutopsius. 90
roughmya
   ample. 46
   Arctic. 46
   Bering 46
roundlake 28
rowelli
   Cerithiopsis 74
   Eremarionta 143
   Vertigo. 127
rubella
   Kurtziella. 98
   Problacmaea. 54
rubellum, Pleurobema 32
rubescens
   Octopus. 152
   Thordisa 118
rubida, Chlamys 26
rubidus, Murex 84

rubra
   Tonicella 149
   Velutina. 80
rubrilineata
   Godiva. 121
   Tricolia 58
rubropicta, Semele 43
rubrum, Gastropteron. 108
rudis
   Pinna 25
   Pododesmus 28
rufa, Placiphorella 150
rufescens, Haliotis 52
rufocincta, Micrarionta 144
rufum, Campeloma. 59
rufus
   Arion 131
   Dendronotus 119
rugatina, Ventricolaria. 45
rugatum, Bittium 73
rugatus, Leptochiton 148
rugeli
   Mesodon 139
   Mesomphix 134
rugifera, Neaeromya 37
rugirima, Glyphoturris 97
rugoderma, Anguispira. 131
rugosa
   Thracia 48
   Triodopsis 141
rugosula, Vertigo 127
rugosum, Trigonostoma 94
rugulosa, Metaxia. 74
ruidosensis, Gastrocopta 126
rumia, Cadlina 116
Rumina 128
rumphiana, Quadrula. 33
Runcina. 109
Runcinidae 109
Rupellaria 45
rupellaria
   Atlantic 45
   cancellate. 45
   hearty 45
   thin 45
rupicola
   Epitonium 76
   Gastrocopta. 126
   Semele. 43
rushii
   Epitonium 76
   Polyschides 51
   Terebra 95

rustica
- Caelatura ... 68
- Fossaria ... 123
- Thais ... 86

rusticana, Succinea ... 133
rusticoides, Columbella ... 87
rustyus, Eubranchus ... 120
rutila, Melanella ... 77
rycia, Polycera ... 116
ryssotitum, Caecum ... 71

S

Sabia ... 78
sabina, Chelidonura ... 108
Sabinella ... 77
sabinii, Colus ... 89
sabinoensis, Sonorella ... 146
saccharina, Tonicella ... 149
Saccharoturris ... 100
sacculifer, Modiolus ... 23
Sagdidae ... 141
sagittatum, Amygdalum ... 23
sagra, Philine ... 108
sagraiana
- Natica ... 82
- Rissoina ... 68

sagrinata, Myrtea ... 35
Sakuraeolis ... 121
salebrosa, Lithasia ... 65
salinasensis
- Ashmunella ... 138
- Disconaias ... 29

salishorum, Pulsellum ... 51
salleana, Hastula ... 95
salleanum, Oxyloma ... 133
sallei, Mytilopsis ... 43
salmonacea, Coryphella ... 120
salmonaceus, Helicodiscus ... 131
salmonea, Halodakra ... 43
salmoneus, Margarites ... 56
salsa, Spurwinkia ... 63
saludensis, Helicodiscus ... 131
salvanium, Teinostoma ... 70
sampsoni, Micromenetus ... 124
sampsonii, Epioblasma ... 30, 156
sanburni, Cryptomastix ... 138
sanctaemarthae, Solecurtus ... 43
sanctus, Aesopus ... 86
sandersoni, Atys ... 109
sandiegensis, Diaulula ... 118
sandshell
- black ... 31
- southern ... 30
- yellow ... 31

sanesia, Yoldiella ... 22
sanguin
- Atlantic ... 42
- gaudy ... 42
- operculate ... 42

sanguinea, Aldisa ... 117
Sanguinolaria ... 42
sanguinolenta, Sanguinolaria ... 42
sanibelensis, Crssispira ... 96
sanjuanense, Bittium ... 73
sanjuanensis
- Barleeia ... 68
- Eubranchus ... 120
- Odostomia ... 103

Sansonia ... 57
santaritana, Sonorella ... 146
santarosana
- Maxwellia ... 84
- Turbonilla ... 106

santarosanus, Antiplanes ... 96
sanus, Mesodon ... 139
sapelona, Okenia ... 115
sapia, Odostomia ... 103
sapidum, Calliostoma ... 55
sapiella, Teinostoma ... 70
sapotilla, Yoldia ... 22
Sarasomia ... 147
sarda, Chama ... 36
sargenti, Somatogyrus ... 62
sargentianus, Mesodon ... 139
sarsii, Oenopota ... 100
sathon, Notogillia ... 61
satura
- Lampsilis ... 31
- Odostomia ... 103

saturna, Limatula ... 26
sawinae, Nitidiscala ... 76
saxatilis, Littorina ... 66
saxicave, little ... 46
Saxicavella ... 46
saxicola, Entodesma ... 47
Saxidomus ... 45
sayana, Oliva ... 92
sayanus, Mesodon ... 139
Sayella ... 104
sayi, Diodora ... 52
scaber, Plectodon ... 49
scabra
- Collisella ... 54
- Lima ... 26

scabricostata, Lepidozona ... 149
scabricula, Chaetopleura ... 149
scabriuscula, Cadlina ... 116

scabrum, Chaetoderma	20
Scaeurgus	152
scalariforme, Buccinum	88
scalariformis	
Cerithidea	72
Coralliophil	86
Pyrgulopsis	62
scalaris	
Oenopota	100
Planorbella	125
Truncatella	69
Scalenostoma	77
scaleshell	31
scalesnail	
parasitic	71
striate	71
scallop	
Antillean	27
Atlantic calico	26
bay	26
Benedict	26
Bering	26
dwarf zigzag	26
fine-ribbed	26
giant rock	26
Greenland	26
Iceland	26
Jordan	26
kelp	27
knobby	26
lined	26
lions-paw	27
Lowe	26
Mildred	26
nucleus	26
ornate	26
Pacific calico	26
paper	26
Randolph	26
Ravenel	27
reddish	26
red-ribbed	26
rough	26
San Diego	27
scaly	26
sea	27
Sigsbee	27
spathate	26
spiny	26
striate	27
thistle	26
Vancouver	27
vitreous	27
wavy	27
weathervane	27
white	26
zigzag	27
Scaphander	107
Scaphandridae	107
Scaphella	93
scaphoides, Adrana	21
Scelidotoma	53
Schizoplax	149
schrammi, Tonicia	150
schroederi, Cerodrillia	96
Schwartziella	68
schwengelae, Eulima	77
Schwengelia	77
Scintillona	37
Scissilabra	70
scissurata, Yoldia	22
scissuratus, Nassarius	91
Scissurella	52
scissurelle	
belt	52
crispate	52
Florida	52
lyre	52
rim	52
Scissurellidae	52
scitulus, Nipponotrophon	85
sclera	
Trachypollia	86
Urosalpinx	86
Sclerodoris	117
Sconsia	82
scopulosum, Sinum	82
scrobiculatus, Ischnochiton	48
scrubsnails	
Appalachian	139
banded	139
Choctaw	139
Florida	139
Hidalgo	139
ridge	139
sandyland	139
striped	139
vagrant	139
white	139
sculptilis, Glyphyalinia	134
sculptum marielinum, Cerion	128
sculpturatus, Oenopota	100
scutata, Pterotrachea	81
Scutopus	20
scutulata, Littorina	66
scutum, Notoacmea	54
Scyllaea	119
Scyllaeidae	119

seacat, warty	112	semen, *Retusa*	109
seahare		semiaspera, *Diplodonta*	36
Atlantic black	112	semicarinata, *Elimia*	64
blue-ring	112	semiinflata, *Crassispira*	96
California	112	semiliratus, *Dendrochiton*	148
California black	112	seminalis, *Fluminicola*	61
net-foot	112	seminuda	
ragged	112	*Atrina*	25
sooty	112	*Boonea*	101
spotted	112	*Crenella*	23
walking	112	*Yoldia*	22
yellow	112	semiplicata, *Anachis*	86
sea-lemon		semipolita, *Graptacme*	50
Monterey	117	*Semirossia*	151
mottled pale	118	semirugosa, *Cyathodonta*	48
Pacific	118	semisculptus, *Circulus*	69
Searlesia	90	semistriata, *Ringicula*	107
seaslug		semistriatum, *Teinostoma*	70
batwing	108	semistriolata, *Graptacme*	50
gray batwing	108	semisulcata	
Hansine	113	*Papyridea*	39
Pacific batwing	108	*Sigatica*	82
secreta, Pallifera	132	senex, *Aspella*	84
secta, Macoma	41	senhousia, *Musculista*	23
securis, Musculium	34	sennottorum, *Conus*	95
sedlaki, Cyphoma	81	sentis, *Chlamys*	26
seftoni, Admete	94	sepioidea, *Sepioteuthis*	151
seguenzae, Verticordia	49	Sepiolidae	151
Seguenzia	56	*Sepioteuthis*	151
seguenzia		septadens, *Paravitrea*	135
California	56	septemvolva, *Polygyra*	139
pygmy	56	septentrionalis	
Seguenziidae	56	*Glycymeris*	25
Seila	74	*Odostomia*	103
selenina, Lacteoluna	141	*Thracia*	48
selinitoides, Discus	131	*Septifer*	24
sellatus, Philomycus	132	sequoicola, *Helminthoglypta*	144
Semele	43	seradens, *Paravitrea*	135
semele		serca, *Elysia*	113
Asian	43	serga, *Kurtziella*	98
Atlantic	43	sericatum, *Buccinum*	88
beautiful	43	sericifilum, *Epitonium*	76
California	43	*Serpulorbis*	72
cancellate	43	serra, *Bittium*	73
clipped	43	serrae, *Turbonilla*	106
contracted	43	serrana, *Alaba*	72
incongruous	43	serrata	
nut	43	*Atrina*	25
purplish	43	*Lepidozona*	149
rock	43	serrei, *Marginellopsis*	94
rose-painted	43	serricata, *Axinopsida*	35
Tellin	43	*Serripes*	39
Semelidae	42	setacea, *Bankia*	47
Semelina	43	setifera, *Limatula*	26

setosa
   Monadenia 144
   Philobrya 25
shagreen
   Alabama 139
   deep-tooth 139
   Magazine Mountain 138
   smooth-lip 138
   tight-gapped 138
sharonae, Marseniopsis 80
sharpii, Lepidochitona 149
shasta, Vespericola 141
sheepnose 31
sheldoni, Hoyia 61
shelledslug
   Guadalupe 132
   Santa Barbara 132
shepardae, Pristiloma 136
shepardiana
   Aclis 76
   Elliptio 29
shimeki
   Discus 131
   Helicodiscus 131
shipworm
   Bartsch 47
   big-ear 47
   blacktip 47
   carinate 47
   feathery 47
   fimbriate 47
   Gould 47
   Knox 47
   malleate 47
   naval 47
   Norway 47
shoulderband
   Big Sur 144
   bigtree 144
   Breckenridge 144
   bronze 143
   Coalinga 144
   Coast Range 144
   Contracosta 143
   Cottonwood 143
   coyote 144
   Cuyama 143
   Cuyamaca 143
   cypress 143
   El Paso 144
   foothill 143
   granite 143
   Greenhorn 143
   Horse Meadows 143
   Kern 143
   Kings 143
   Laguna 144
   Lowe Canyon 144
   Mendocino 143
   Merced Canyon 143
   mimic 143
   Mohave 143
   Morris Canyon 143
   Morro 144
   Oregon 143
   Paiute 144
   Panamint 143
   Peninsular Range 144
   Pine Valley 143
   Pinnacles 143
   Point Pinos 143
   Point Reyes 144
   redwood 144
   San Juaquin 143
   San Miguel 143
   Santa Cruz 143
   sidewinder 143
   Sierra 143
   silky 143
   Soledad 143
   Sonoma 144
   southern 144
   surf 143
   Sweetwater 143
   Tehachapi 143
   Transverse Range 144
   Tulare 144
   Victorville 144
   Yosemite 144
   yucca 143
showalteri
   Elimia 64
   Lepyrium 61
   Pleurocera 66
showalterii, Leptoxis 65, 157
shuttleworthianus, Callistochiton 148
shuyakensis, Turbonilla 106
sicarius, Solen 40
sicula, Emarginula 53
sideband
   keeled 144
   Klamath 144
   Mariposa 144
   Pacific 144
   redwood 144
   Shasta 144
   Sierra 144
   Tuolumne 144

sideralis, Berthella	114
sierranus, Vespericola	141
Sigatica	82
signa, Cerithiopsis	74
signae, Turbonilla	106
signatum, Cyphoma	81
significans	
Bartschia	87
Paravitrea	135
sigsbeei, Pectinella	27
silicula, Juga	65
Siliqua	40
siliqua, Cyrtodaria	46
Siliquaria	72
Siliquariidae	72
siliquoidea, Lampsilis	31
sillana, Odostomia	103
sillimani, Oxyloma	133
siltsnail	
alligator	61
armored	63
Comal	60
conical	63
creek	60
crystal	60
Enterprise	60
globe	60
hyacinth	60
Ichetucknee	60
midland	60
ponderous	60
pumpkin	63
pygmy	60
Rock Springs	60
satyr	61
Seminole	60
Wekiwa	60
simbla, Margarites	56
similans, Helminthoglypta	144
similaris, Bradybaena	142
simile	
Sphaerium	34
Stenotrema	140
similis	
Gastrocopta	127
Tellina	42
simmonsi, Sonorella	146
simnia	
Keys	81
one-tooth	81
Pacific	81
pear	81
Santa Barbara	81
sea-whip	81
slender	81
West Indian	80
Simnialena	81
simplex	
Anomia	27
Columella	126
Elimia	64
Mohnia	89
Oenopota	100
Thyasira	36
Volutopsius	90
Simpsonaias	33
simpsoni	
Cerodrillia	96
Paravitrea	135
Pitar	44
Polygyra	139
Turbonilla	106
simpsonianus, Medionidus	31
simulatum, Buccinum	88
sincera, Valvata	58
Sinezona	52
singleyana, Euglandina	129
singleyanus, Helicodiscus	131
sinistrum, Busycon	90
sinuata	
Mopalia	150
Philine	108
sinudentata, Lepidozona	149
Sinum	82
sinuosa, Chama	36
Siphonaria	126
Siphonariidae	126
Siphonodentaliidae	51
Siphonodentalium	51
Siraius	117
Siratus	85
sitiens, Sonorella	146
sitkana, Littorina	67
sitkensis, Tonicella	149
Skenea	57
skenea	
Alaska	69
beaded	57
California	57
Carmel	57
Coronado	57
flat	69
Skeneidae	57
Skeneopsidae	69
Skeneopsis	69
skidegatensis, Odostomia	103
skinneri, Physa	123
skomma, Crenella	23

slabshell
Altamaha 29
Carolina 29
Chipola 29
Roanoke 29
slippersnail
cheeky 79
common Atlantic 79
convex 79
eastern white 79
great 79
greedy 79
hooked 79
onyx 79
ridged 79
round 79
spiny 79
spotted 79
western white 79
white 79
slitmouth
Alabama 140
Apalachicola 140
Appalachian 140
Bear Creek 140
bristled 139
carinate 140
Chattanooga 140
Cohutta 140
Doaks Creek 140
fringed 139
Great Smoky 140
hairy 140
highland 139
inland 140
Kentucky 139
Missouri 140
Monte Sano 140
Oachita 140
Ozark 140
pygmy 149
Rich Mountain 140
ridge-and-valley 140
ridge-lip 140
Sequatchie 140
Talladega 140
slitshell
excised 65, 157
pagoda 65, 157
pyramid 65, 157
ribbed 65, 157
round 65, 157
striate 65, 157
slitsnail, lovely 52
sloani, Leidyula 147
sloati, Siliqua 40
sloatianus, Elliptoideus 29
slug
Atlantic sidegill 114
Atlantic umbrella 114
black 132
California sidegill 114
earshell 137
greenhouse 137
Hemphill 132
lettuce 113
marsh 137
meadow 137
salamander 143
sheathed 132
spotted 132
tree 137
vine 137
yellow umbrella 114
smaragda, Phyllaplysia 112
Smaragdia 58
smirna, Acteocina 107
smirnia, Neptunea 89
smithi
Aplysiopsis 113
Margarites 56
Mesodon 139
Neoplanorbis 125
Punctum 130
Vitrinella 70
smithii
Astarte 39
Carditopsis 38
Diplothyra 46
Odostomia 103
Olssonella 94
snaggletooth
Apache 126
armed 126
bark 126
bluegrass 126
bottleneck 126
comb 126
crested 126
cross 126
Erie 126
Great Lakes 127
Gulf Coast 126
lambda 126
montane 126
plains 126
Rio Grand 126
Ruidoso 126

shortneck	126	*Solariella*	56
slim	126	*Solariorbis*	70
sluice	126	solastra, *Succinea*	133
Sonoran	126	Solecurtidae	43
tapered	126	*Solecurtus*	43
white	127	solecurtus	
wing	126	corrugate	43
snail		St. Martha	43
atom	69	solemi, *Primovula*	81
badwater	68	*Solemya*	22
boulder	65	Solemyidae	22
brittlestar	77	*Solen*	40
Caribbean bivalved	113	Solenidae	40
chink	66	soleniformis, *Papyridea*	39
chocolate-band	146	solenogaster	
corncob	131	corkscrew	20
decollate	128	elongate	20
giant African	129	glisten-worm	20
herald	125	jawed	20
milk	146	leaf-spiculed	20
sargassum	73	mole	20
sculpin	63	psammon	20
seagrape	144	tailed	20
spruce	141	solenum, *Buccinum*	88
teardrop	62	solida	
Vancouver	141	*Bulla*	109
watercress	61	*Strictispira*	100
snakeskin-snail		solidissima, *Spisula*	40
cancellate	100	solidus	
gray	100	*Oenopota*	100
snuffbox	30	*Vorticifex*	125
socorroensis, *Oreohelix*	142	solitaria, *Haminoea*	109
soelneri, *Triodopsis*	141	*Somatogyrus*	62
softshell	45	sombrerensis, *Linga*	35
boring	45	sonoma, *Helminthoglypta*	144
California	45	sonomaensis, *Fossaria*	123
deep	45	sonorana, *Pupilla*	127
false	45	*Sonorelix*	145
Siberia	45	*Sonorella*	145
subovate	45	sorenseni, *Haliotis*	52
truncate	45	soror, *Diplodonta*	36
solandri, Trivia	80	souleyeti, *Protatlanta*	81
solarelle		souverbii, *Lobiger*	113
channeled	56	sowerbii	
fine-groove	56	*Lucapina*	53
intermediate	56	*Typhis*	86
lamellose	56	sowerbyi, *Episiphon*	50
look-around	56	soyoae, *Scissurella*	52
lovely	56	spadicea, *Cypraea*	80
naked	56	spargana, *Nuculana*	22
obscure	56	sparsa	
smooth	56	*Anachis*	86
varicose	56	*Cadlina*	116
wrinkled	56	*Quadrula*	33, 154

speciosa, Eucrassatella 39
spectabilis
  Corolla 111
  Mopalia 150
  Polyschides 51
spectaclecase 29
  little 33
spectralis, Glycymeris 25
specus, Glyphyalinia 134
Speleodiscoides 131
spelta capitia, Neosimnia 81
spelunca, Physella 124
Spengleria 46
sphaericosa, Adontorhina 35
Sphaeriidae 34
Sphaerium 34
Sphenia 45
sphenia
  Antillean 45
  ovoid 45
  Pacific 45
  tumid 45
Spheniopsidae 46
sphoni, Pseudosimnia 81
Spiculata 81
spike 29
  Alabama 29
  Atlantic 29
  box 29
  brother 29
  Cape Fear 29
  Carolina 29
  delicate 29
  false 33
  flat 29
  shiny 29
  variable 29
  Waccamaw 29
  winged 29
Spilochlamys 63
spinctostoma, Texadina 63
spindle
  apricot 91
  brown 91
  California false 93
  graceful 91
  ligate false 93
  lined egg 81
  modest 91
  Montery false 93
  ornamented 91
  painted 91
  Santa Barbara 91
  slender 91

stout 91
turnip 91
Yucatan 91
spindle-bubble
  California 109
  cylindrical 109
  Panamic 109
  southern 110
  spined 110
  spineless 110
  Texas 110
spinosa, Elliptio 29
spinosum
  Crucibulum 79
  Stenotrema 140
spinosus, Pyrgophorus 62
spinymussel
  Altamaha 29
  James 32, 154
  Tar 29, 154
spiralis, Antroselatus 60
spirata
  Acanthina 84
  Vermicularia 72
spiratus, Busycotypus 90
Spiraxidae 129
spirellum, Speleodiscoides 131
Spiroglyptus 72
Spirolaxis 74
Spiromoelleria 57
Spirula 151
spirula, Spirula 151
Spirulidae 151
Spisula 40
spitzbergensis, Colus 89
splendens, Calliostoma 55
splendida
  Lampsilis 31
  Leptadrillia 98
Splendrillia 100
Spondylidae 27
Spondylus 27
spongiosa, Opalia 76
spongotheras, Tylodinella 114
spoonclam
  Aleutian 48
  angled 48
  delicate 48
  flat 48
  fragile 48
  Lea 48
  paper 48
  related 48
  round 48

sulcate	48
unequal	48
sportella	
Corbula	38
elliptical	38
Pilsbry	38
square	38
textured	38
tube	38
sportella, Haplotrema	129
Sportellidae	38
spot	
brown	130
conical	130
glass	130
lamellate	130
ribbed	130
small	130
striate	130
spreadboroughi, Odostomia	103
springsnail	
Bear Lake	61
Blue Ridge	61
Crooked Creek	61
desert	61
Harney Lake	61
hidden	61
Idaho	61
Jackson Lake	61
Oasis Valley	61
Socorro	61
Yaqui	61
sprite	
bugle	124
button	124
disc	124
marsh	124
penny	124
sharp	125
shoal	124
umbilicate	125
sprouli, Ashmunella	138
spurca acicularis, Cypraea	80
Spurilla	121
spurilla	
frosted	121
neapolitan	121
red-tentacle	121
Spurillidae	121
spurius, Conus	95
Spurwinkia	63
squalida	
Ceratozona	149
Physella	124
squama, Siliqua	40
squamata, Siliquaria	72
squamifera, Tellina	42
squamiger, Serpulorbis	72
squamula, Anomia	27
squamulifera, Ocenebra	85
squawfoot	33
squid	
arrow	151
Atlantic bird	152
Atlantic brief	151
Berry armhook	151
bigeye	151
boreal armhook	151
boreal clubhook	151
Brown Bear armhook	151
California market	151
California two-spot	152
Caribbean reef	151
Caribbean two-spot	152
clawed armhook	151
common clubhook	151
dart	151
diamondback	152
flying	152
jumbo	151
longfin	151
luminous flying	152
magistrate armhook	151
northern shortfin	152
orangeback	152
Oregon armhook	151
Panama brief	151
purpleback	152
ram's horn	151
robust clubhook	151
Roper	151
sharptail shortfin	152
southern shortfin	152
webbed	152
squillaclam	37
squillina, Parabornia	37
stageri, Helminthoglypta	144
stagnalis, Lymnaea	123
Stagnicola	123
staminea, Protothaca	45
stancyki, Ersilia	77
stapes, Quadrula	33, 154
staphylinus, Boreotrophon	84
staphylophera, Chaetopleura	149
starcki, Dimyella	27
starsnail	
American	57
carved	57

green 57
longspine 57
West Indian 57
star-turris
elegant 96
lesser 96
stauberi, Lomanotus 119
stearnsi
Conus 95
Facelina 121
Pristiloma 136
Triphora 75
Vitrinella 71
stearnsiana
Fontelicella 61
Megasurcula 98
Plesarionta 144
stearnsii
Aesopus 86
Arctomelon 93
Compressidens 51
Marsenina 80
Tachyrhynchus 71
stefanssoni, Volutopsius 90
stegaria, Cyprogenia 29
stegeri
Daphnella 97
Fusinus 91
Olivella 92
steinbergae, Doridella 114
steinstansana, Elliptio 29, 154
stejnegeri, Cerithiopsis 74
stellata
Agathotoma 95
Stellatoma 100
Stellatoma 100
stelleri
Cryptochiton 150
Turbonilla 106
Stenophysa 124
Stenoplax 149
stenoschizum, Pseudantalis 50
Stenosemus 149
Stenotrema 139
stenotrema, Stenotrema 140
stephensae
Cerithiopsis 74
Psephidia 45
Sterkia 127
sterkii
Guppya 133
Vertigo 127
stewardsoni, Epioblasma 30, 156
Sthenorytis 76

sticta, Pseudomelatoma 100
Stigmaulax 82
stilesi, Neptunea 89
Stiliger 113
stiliger
blue 113
brown-streak 113
dusky 113
Stiligeridae 113
stillmani, Turbonilla 106
stimpsoni
Beringius 87
Colus 89
Conus 95
Cuthona 121
Gastrochaena 46
Pelseneeria 77
Placiphorella 150
Thracia 48
Transennella 45
Turbonilla 106
stimpsonii
Pazionotus 85
Truncatella 69
Stiobia 63
stirrupshell 33, 154
stiversiana, Helminthoglypta 144
stomatella
false 56
painted false 56
Stosicia 68
straminea
claibornensis, Lampsillis 31
straminea, Lampsillis 31
Mitra 92
strawberry-cockle
Atlantic 39
Guppy 39
western 39
streckeri, Lampsilis 31
strengi, Somatogyrus 62
Streptaxidae 129
stretchiana, Catinella 133
striata
Aclis 76
Alcadia 58
Bulla 109
Cardiomya 49
Cochliolepis 71
Hiatella 46
Hyalocylis 110
Lyonsia 48
Martesia 47
Sconsia 82

striate
- black 136
- fine-ribbed 136
- median 136
- northwest 136
- ribbed 136

striatinum, Sphaerium 34
striatissimum, Buccinum 88
striatoides, Acanthotrophon 84
striatula
- *Couthouyella* 75
- *Elimia* 64
- *Mysella* 37

striatum
- *Crucibulum* 79
- *Palliolum* 27

Striatura 136
Strictispira 100
strigata, Succinea 133
Strigilla 41

strigilla
- false 41
- Gabb 41
- large 41
- ovate 41
- pea 41
- Surinam 41
- white 41

strigillatum fucanum, Buccinum 88
strigillinus, Circomphalus 44

strigosa
- *Elimia* 64
- *Oreohelix* 142

striolata
- *Crepidula* 79
- *Trichia* 146

striolatus, Ischnochiton 148
striosa, Rissoina 68
Strobilops 128
Strobilopsidae 128
strodeanum, Pleurobema 32
Strombidae 78
Strombiformis 77
Strombus 78

strongi
- *Engina* 89
- *Odostomia* 103
- *Pleurobranchus* 114
- *Turbonilla* 106

strongyloides, Anguispira 131
Strophitus 33
stuarti, Boreotrophon 84
stultorum, Tivela 45
stygium, Carychium 125

stygius, Amnicola 60
Stygopyrgus 63
styliformis, Turbonilla 106
stylina, Turbonilla 106
Styliola 110
Stylocheilus 112
styria, Vexillum 93
Suavodrillia 100
subaequilatera, Nuculana 22

subangulata
- *Lampsilis* 31
- *Urosalpinx* 86
- *Yoldiella* 22

subauriculata, Limatula 26
subcancellata, Ervilia 40

subcarinata
- *Eulimostraca* 77
- *Lioplax* 59

subclausa, Polygyra 139
subdiaphana, Cooperella 45
subdiaphanus, Compsomyax 44
subeffusum, Oxyloma 133
subfuscus, Arion 131
subglacialis, Cuspidaria 49

subglobosa
- *Diplodonta* 36
- *Odostomia* 103

subglobosus, Birgella 60
submarmorea, Tonicella 149
subobsoleta, Glycymeris 25
suborbicularis, Kellia 37
suborbiculata, Anodonta 28
subornata, Elysia 113

subovata
- *Lima* 26
- *Paramya* 45

subpalliatus, Mesodon 139
subplana, Lexingtonia 31
subplanus, Mesomphix 134
subpurpureus, Viviparus 59
subramosus, Dendronotus 119
subrosacea, Coryphella 120
subrostrata, Ligumia 31

subrotunda
- *Fusconaia* 30
- *Lanx* 123
- *Obovaria* 31

subrudis, Oreohelix 142
subrufus, Hipponix 78
subrupicola, Pristiloma 136
Subsimnia 81
substriata, Tricolia 58
substriatum, Laevicardium 39
substriatus, Somatogyrus 62

subtentum, Ptychobranchus............. 32
subtenuis, Barleeia.................... 68
subteres, Tagelus..................... 43
subtilis
    Glycymeris........................ 25
    Paravitrea........................ 135
subtrigona, Halodakra................. 43
subtrigonus, Cystiscus................. 93
subtriplicata, Volvarina................ 94
subtruncata, Barnea................... 46
subtruncatum, Pisidium................ 34
subturrita
    Anachis........................... 87
    Odostomia........................ 103
subturritus, Aesopus................... 86
subula, Styliola....................... 110
subulata, Turbonilla................... 106
Subulina............................. 128
Subulinidae.......................... 128
subvexus, Strophitus.................. 33
subviridis
    Lasaea............................ 37
    Lasmigona........................ 31
subvitreus, Oenopota................. 100
succincta, Lirularia................... 55
Succinea....................... 133, 155
succinea
    Assiminea........................ 68
    Haminoea........................ 109
    Lacuna........................... 66
    Marinula......................... 122
Succineidae.................... 132, 155
succisa, Thyasira..................... 36
succissa, Fusconaia................... 30
suffusa
    Lucapina......................... 53
    Trivia............................ 80
sugarspoon....................... 29, 156
sulcata
    delicata, Epioblasma.............. 154
    Halistrepta....................... 48
    Limopsis......................... 24
    Oocorys.......................... 83
    Retusa........................... 109
    Stigmaulax....................... 82
sulcidentata, Eupleura................. 84
Sulcoretusa........................... 109
sulcosa, Odostomia................... 103
sulculosa, Lioplax.................... 59
sumneri, Turbonilla................... 106
sundial
    Atlantic cylinder.................. 74
    beaded........................... 74
    channeled........................ 74
    common.......................... 74
    keeled........................... 74
    noduled.......................... 74
    smooth........................... 74
sunsetclam
    California........................ 42
    painted........................... 42
superba, Gaza........................ 55
supercoil
    amber............................ 135
    barred............................ 135
    brown............................ 135
    buff.............................. 135
    Caneyfork........................ 135
    Cherokee......................... 135
    club.............................. 135
    comb............................ 135
    crowned......................... 135
    dentate........................... 135
    dimple........................... 135
    domed........................... 135
    funnel............................ 135
    gem.............................. 135
    glossy............................ 135
    goddess.......................... 135
    gray.............................. 135
    high mountain.................... 135
    hunted........................... 135
    lamellate......................... 135
    Mirey Ridge...................... 135
    Natural Bridge................... 135
    open............................. 135
    pearl............................. 135
    Ramp Cove...................... 135
    roan.............................. 135
    round............................ 135
    sculpted.......................... 135
    sharp............................ 135
    shrew............................ 135
    sidelong.......................... 135
    slender........................... 135
    spirit............................. 135
    teasing........................... 135
    tight.............................. 135
    translucent....................... 135
    triangular........................ 135
    variable.......................... 135
    white-foot........................ 135
superstitionis, Sonorella............... 146
supinum, Pisidium................... 34
suppressus
    Circulus.......................... 69
    Ventridens....................... 136
supragranosum, Calliostoma........... 55

supravallatum, Teinostoma	70	*Tagelus*	43
surfclam		tagelus	
Arctic	40	California	43
Atlantic	40	lesser	43
bent	40	neighbor	43
California	40	purplish	43
Clery	40	stout	43
dish	40	taildropper	
dwarf	40	banded	132
flat	40	blue-gray	132
fragile	40	mottled	132
hatchet	40	northern	132
Hemphill	40	papillose	132
hooked	40	reticulate	132
winged	40	scarletback	132
surinamensis		smoky	132
Cypraea	80	yellow-bordered	132
Strigilla	41	*taitiana, Elimia*	64
susannae, Nesovitrea	135	*taitianum, Pleurobema*	32, 154
swainsonii antillensis, Mitra	92	*takanosimensis, Aeolidiella*	121
swani, Turbonilla	106	*talma, Turbonilla*	106
swanii, Mopalia	150	*talpoideus, Limifossor*	20
swedmarki, Meiomenia	20	*talquinensis, Lioplax*	59
swiftiana, Corbula	46	talussnail	
swopei, Oreohelix	142	Agua Caliente	146
sybaritica		Agua Dulce	145
Problacmaea	54	Animas	145
Tellina	42	Apache	145
sybariticum, Laevicardium	39	Baboquivari	145
sykesi, Vexillum	93	Bear Canyon	145
sylvaticus, Arion	131	Big Emigrant	146
symmetrica, Elimia	64	Black Mountain	146
Symplectoteuthis	152	Blue	145
Synaptocochlea	56	Chiricahua	146
syngenes, Pupilla	127	Clark Peak	145
syrtensis, Plicifusus	90	Dona Ana	146
		Dos Cabezas	145
T		Doubtful Canyon	146
		Dragoon	145
tabula, Ensitellops	38	Empire Mountain	145
tabulata, Neptunea	89	Escabrosa	145
tabulatus, Oenopota	100	evening	146
Tachyrhynchus	71	Franklin Mountain	145
tacomaensis		Galiuro	145
Melanella	77	Garden Canyon	145
Odostomia	103	Grand Canyon	145
taeniata		Happy Valley	145
Leptoxis	65	Horseshoe Canyon	145
Praticolella	139	Huachuca	145
Villosa	33	Kitt Peak	146
taeniolata		Las Guijas	146
Metaxia	75	little	146
Volvarina	94	Madera	145
tageliformis, Macoma	41	Mik Ranch	145

mimic	145	*tappaniana, Gastrocopta*	127
Mustang	146	*tarda, Pleurobranchaea*	114
New Mexico	145	*Tarebia*	63
Oak Creek	145	*Taringa*	118
Organ Mountain	146	taringa, dusky brown	118
Papago	145	*tartarea, Fontigens*	61
Picacho	146	*taurina, Terebra*	95
Pinaleno	145	*taylori*	
Portal	146	*Hadoceras*	61
Posta Quemada	146	*Musculus*	24
pungent	146	*Phyllaplysia*	112
pygmy	145	*Turbonilla*	106
Quartzite Hill	145	*tazewelliana, Fossaria*	123
Ramsey Canyon	145	*techella, Fossaria*	123
Richinbar	145	*Tectarius*	67
Rincon	145	*Tectininus*	67
Roosevelt	146	*Tectonatica*	82
Rosemont	146	*tectum, Lithopoma*	57
San Xavier	145	*Tectura*	54
Sanford	146	*tedbayeri, Natica*	82
Santa Catalina	146	*Tegula*	56
Santa Rita	146	tegula	
Sierra Ancha	145	banded	56
Sonoran	145	black	56
Squaw Park	145	brown	56
St. Francis	145	Caribbean	56
Stronghold Canyon	145	dusky	56
Superstition Mountain	146	gilded	56
Table Top	146	green-base	56
Tollhouse Canyon	145	Monterey	56
Tortolita	146	queen	56
Total Wreck	145	silky	56
Walnut Canyon	145	speckled	56
Wet Canyon	145	West Indian	56
Whetstone	145	*tegula, Nassarius*	91
tampaensis		*tehamana, Trilobopsis*	141
Bermudaclis	77	*Teinostoma*	70
Pyrgospira	100	*tellea, Polystira*	100
Tellina	42	*Tellidora*	41
Urosalpinx	86	*tellimyalis, Rupellaria*	45
tampicoensis, Cyrtonaias	29	tellin	
tanna, Cerberilla	121	alternate	41
tanneri, Diodora	52	American	41
tannum, Campeloma	59	angulate	41
tantalus, Holospira	129	apricot	42
tantilla		bloodless	42
Lucidella	58	Bodega	41
Paravitrea	135	Cande	41
Transennella	45	candystick	42
tantillus, Neoplanorbis	125	Carpenter	41
tanya, Sclerodoris	117	consobrine	41
Tapes	45	crenulate	42
taphria, Nuculana	22	crystal	41
taphrium, Antalis	50	cuneate	42

delicate	42
favored	42
Guilding	42
Ida	42
Jutting	42
many-colored	42
Martinique	42
oval	42
perfect	42
plain	42
pure	42
rainbow	42
rose-petal	42
salmon	42
Say	42
shiny	41
slandered	42
smooth	42
speckled	42
striate	41
sunrise	42
sybaritic	42
Tampa	42
tinted	41
Vespucci	42
watermelon	42
western	40
white-crest	41
Tellina	41
Tellinidae	40
tellinoides	
Adrana	21
Cumingia	43
telopia disa, Taringa	118
Tenaturris	100
tenax, Somatogyrus	62
tenella	
Onchidoris	115
Tellina	42
Tenellia	121
tenellum, Buccinum	88
tenera	
Barbatia	24
Semirossia	151
tenerrima, Protothaca	45
tenerum	
Periploma	48
Trigonostoma	94
tennesseensis	
Somatogyrus	62
Triodopsis	141
tenta, Macoma	41
tentaculata, Bithynia	59
tenuicula, Turbonilla	106
tenuiliratum, Cymatium	83
tenuiliratus cymatus, Oenopota	100
tenuipes, Littoridinops	61
tenuis	
Aclis	76
Cyclinella	44
Euspira	82
Nucula	21
Planorbella	125
tenuisculpta	
Finella	73
Odostomia	103
Parvilucina	35
tenuissimus, Oenopota	100
tenuistriata, Oreohelix	142
tenuisulcata, Nuculana	22
Terebra	95
Terebridae	95
Teredinidae	47
Teredo	47
Teredora	47
teres	
Elimia	64
Lampsilis	31
Odostomia	103
Tergipedidae	120
Tergipes	121
tergipes, Tergipes	121
terminalis, Vitrinella	71
ternaria, Paravitrea	135
teskeyae, Vertigo	127
Teskeyostrea	28
tessellata, Nerita	58
Testacella	137
Testacellidae	137
testiculus, Cypraecassis	82
testudinalis, Notoacmea	54
Tethyidae	119
tetrabranchiata, Prionoglossa	111
tetralasmus, Uniomerus	33
tetraquetra, Tochuina	118
tetrodon	
Ashmunella	138
Polyschides	51
Texadina	63
texana	
Cochliopa	60
Humboldtiana	144
Tellina	42
Terebra	95
Vitrinella	71
texanus, Microceramus	130
texasensis, Toxolasma	33

texasiana
- *Aligena* 36
- *Euglandina* 129
- *Polygyra* 139
- *Volvulella* 110

texasianius, Donax 42
texasianum, Agriopoma 44
texasianus, Strobilops 128
textile, Caecum 71
textilis, Turbonilla 106
Thais 85
Thala 92
thala, beaded 92
thalassicola, Tricolia 58
thalassinus, Cyclopecten 27
thalia, Odostomia 103
thamnoporus, Dendrochiton 148

thea
- *Cerodrillia* 96
- *Odostomia* 103

Theba 146
theiocrenetum, Aphaostracon 60
theloides, Ventridens 136
Theora 43
thermimontis, Helminthoglypta 144

thersites
- *Balcis* 77
- *Siphonaria* 126

Thiaridae 63
Thliptodon 111
Thliptodontidae 111

thomasi
- *Nuttallina* 149
- *Vitrinella* 71

thompsoni
- *Akera* 111
- *Eudolium* 83
- *Olivella* 92

thompsoniana, Ashmunella 138

thordis
- red 118
- two-spot 118

Thordisa 118

thorn
- Appalachian 125
- cave 125
- file 125
- floodplain 125
- ice 125
- obese 125
- southern 125
- western 125

thornmouth, foliate 84

thorny-oyster
- Atlantic 27
- digitate 27

Thracia 48

thracia
- Adams 48
- Arctic 48
- Bering 48
- Challis 48
- Conrad 48
- Corbula 48
- distorted 48
- kidneybean 48
- Morrison 48
- northern 48
- rugose 48
- short 48
- sloping 48
- Stimpson 48
- trapezoid 48

thraciaeformis, Yoldia 22

thracid
- doubtful 48
- hairy 48
- Hemphill 48
- wavy 48

Thraciidae 48

threeband
- agave 144
- boulder slide 144
- Capote 144
- Chisos 144
- Davis Mountain 144
- Mitre Peak 144
- Mt. Livermore 144
- northern 144
- San Carlos 144
- Stockton Plateau 144

threeridge 28
- fat 28

threetooth
- Alabama 140
- Atlantic 140
- baffled 140
- budded 141
- buttress 141
- Cape Fear 141
- Cheat 141, 155
- Claiborne 140
- coiled 141
- dished 141
- dwarf 140
- glossy 140
- Hanging Rock 140

magnolia	140
mimic	140
northern	141
nubbin	140
Ozark	140
pineywood	140
pinhole	140
Pittsylvania	140
Post Oak	140
rivercliff	140
Santee	140
Spruce Knob	140
Texas	141
Thyasira	35
Thyasiridae	35
Thyonicola	78
thyroidus, Mesodon	139
Thysanophora	141
Thysanophoridae	141
Thysanoteuthidae	152
Thysanoteuthis	152
tiara, Paravitrea	135
tightcoil	
blackfoot	135
broadwhorl	135
cedar	135
crowned	136
denticulate	136
island	136
minute	136
northern	135
riparian	136
shiny	136
southern	136
striate	136
thinlip	135
waxy	135
tigrina, Armina	119
timessus, Heilprinia	91
timetus, Colus	89
tina, Cuthona	121
tincta	
Barleeia	68
Nitidiscala	76
Philine	108
Pisania	90
tinctum, Nemocardium	39
titubans, Melanella	77
Tivela	45
tivela	
Abaco	45
Florida	45
trigonal	45
tochui, giant orange	118

Tochuina	118
todseni	
Ashmunella	138
Sonorella	146
togatus, Philomycus	132
tollini, Epitonium	76
tolmiei, Polyschides	51
toma, Paravitrea	135
tongue, flamingo	81
Tonicella	149
Tonicia	150
Tonna	83
Tonnidae	83
tooth, bleeding	58
toothshell	
aberrant	51
Agassiz	51
California	51
Carolina	51
compressed	51
elongate	51
flattened	51
fourslit	51
fourtooth	51
fusiform	51
grand	51
Greenlaw	51
Hepburn	51
inflated	51
kingly	51
little	51
lobate	51
Mayor	51
Miami	51
nineside	51
notable	51
Rush	51
Salish	51
Stearns	51
tetrodon	51
Tolmie	51
Verrill	51
Watson	51
westerly	51
top, boreal	128
topsnail	
beautiful	55
blue	55
boreal	55
channeled	55
chocolate-line	55
depressed	55
enchanting	55
gem	55

glorious ... 55
granulose ... 55
Keys ... 54
mahogany ... 55
mottled ... 55
psyche ... 55
purple-ring ... 55
red-spot ... 55
rosy ... 55
rough ... 55
sculptured ... 55
silvery ... 55
spindle ... 55
splendid ... 55
tricolor ... 55
two-tooth ... 56
variegate ... 55
West Indian ... 55
Torellia ... 79
toreta, Peristichia ... 103
tornata, Odostomia ... 103
Tornidae ... 71
Tornus ... 71
toroensis, Rissoa ... 68
torosa, Pseudomelatoma ... 100
torquata, Turbonilla ... 106
torticula, Volvarina ... 94
tortile, Caecum ... 71
tortillita, Sonorella ... 146
tortugaensis, Rossia ... 151
tortuganum, Cyclostrema ... 57
torulosa
  gubernaculum, Epioblasma ... 30, 154
  rangiana, Epioblasma ... 30
  torulosa, Epioblasma ... 30, 154
totteni
  Buccinum ... 88
  Hydrobia ... 61
townsendiana, Allogona ... 137
Toxolasma ... 33, 154
toyatani, Turbonilla ... 106
trabalis, Villosa ... 33, 154
tracheia, Ocenebra ... 85
trachis, Odostomia ... 103
trachisma, Alvania ... 67
Trachycardium ... 39
trachypepla, Trilobopsis ... 141
Trachypollia ... 86
Tralia ... 122
trampsnail, Asian ... 142
Transennella ... 45
transennella
  confusing ... 45
  Conrad ... 45

Cuba ... 45
purple ... 45
Stimpson ... 45
transfuga, Haplotrema ... 129
transitorius, Cadulus ... 51
transversa, Anadara ... 24
transversum
  Musculium ... 34
  Nemocardium ... 39
Trapania ... 115
Trapeziidae ... 43
trapezoides
  Thracia ... 48
  Verticordia ... 49
traski
  Helminthoglypta ... 144
  Physella ... 124
  Planorbella ... 125
  Stagnicola ... 123
traskii, Acteon ... 107
tree-oyster, flat ... 25
treesnail
  banded ... 130
  Florida ... 130
  Florida Keys ... 130
  lined ... 130
  manatee ... 130
  master ... 130
  Stock Island ... 130, 155
Tremoctopodidae ... 152
Tremoctopus ... 152
tremperi
  Odostomia ... 103
  Turbonilla ... 106
Tresus ... 40
triacantha, Peracle ... 110
trialata, Pteropurpura ... 85
triangularis
  Collisella ... 54
  Pterotyphis ... 85
triangulatus, Boreotrophon ... 84
tricarinata
  Arene ... 57
  Valvata ... 58
Trichia ... 146
Trichotropidae ... 79
Trichotropis ... 79
Tricolia ... 58
tricolor
  Calliostoma ... 55
  Eubranchus ... 120
  Polycera ... 116
Tridachia ... 113

tridens
    Helicodiscus 131
    Paravitrea 135
tridentata
    Cavolinia 110
    Pleuromeris 38
    Triodopsis 141
    Turbonilla 106
    Vertigo 128
tridentatum, Cerion 128
Tridonta 39
tridonta
    Alaska 39
    Arctic 39
    Bennett 39
    boreal 39
    Filatova 39
    Montagu 39
    Rolland 39
    varnished 39
trifidus, Ischnochiton 148
trigonalis, Isorobitella 37
Trigoniocardia 40
Trigonostoma 94
trilineata
    Coryphella 120
    Leptoxis 65
    Pandora 48
Trilobopsis 141
trimatris, Praticolella 139
Trimusculidae 126
Trimusculus 126
Triodopsis 140, 155
triodus, Helicodiscus 131
Triopha 115
triopha
    grand 115
    maculated 115
    sea-clown 115
Triophidae 115
Triphora 75
triphora
    beautiful 75
    black-line 75
    mottled 75
    San Pedro 75
    St. Thomas 75
    white Atlantic 75
Triphoridae 74
Triptychus 104
triquetra
    Epioblasma 30
    Mysella 37
    Parastarte 44

triseriata, Euspira 82
trisinuata, Thyasira 36
trispinosa, Diacria 110
Tritogonia 33
triton
    angular 83
    arrow dwarf 90
    Atlantic trumpet 83
    doghead 83
    dwarf 83
    frilly dwarf 84
    giant 83
    goldmouth 83
    hairy 83
    knobbed 83
    lip 83
    miniature trumpet 90
    obscure dwarf 90
    Oregon 83
    ringed 83
    slender 83
    twisted 83
Tritonia 118
tritonia
    diamondback 118
    rosy 118
Tritoniidae 118
tritonis variegata, Charonia 83
Trivia 80
trivia
    Antilles 80
    California 80
    coffeebean 80
    fourspot 80
    pink 80
    snowy 80
    white 80
Triviidae 80
trivittata, Ilyanassa 91
trivolvis, Planorbella 125
Trochidae 54
trochiformis
    Limacina 110
    Pleurocera 66
Trochoidea 146
trochulus, Euconulus 133
troglodytes
    Monadenia 144
    Sabinella 77
trombinus, Colus 89
troostiana
    Elimia 64
    Polygyra 139
trophina, Coryphella 120

trophius, Colus 89
trophon
  Avalon 84
  bobtail 84
  carrot 84
  clathrate 84
  corded 84
  crown 85
  elegant 84
  girdled 84
  grooved 84
  Kamchatka 86
  knobbed 84
  ribbed 84
  rotund 84
  sandpaper 85
  scallion 84
  spiny 85
  tabulate 84
  threaded 84
  triangular 84
  wavy 84
  white-spine 84
  winged 84
trophonium, Vexillum 93
trophonius, Volutopsius 90
Trophonopsis 86
troschelianum, Pleurobema 32
trosti, Poromya 49
trothis, Somatogyrus 62
trumpet, nymph 61
truncata
  Barnea 46
  Cerithiopsis 74
  Mya 45
  Planorbella 125
  Truncilla 33
Truncatella 69
truncatella
  beautiful 69
  California 69
  Caribbean 69
  ladder 69
Truncatellidae 69
truncatula, Fossaria 123
truncatus, Boreotrophon 84
Truncilla 33
tryoni
  Murex 84
  Somatogyrus 62
  Vitrinella 71
  Xerarionta 146
Tryonia 63

tryonia
  desert 63
  devil 63
  grated 63
  mimic 63
  Phantom 63
tryoniana, Sonorella 146
tschuktschica, Adalaria 115
tuba, hollow 126
tuba, Chaenaxis 126
tuber, Lithopoma 57
tubercularis floridana, Cerithiopsis 74
tuberculata
  Cyclonaias 29
  Nodilittorina 67
  Sansonia 57
tuberculatus
  Chiton 150
  Melanoides 63
tuberculosa, Emarginula 53
tuberosa
  Cassis 81
  Mitrella 87
  Quadrula 33
tuca, Elysia 113
tudiculata, Helminthoglypta 144
tularensis, Helminthoglypta 144
tulip
  banded 91
  true 91
  yellow 91
tulipa, Fasciolaria 91
tulotoma 59
tumens, Hipponix 78
tumida
  Cerithiopsis 74
  Mysella 37
  Sphenia 45
tumidum, Bittium 73
tun
  Atlantic partridge 83
  giant 83
  mottled 83
  straw 83
tunicata, Katharina 150
tuolumnea, Monadenia 144
tupala, Berthella 114
turban
  channeled 58
  chestnut 58
  filose 58
  red 57
  wavy 57
turbinella, Engina 89

Turbinellidae 93
Turbinidae 57
turbiniformis, Fluminicola 61
turbinum, Calliostoma 55
Turbo 58
Turbonilla 104
Turcica 56
turgida, Spilochlamys 63
turgidula, Epioblasma 30, 154
turnerae, Penitella 47
turretsnail
    boring 71
    eastern 72
    eroded 71
    milky 71
    needle 72
    symmetrical 72
    variegate 72
turricula
    Odostomia 103
    Oenopota 100
turriculata, Atlanta 81
turriculus, Melanoides 63
Turridae 95
turristhomae, Triphora 75
turrita, Aclis 76
Turritella 71
turritella, Fontigens 61
Turritellidae 71
Turritellopsis 72
turton 38
    minute 38
    western 38
turtoni, Beringius 87
Turtonia 38
Turtoniidae 38
tuskshell
    American 50
    annulate 50
    Antillean 50
    banded 50
    Berry 50
    costate 50
    Dall 50
    gorgeous 50
    Gould 50
    green 50
    half-smooth 50
    hexagon 50
    ivory 50
    meridian 50
    narrow-slit 50
    occidental 50
    ovate 50
    Pilsbry 50
    reed 50
    reticulate 50
    scratched 50
    slender 50
    stained 50
    straight 50
    thistle 50
    trench 50
    turned 50
    wampum 50
    Watson 50
    waxy 50
    yellow 50
Tylodina 114
Tylodinella 114
Tylodinidae 114
Typhis 86
typhis, frilly 86
Typhlomangelia 100
typica, Rupellaria 45

U

uberinus, Polinices 82
Udosarx 142
uinta, Oreohelix 142
ulla, Ascobulla 112
ultima, Humboldtiana 144
ultramontanum, Pisidium 34
umbilicaris, Paravitrea 135
umbilicata
    Clappia 60, 157
    Glyphyalinia 134
    Helminthoglypta 144
    Leptoxis 65
    Retusa 109
umbilicatellus, Promenetus 125
umbilicatus, Neoplanorbis 125
Umbraculidae 114
Umbraculum 114
umbraculum, Umbraculum 114
umnaka, Cyclocardia 38
unalashkensis, Admete 94
unalaskensis, Odostomia 103
uncialis, Pleurocera 66
unciferum, Stenotrema 140
uncinata, Cavolinia 110
undata
    Amphissa 86
    Astarte 39
    Glycymeris 25
    Velutina 80
undatella, Chione 44

undatum, Buccinum 88
undatus
    Hyalopecten 27
    Micromelo........................ 107
underwoodae, Aclis 76
undocarulea, Tonicella................ 149
undosum, Lithopoma 57
undulata
    Alasmidonta 28
    Raeta............................ 40
undulatus, Strophitus 33
Unela 110
ungaricus, Capulus.................... 78
Ungulinidae........................... 36
unicirrhus, Scaeurgus................. 152
unicolor
    Obovaria......................... 31
    Succinea 133
unicorn
    angular........................... 84
    checkered 84
    dark............................. 84
    spotted........................... 84
unidentata, Odostomia................. 103
unifasciata
    Lacuna 66
    Pyramidella...................... 104
unifasciatum, Epitonium 76
unilirata, Turbonilla.................. 106
Uniomerus 33
Unionidae 28, 154, 156
uniplicata, Simnialena 81
unisulcatus, Pedipes................... 122
urbana, Succinea 133
urocoptid
    Florida........................... 130
    pontiff 130
    Texas............................ 130
    truncate 129
Urocoptidae 129
Urosalpinx 86
ursabrunae, Gonatus 151
utahensis
    Physella.......................... 124
    Stagnicola 123
    Valvata 58
uva, Doto 119
uvulifera, Polygyra................... 139
uzenensis, Mya....................... 45

V

vaccaria, Aplysia 112
vadosus, Boreacola 37

vagans, Catinella 133
vaginata, Lacuna 66
vahlii, Margarites..................... 56
valentianus, Limax.................... 137
vallei, Amphithalamus 68
vallicolens, Dentalium 50
Vallonia.............................. 128
vallonia
    costate........................... 128
    indecisive........................ 128
    Iroquois.......................... 128
    lovely............................ 128
    multirib 128
    silky............................. 128
    thin-lip.......................... 128
    trumpet 128
vallonia, Torellia..................... 79
Valloniidae 128
Valvata 58
valvata
    desert............................ 58
    emerald 59
    European stream................. 58
    flanged.......................... 59
    fringed 58
    glossy............................ 58
    mossy 58
    purplecap........................ 58
    rams-horn 58
    threeridge 58
    two-ridge 58
Valvatidae............................ 58
valvulus, Solemya.................... 22
vanattae, Prophysaon................. 132
vanattai, Glyphyalinia 134
vancouverense
    Bittium.......................... 73
    Haplotrema...................... 129
vancouverensis
    Delectopecten 27
    Fossaria.......................... 123
    Hermaea 113
    Limatula 26
    Odostomia....................... 103
    Okenia........................... 115
vanhyningi
    Cerithiopsis 74
    Cincinnatia 60
    Pseudosimnia 81
vanhyningiana, Elimia 64
Vanikoro............................. 78
vanikoro, West Indian 78
Vanikoroidae.......................... 78
vannostrandi, Triodopsis 141

vanuxemensis
    umbrans, *Villosa* 33
    vanuxemensis, *Villosa* 33
vanuxemiana, Elimia 64
varia, Pallifera 132
variabile, Pisidium 34
variabilis
    Arene 57
    Donax 42
    Paravitrea 135
varians
    Elimia 65, 157
    Hemitrochus 144
    Petaloconchus 72
variata, Elimia 65
varicifera, Ashmunella 138
Varicorbula 46
varicorbula, oval 46
varicosa
    Alasmidonta 28
    Solariella 56
varidens, Paravitrea 135
variegata
    Diodora 52
    Kurtziella 98
    Lacuna 66
    Turritella 72
variegatum, Calliostoma 55
variolosa, Bothriopupa 126
varium, Bittium 73
vase, Caribbean 93
Vasum 93
vaughaniana, Villosa 33
velata, Placiphorella 150
veliei, Volvarina 94
velox, Trapania 115
velum, Solemya 22
Velutina 80
velutina, Velutina 80
Veneridae 44
venezuelensis, Diplodonta 36
ventilabrum, Tenellia 121
Ventricolaria 45
ventricosa
    Cyclocardia 38
    Neptunea 89
    Pyramidella 104
    Vertigo 128
ventricosum, Pisidium 34
ventricosus, Colus 89
Ventridens 136
venus
    beaded 44
    California 44

Clench 44
cross-barred 44
downy 44
empress 44
Ford 44
frilled 44
glory-of-the-seas 44
imperial 44
Kennerley 44
king 44
lady-in-waiting 44
lamellar 44
lily 44
milky 44
princess 44
queen 45
rigid 45
smooth 44
sunray 44
Texas 44
white 44
Venustaconcha 33
venustula, Arene 57
venustus, Philomycus 132
verae, Naticarius 82
verkruzeni, Plicifusus 90
Vermetidae 72
Vermicularia 72
vermiculata, Eobania 146
vernicosa, Tridonta 39
vernicosus, Musculus 24
veronica, Cuthona 121
Veronicellidae 147
verrilli
    Cerodrillia 96
    Diplodonta 36
    Pulsellum 51
verrilliana, Nuculana 22
verrillii
    Cylichna 108
    Turbonilla 106
verrucicornis, Berghia 121
verrucosa
    Doris 117
    Lithasia 65
    Rufibranchialis 120
    Tritogonia 33
versicolor
    Amphissa 86
    Nerita 58
    Tellina 42
verticord
    even-ribbed 49
    Fischer 49

flexed	49
ornate	49
sculptureless	49
Seguenza	49
sharp-ribbed	49
trapezoid	49
Wood	49
Verticordia	49
Verticordiidae	49
Vertigo	127
vertigo	
Alabama	127
blade	127
bluff	127
capital	127
chestnut	127
Columbia	127
cone	127
Conecuh	127
crested	127
cross	127
cupped	127
cylindrical	127
deep-throat	127
delicate	127
five-tooth	128
heart	127
honey	128
horseshoe	127
Idaho	127
Keys	127
mitered	127
Monte Sano	128
olive	127
ovate	127
Pacific	127
palmetto	127
ribbed	127
rotund	127
six-whorl	127
smallmouth	127
smalltooth	127
striate	127
swamp	127
tapered	127
threaded	127
tundra	127
variable	127
verum, Pleurobema	32
vesica, Vesicomya	43
Vesicomya	43
vesicomya, fat	43
Vesicomyidae	43
vesicula, Haminoea	109
Vespericola	141
vespertilium, Gastropteron	108
vespertina, Sonorella	146
vespuciana	
Nannodiella	98
Tellina	42
vestale, Microstelma	67
vestita	
Amicula	149
Pleurocera	66
Torellia	79
vestitum, Caecum	71
vetula, Inodrillia	97
vexativa, Turbonilla	106
Vexillum	92
viabrunnea, Bathytoma	96
viaregis, Cerion	128
vibex	
Nassarius	91
Polystira	100
Villosa	33
vicola, Odostomia	103
victoriana, Mangelia	98
viennaensis, Elimia	65
villepini, Conus	95
Villosa	33, 154
villosa, Villosa	34
villosior, Asthenothaerus	48
vincta	
Lacuna	66
Odostomia	103
vinosa	
Neptunea	90
Physella	124
Terebra	95
violaceus	
Oenopota	100
Tremoctopus	152
virens	
Cuthona	121
Fluminicola	61
Plicifusus	90
Valvata	59
virescens	
Adalaria	115
Haminoea	109
Lampsilis	31, 154
virga, Turbonilla	106
virgata	
Leptoxis	65
Physella	124
Turbonilla	106
virginea	
Neritina	58

Physella......... 124
virginiana, Marginella......... 94
virginica
   Cerithiopsis......... 74
   Crassostrea......... 28
   Elimia......... 65
   Glyphyalinia......... 134
   Odostomia......... 103
   Turbonilla......... 106
virginicus
   Philomycus......... 132
   Polygyriscus......... 131, 155
   Somatogyrus......... 62
   Ventridens......... 136
virgo, Turbonilla......... 106
virgula, Creseis......... 110
viridaria, Turbonilla......... 106
viride, Liocyma......... 44
viridis
   Alasmidonta......... 28
   Axinopsida......... 35
   Phyllobranchillus......... 112
   Smaragdia......... 58
   Solen......... 40
Viridrillia......... 100
viridula, Diodora......... 53
viridum, Buccinum......... 88
virilis, Sonorella......... 146
vitellina, Erato......... 80
Vitrea......... 136
vitrea, Pteria......... 25
vitreum
   Dacrydium......... 23
   Punctum......... 130
vitreus, Delectopecten......... 27
Vitricythara......... 101
Vitrina......... 136
Vitrinella......... 70
vitrinella
   Biscayne......... 70
   cupola......... 69
   Florida......... 70
   fringed......... 70
   hairy......... 70
   helix......... 70
   interrupted......... 70
   premature......... 70
   San Diego......... 71
   splitlip......... 70
   suppressed......... 69
   terminal......... 71
   Texas......... 71
   threaded......... 69
   two-tail......... 70

   upright......... 70
Vitrinellidae......... 69
Vitrinidae......... 136
Vitrinorbis......... 71
Vitrizonites......... 136
vittata, Leptoxis......... 65, 157
Viviparidae......... 59
Viviparus......... 59
vokesae, Pteropurpura......... 85
volcano, Fissurella......... 53
volusiae, Ventridens......... 136
volute
   Alaska......... 93
   banded......... 93
Volutharpa......... 90
Volutidae......... 93
Volutomitra......... 93
Volutomitridae......... 93
Volutopsius......... 90
Volva......... 81
volva striata, Volva......... 81
Volvarina......... 94
Volvulella......... 109
vortex, Oreohelix......... 142
vorticifer, Margarites......... 56
Vorticifex......... 125
vossi, Stiliger......... 113
voyanum, Haplotrema......... 129
voyi, Antiplanes......... 96
vulgaris, Octopus......... 152
vulgata, Triodopsis......... 141
vulgatus, Mesomphix......... 134
vultuosa, Triodopsis......... 141

W

waccamawensis, Elliptio......... 29
walallensis, Haliotis......... 52
waldense, Stenotrema......... 140
walker
   brown......... 63
   marsh......... 63
   Pacific......... 63
   robust......... 63
   slender......... 63
walkeri
   Amnicola......... 60
   Ashmunella......... 138
   Ferrissia......... 125
   Gyrotoma......... 65, 157
   Medionidus......... 31
   Pisidium......... 34
   Pleurocera......... 66

Sonorella	146
walkeriana, Helminthoglypta	144
walkerianus	
Somatogyrus	62
Stagnicola	123
waltoni	
Elliptio	29
Helminthoglypta	144
Oreohelix	142
Sonorella	146
wandae, Catinella	133
wandoense, Vexillum	93
wardiana, Pandora	48
warta, Dendrodoris	117
wartyback	32
purple	29
threehorn	31
white	31, 154
wascoense, Pristiloma	136
washboard	31
round	31
washingtona, Xylophaga	47
washingtonia, Odostomia	103
watermani, Olivella	92
watsoni	
Polyschides	51
Rhabdus	50
Scaphander	108
weberi	
Caducifer	88
Fontigens	61
Parviturbo	57
Teskeyostrea	28
wedge	
Arkansas	140
bladetooth	140
blunt	140
sharp	140
velvet	140
wedgeclam	
Arctic	40
gilded	40
wedgemussel, dwarf	28
wekiwae, Cincinnatia	60
weldi, Turbonilla	106
wellsi, Tritonia	118
wentletrap	
angulate	75
Atlantic	76
bladed	75
boreal	76
brown-band	76
California	76
cancellate	76
chalky	75
coarse	76
concave	76
dwarf	76
fine-mesh	76
fine-ribbed	76
frosted	76
Greenland	75
humble	76
lamellose	76
many-rib	76
money	76
Monterey	76
mottled	75
New England	76
noble	76
one-band	76
pitted	76
polished	75
reticulate	75
rough	75
scalloped	76
semismooth	75
silky	76
slender	75
smooth	75
sparse	76
spongy	76
tabulate	76
tinted	76
tooth-rib	75
tower	75
well-threaded	75
wide-coil	75
wrinkle-rib	75
wetherbyi	
Megapallifera	132
Mesodon	139
Notogillia	61
wetmorei, Diodora	53
wheatleyi	
Glyphyalinia	134
Mesodon	139
wheeleri	
Arkansia	29
Somatogyrus	62
Vertigo	128
whelk	
Aleut	87
angular	87
arctic	90
basket	89
berry	88
bluish	88

broad-cord	90
brown	90
bruised	89
Carolina	89
carved	88
chalky	90
channeled	90
chestnut	88
corded	89
crude	88
dire	90
egg	89
elongate	90
fat	89
finch	88
fragile	90
frilled	90
garter	89
glacial	88
globose	90
gray	90
green	90
hairy	89
hammered	87
helmet	89
Iceland	89
inflated	89
knobbed	90
Kodiak	88
ladder	88
left-hand	90
lightning	90
lirate	88
lyre	89
noble	89
Norway	90
nut	89
oblique	89
Okhotsk	88
painted	88
paper	90
plump	89
polar	88
Pribilof	89
prickly	90
pygmy	89
regular	90
rosy	89
rotund	90
Santa Barbara	88
seahorse	90
shady	89
shingled	87
shouldered	90
shrew	89
silky	88
simple	90
sinuous	88
snow	90
splendid	90
strombiform	90
swollen	88
tabled	89
thick-cord	87
thick-ribbed	89
thin	88
thin-ribbed	89
threaded	90
tulip	90
turban	88
ventricose	89
volute	90
wandering	88
warped	90
waved	88
wayward	88
wine	90
wrinkle	89
whetstonensis, Holospira	129
whiteavesi, Turbonilla	106
whitei, Odostomia	103
whitelip	
big-tooth	140
coastal	140
Kaw	140
Ozark	140
southeastern	140
striped	140
Tulsa	140
western	140
wichitorum, Euchemotrema	138
wickhami, Turbonilla	107
willcoxi, Aplysia	112
willetti	
Antiplanes	96
Astarte	39
Cerithiopsis	74
Lepidozona	149
Odostomia	103
Scaphander	108
Suavodrillia	100
Turbonilla	107
williami, Viridrillia	100
Williamia	126
williamsoni, Vitrinella	71
wilsoni, Succinea	133
wing, turkey	24

wing-oyster
   Atlantic 25
   glassy 25
   scaly 25
winkleyi, Odostomia 103
winnebagoensis, Valvata 59
wolfsnail
   glossy 129
   rosey 129
   striate 129
Woodbridgea 109
woodeater
   Atlantic 47
   Washington 47
woodianus, Oenopota 100
woodii, Verticordia 49
woodlandsnail
   angulate 137
   Animas Peak 137
   Barefoot 137
   Big Hatchet 137
   Black Range 137
   Blue Mountain 137
   Boulder Canyon 137
   Capitan 137
   Cave Creek 137
   Chiricahua 137
   Cook's Peak 137
   Dry Creek 138
   Florida Mountain 138
   Franklin Mountain 137
   Goat Cave 137
   Goat Mountain 137
   Guadalupe 137
   Hacheta Grande 137
   Hell's Canyon 138
   Huachuca 137
   Iron Creek 137
   Jemez 137
   Maple Canyon 138
   Miller Canyon 138
   Mogollon 137
   Mount Riley 138
   Organ Mountain 137
   Pine Springs 137
   Reed's Mountain 137
   Salinas Peak 138
   San Andreas 137
   Sangre de Cristo 138
   Sawtooth Mountain 137
   Sierra Blanca 138
   Silver Creek 137
   Whitetail 137
   Whitewater Creek 137

woodringi, Splendrillia 100
woodruffi, Stagnicola 123
woodworthi, Admete 94
worki, Anisodoris 118
wormsnail
   California 72
   compact 72
   decussate 72
   erect 72
   flat 72
   Florida 72
   irregular 72
   Monterey 72
   northern 72
   ringed 72
   scaled 72
   slit 72
   variable 72
   West Indian 72
wrighti, Physella 124
wrightiana, Alasmidonta 28, 156

X

xanthenes, Sonorella 146
xanthias, Alvania 67
Xenophora 79
Xenophoridae 79
Xerarionta 146
Xylophaga 47
xynoelictum, Aphaostracon 60
xystrum, Sulcoretusa 109

Y

yatesi, Ammonitella 141
yavapai, Oreohelix 142
Yoldia 22
yoldia
   almond 22
   broad 22
   Cooper 22
   crisscrossed 22
   different 22
   file 22
   frigid 22
   halfsmooth 22
   inconspicuous 22
   inflated 22
   intermediate 22
   iridescent 22
   lenticulate 22
   northern 22
   oval 22

rainbow 22
regular 22
sanesia........................... 22
short 22
sibling 22
subangular....................... 22
witness.......................... 22
Yoldiella 22
yoldiformis, Macoma................ 41
youngi, Odostomia.................. 103
yucatanensis, Holospira.............. 129
yucatecana, Mathilda................ 74
yucatecanum, Calliostoma............ 55

Z

Zachrysia 141
zachrysia
   garden 141
   golden 141
Zacoleus......................... 132

zaletus, Mesodon 139
Zebina........................... 68
zebra
   Aplysiopsis 113
   Arca........................... 24
   Cypraea........................ 80
zetlandica, Aldisa................... 117
ziczac
   Littorina 67
   Pecten......................... 27
zionis, Physella..................... 124
Zirfaea 47
zonalis, Batillaria................... 72
zonata, Limapontia 113
zonatus, Pitar...................... 44
Zonites 130
Zonitidae......................... 134
Zonitoides........................ 136
zoobotryon, Okenia 115
Zoogenetes....................... 128
zosterae, Polycera 116

PORTFOLIO OF MOLLUSK DIVERSITY

Atlantic thorny-oyster, *Spondylus americanus*

Photographs not otherwise credited are from the
"Shell Desk Diary," 1984–1987, courtesy of the Shell Oil Company

Pacific pygmy octopus, *Octopus digueti*

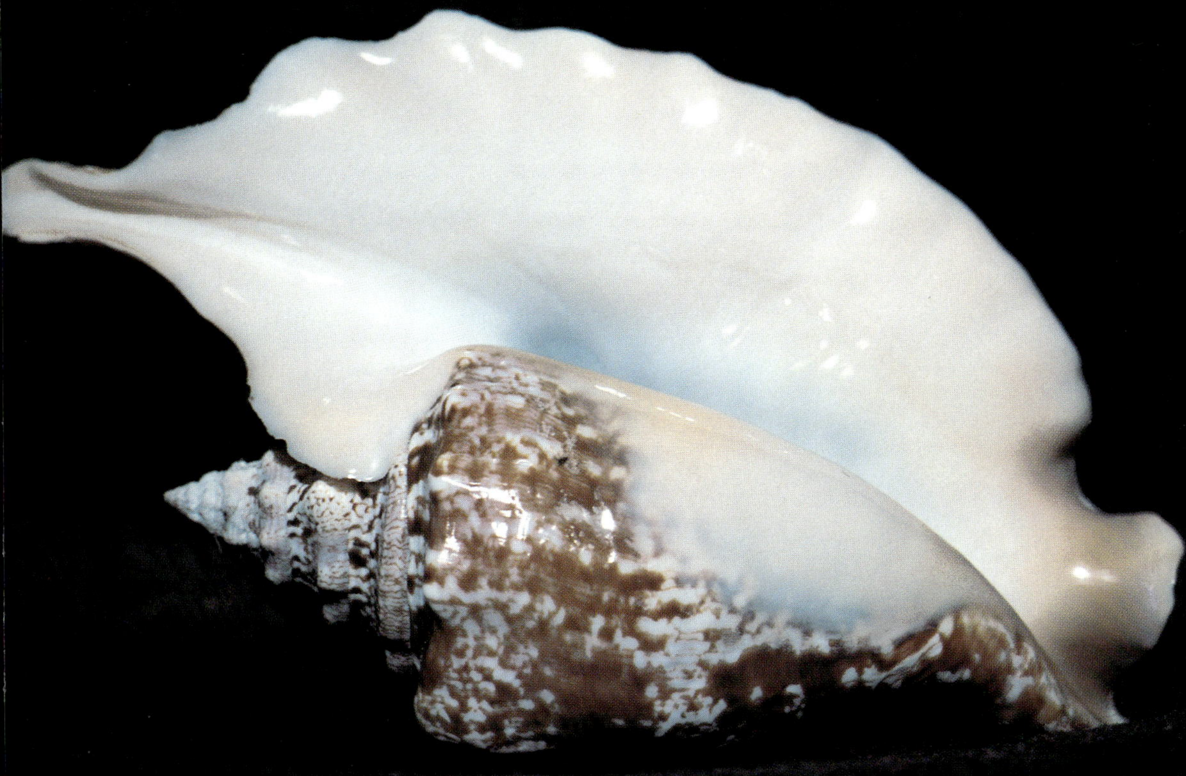

1

2

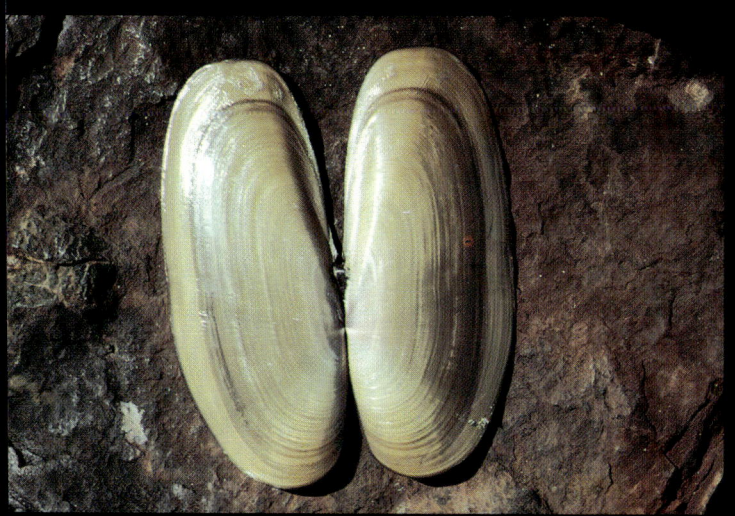

Roostertail conch, *Strombus gallus*

Pacific razor, *Siliqua patula*

Dennison morum, *Morum dennisoni*

3

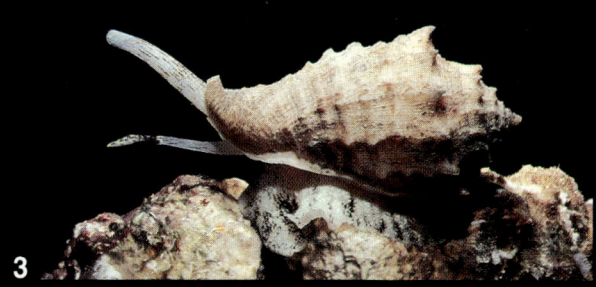

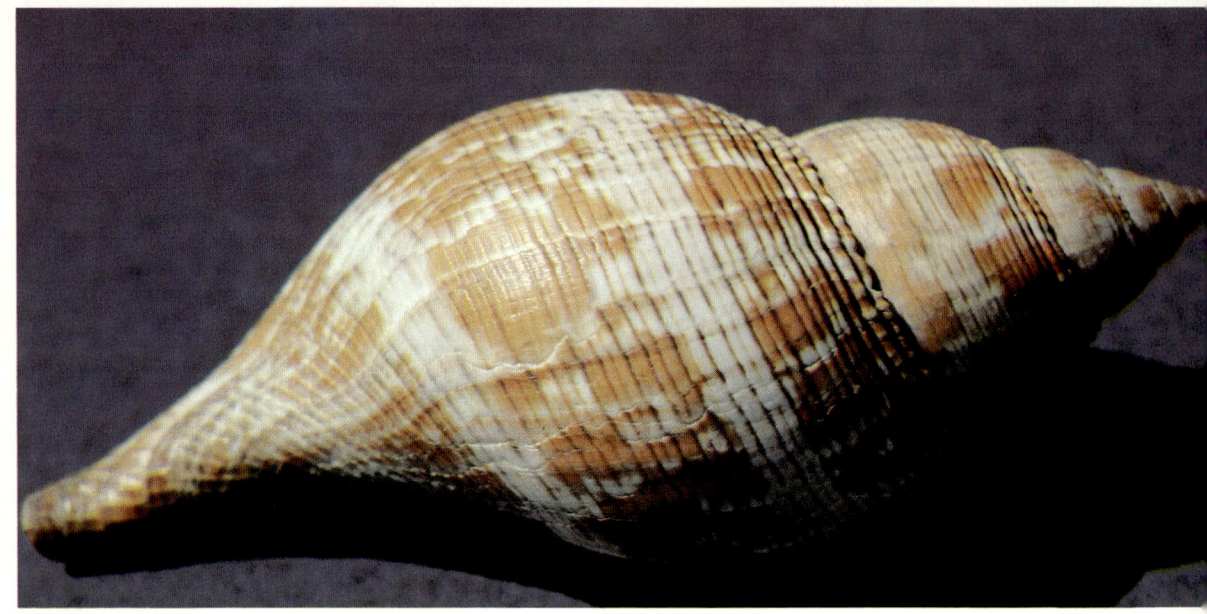

1 True tulip, *Fasciolaria tulipa* 1

2 Lions-paw scallop, *Nodipecten nodosus*

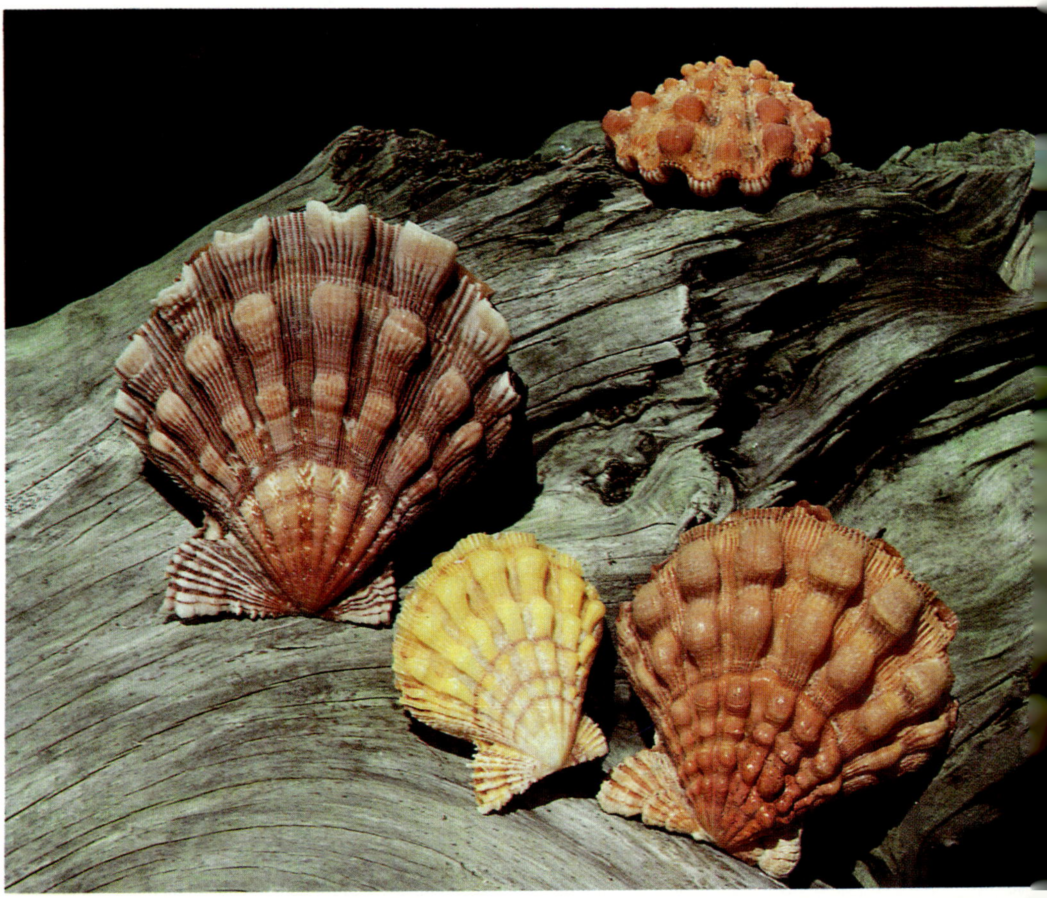

2

1 2

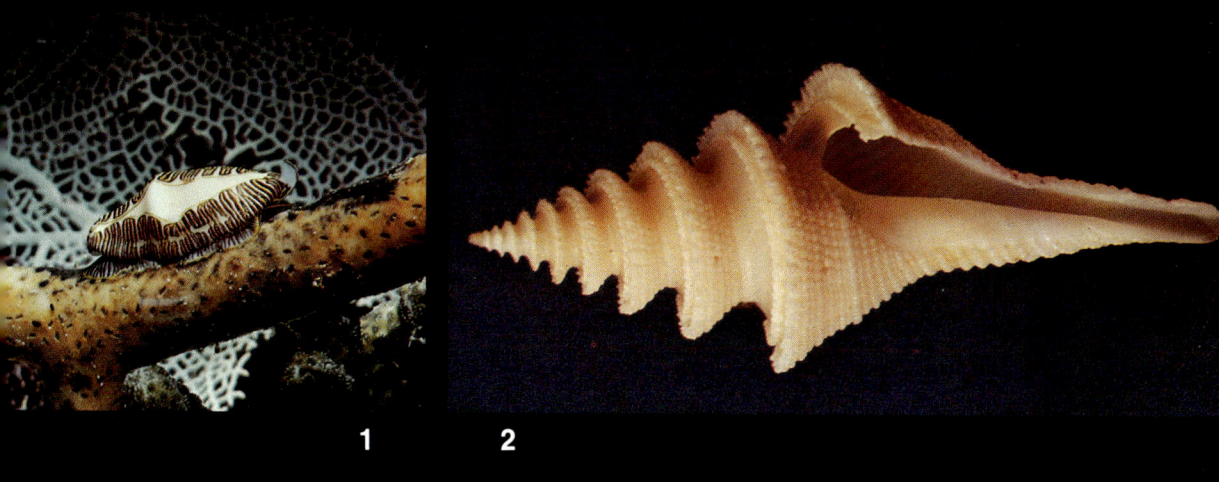

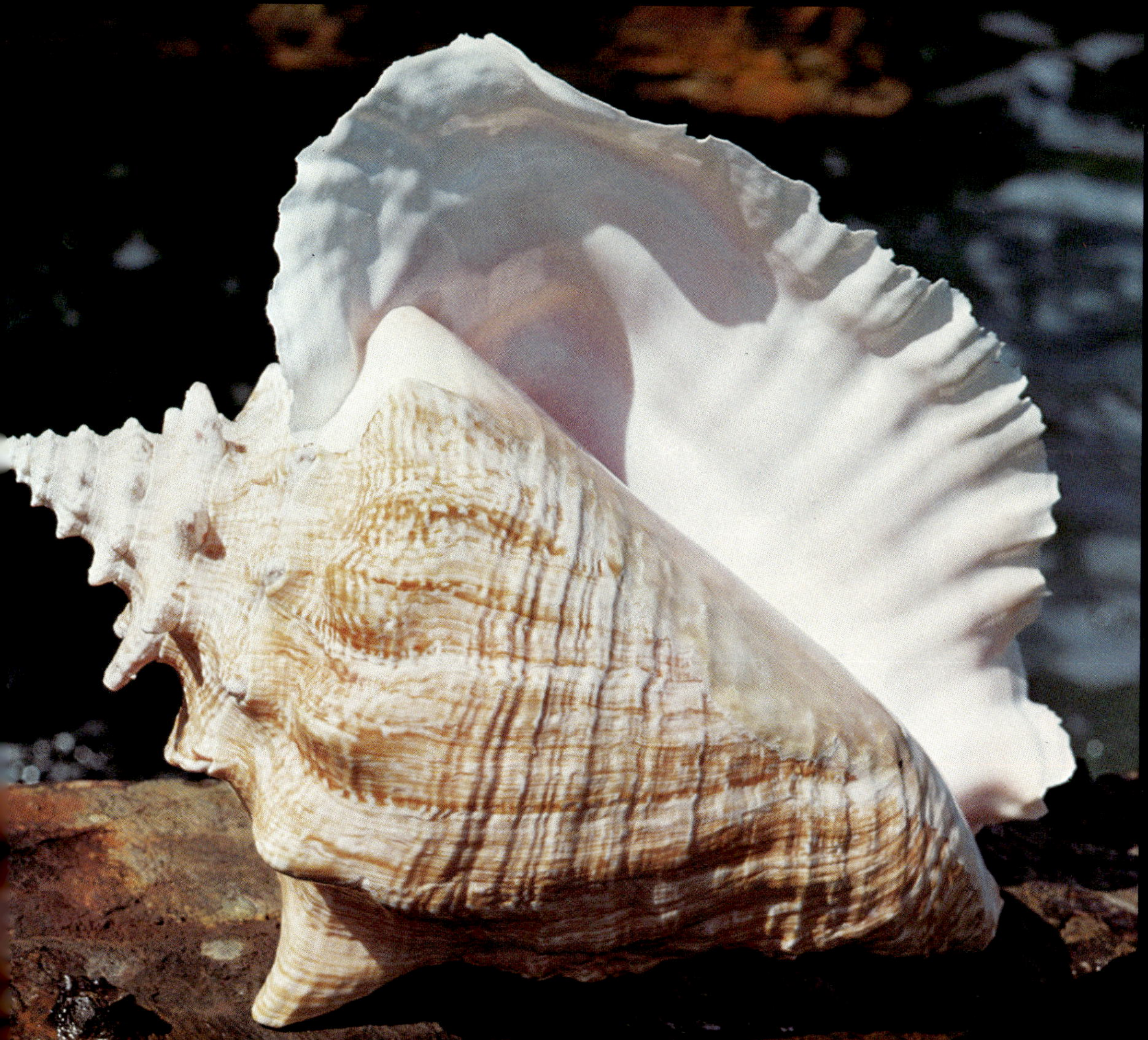

3

1 Fingerprint cyphoma, *Cyphoma signatum*

2 Elegant star-turris, *Cochlespira elegans*

3 Queen conch, *Strombus gigas*

1. Knobby argonaut, *Argonauta nodosus*
2. West Indian green chiton, *Chiton tuberculatus*
3. Rough fileclam, *Lima scabra*
4. Chestnut cowrie, *Cypraea spadicea*

1

2

3

4

1 Atlantic trumpet triton, *Charonia tritonis variegata*
2 Green abalone, *Haliotis fulgens*
3 Reticulate cowrie-helmet, *Cypraecassis testiculus*
4 Caribbean helmet, *Cassis tuberosa*
5 *Octopus* species

1 Amber penshell, *Pinna carnea*
2 Blue mussel, *Mytilus edulis*
3 Crown conch, *Melongena corona*
4 Pacific calico scallop, *Argopecten circularis*

1
Glisten-worm solenogaster, *Chaetoderma nitidulum* (Robert Robertson)
2
Occidental tuskshell, *Antalis entale occidentale* (Terrance M. Gosliner)
3
Northern red chiton, *Tonicella rubra* (Terrance M. Gosliner)

1 Pacific bananaslug, *Ariolimax columbianus* (Arthur E. Bogan)
2 Florida leatherleaf, *Leidyula floridana* (Florida Division of Plant Industry)
3 Three-stripe doris, *Chromodoris macfarlandi* (Terrance M. Gosliner)

Rosey wolfsnail, *Euglandina rosea* (F. H. Roberts)

Stock Island treesnail, *Orthalicus reses reses* (Jane E. Deisler)

1 Black sandshell, *Ligumia recta* (James D. Williams)

2 Arrow squid, *Loligo pleii* (Clyde F. E. Roper)

1

2

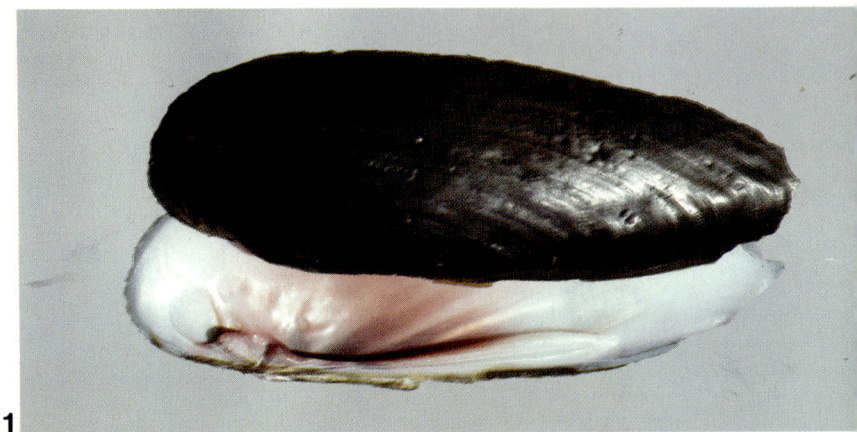

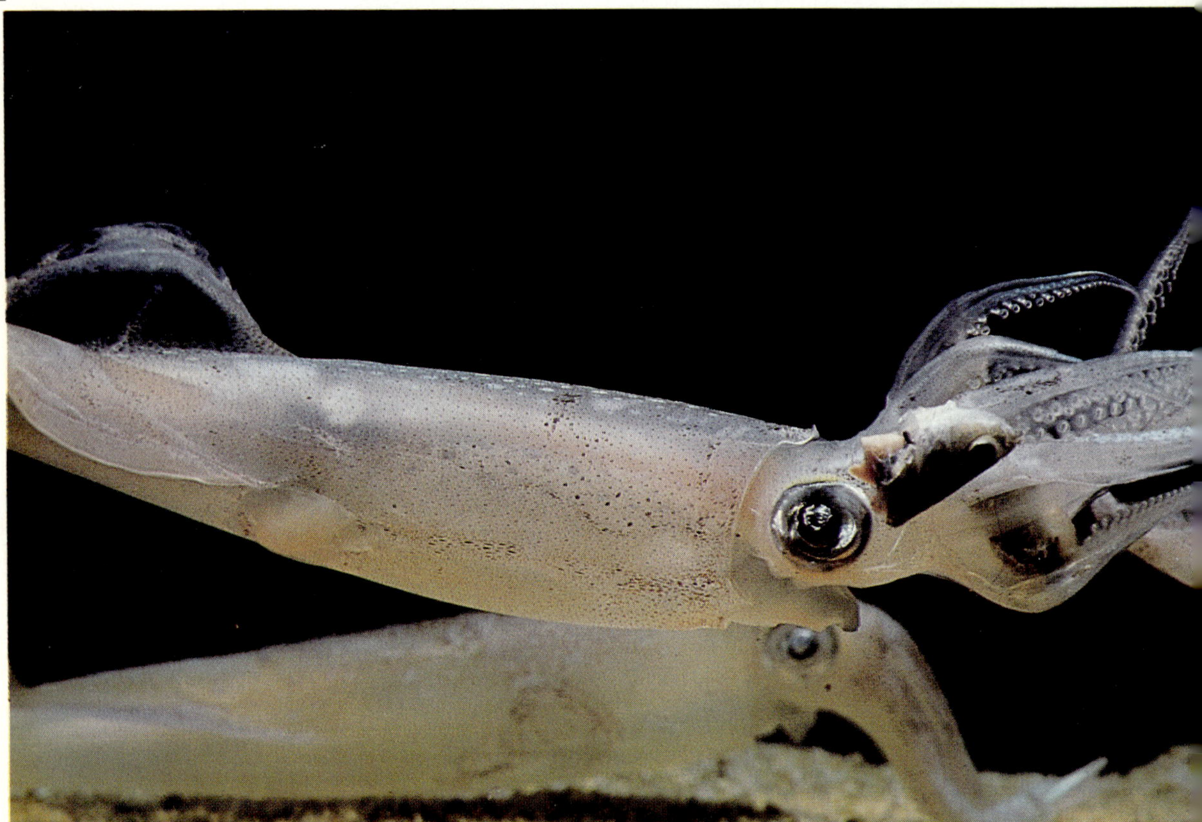